Harvard Studies in Business History XXX

Edited by Alfred D. Chandler, Jr.
Isidor Straus Professor of Business History
Graduate School of Business Administration
George F. Baker Foundation
Harvard University

The British Shipbuilding Industry • 1870-1914

SIDNEY POLLARD

and

PAUL ROBERTSON

Harvard University Press
Cambridge, Massachusetts
and London, England 1979

Printed in the United States of America

Library of Congress Cataloging in Publication Data

Pollard, Sidney.
The British shipbuilding industry, 1870-1914.
(Harvard studies in business history; 30)
Bibliography: p.
Includes index.
1. Ship-building—Great Britain—History.
2. Shipyards—Great Britain—History.
I. Robertson, Paul, 1944- joint author.
II. Title. III. Series.
VM299.7.G7P64 338.4'7'6238200941 78-12500
ISBN 0-674-08287-7

Editor's Introduction

THIS STUDY fits in admirably with the enlarged scope of the *Harvard Studies in Business History*, for it deals with a group of firms in a non-American setting. Professors Pollard and Robertson provide a composite history of the business enterprises that made up the British shipbuilding industry by focusing on the years when the industry was the largest in the world and when it was completing its fundamental transformation from wooden sailing ships to ones made of steel and powered by steam. The book's value lies in the authors' ability to combine business and industrial history. They consider the actions of the entrepreneurs and managers who made the basic decisions about the policies of these firms, and they provide the statistical and other data sought by scholars concerned with changing industrial organization.

This book thus provides rich information on how the businessmen in charge of these enterprises adopted new forms of technology—changes that involved the transformation of the theory and practice of naval architecture and marine engineering. It delineates ways in which the entrepreneurs defined their relations with their suppliers and with their customers, both private and public. The latter were particularly important because the British shipbuilders' shipyards supplied not only the British Navy but also the navies of Japan, Italy, Brazil, and other lesser powers. The book reviews how these entrepreneurs designed and then redesigned the internal organization of their yards, adopted new and increasingly specialized machinery, furnaces, and foundries, and constantly adjusted the coordination between men and machines. The authors have devoted pages to the recruitment and training of labor and to the collective bargaining and other negotiations carried on between owners and managers and powerful labor unions, particularly those of the shipwrights, the engineers, and the boilermakers. Finally, this history pays close attention to the ways in which the firms were financed, in the provision of both long-term permanent capital and short-term working capital.

While covering the many aspects of business history, the authors in no way ignore the more standard items on the agenda of an industrial history. They describe and analyze the patterns of competition among firms, the ways in which they continued to expand, and the impact of mergers—both vertical and horizontal—on the size of the firms and the structure of the industry. Changes in productivity and the impact of the industry on the national and international economy are elaborated. By successfully combining industrial and business history, Professors Pollard and Robertson remind us that a precise demarcation between business and industrial history is artificial and that the combination of both approaches enhances our understanding of modern economic institutions more effectively than either applied alone.

Alfred D. Chandler, Jr.

Acknowledgments

In the course of preparing this book, we have become indebted to many persons and institutions. Sidney Pollard wishes to acknowledge the help he received from his thesis advisors, the late Professor T. S. Ashton and Mr. H. L. Beales of the London School of Economics. Paul Robertson received special inspiration from Eric E. Lampard of the State University of New York at Stony Brook, as well as from Morton Rothstein, Ralph Andreano, and Nathan Rosenberg, his advisors at the University of Wisconsin-Madison. Professors S. G. Checkland, P. L. Payne, Rondo Cameron, Sidney Burrell, and Saul Engelbourg, and Mr. A. Slaven all read parts of the work in earlier drafts and made useful suggestions. Mr. Richard Dell, the Glasgow City Archivist, was most helpful, and Mr. C. R. Connell kindly allowed us to use the records of Charles Connell and Company. Among the institutions that have aided us are the Scottish Record Office, the Glasgow University, and Mitchell Libraries; the libraries of the London School of Economics, the Massachusetts Institute of Technology, Boston University, and the British Museum; the Barrow-in-Furness, Deptford, and Poplar Public Libraries; the Widener Library of Harvard College; and the Baker Library of the Harvard University Graduate School of Business Administration. We also wish to acknowledge our debt to the staffs of these institutions, and to those of Lloyd's Register, *Syren and Shipping*, the Institution of Naval Architects, and the Institute of Marine Engineers. Typing and copying services were provided by the Bologna Center of the Johns Hopkins University School of Advanced International Studies, and the revised draft was typed by Ms. P. McFadyen of the University of Melbourne. Mrs. Jenny Soukup provided valuable help in correcting the proofs.

Research grants from the American Philosophical Society and the Graduate School and Department of History of Boston University greatly eased our efforts, as did the Research Fellowship that the University of Glasgow awarded to Paul Robertson in the summer of 1974 to allow him to complete his portion of the archival research.

We also wish to thank the editors of the *Economic History Review*, the *Journal of Economic History*, *Business History*, and the *International Review of Social History* for permission to reprint extracts from the following articles which appeared originally in their journals. Articles by Sidney Pollard: "Barrow-in-Furness and the Seventh Duke of Devonshire," *Economic History Review* 2nd ser., 8 (1955); "British and World Shipbuilding, 1890-1914: A Study in Comparative Costs," *Journal of Economic History* 17 (1957); The Decline of Shipbuilding on the Thames," *Economic History Review* 2nd ser., 3 (1950); "*Laissez-Faire* and Shipbuilding," *Economic History Review* 2nd ser., 5 (1952). Articles by Paul Robertson: "Demarcation Disputes in British Shipbuilding before 1914," *International Review of Social History* 20 (1975); "Shipping and Shipbuilding: The Case of William Denny and Brothers," *Business History* 16, no. 1 (1974); "Technical Education in the British Shipbuilding and Marine Engineering Industries, 1863-1914," *Economic History Review* 2nd ser., 27 (1974).

Contents

	Introduction	1
1	The Evolution of the Steel Steamship	9
2	Fluctuations in Shipbuilding Output	25
3	Regional Shipbuilding	49
4	The Firms	70
5	Industrial Organization	88
6	The Development of the Shipyard	108
7	Scientific Shipbuilding and the Evolution of Technical Education	130
8	Labor and Labor Relations	151
9	Wages, Employment, and Productivity	173
10	The Influence of the State	201
11	Conclusion	230
	Appendix A: A Note on Tonnage Measurements	237
	Appendix B: Statistical Tables	239
	Notes	255
	Abbreviations	285
	Selected Bibliography	286
	Index	304

TABLES

2.1 The value of British shipbuilding output, 1847-1913. *32*

2.2 The economic weight of the shipbuilding industry. *34*

2.3 Value of merchant hulls and of total merchant output as a proportion of Gross Domestic Product, 1886-1913. *35*

2.4 Aggregate wages in shipbuilding as a proportion of total British wages from employment, 1871-1911. *36*

2.5 British tonnage built for foreign owners and total tonnage built in German and American yards, 1892-1913. *44*

2.6 The British percentage of the world's shipbuilding, 1892-1914. *45*

3.1 Locations of important shipbuilders. *51-52*

3.2 Wage indices of eight maritime districts, shipbuilding and engineering (all districts, 1883 = 100). *53*

3.3 Average tonnage of the vessels built in the principal districts (private yards only). *59*

4.1 Fixed and variable capital of some shipbuilding firms. *79*

4.2 Capitalization of selected British shipbuilding firms, 1900-1901. *80*

4.3 Capital of selected shipbuilding companies, 1900-1912. *81*

4.4 Net profits, capital values, and rates of return for the shipbuilding and marine engineering operations of John Brown and Company, and for the Elswick and Walker Shipyards and the engine works of Sir W. G. Armstrong, Whitworth and Company, 1900-1912. *83*

6.1 Largest Atlantic steamers built at Clydebank, in successive periods, 1881-1913. *110*

6.2 Capacity and output of 20 of the leading shipbuilding firms, 1903-1913. *118*

6.3 Value of shipbuilding machinery by trade in the Clydebank yard of John Brown and Company, 1899 and 1910. *127*
6.4 Capital per shipyard worker, selected trades, John Brown and Company, 1910. *127*
8.1 The percentage of skilled labor in shipyards. *153*
8.2 Employment in the Royal Naval Dockyards. *169*
8.3 Earnings in private shipbuilding firms and in the Royal Naval Dockyards. *170*
9.1 Indices of average wages and earnings of workers in shipbuilding and engineering on the Clyde and at the Leven Shipyard, 1851-1914 (1860 = 100). *175-176*
9.2 Percentages of work performed, by value, by each trade in each quarter of a vessel's construction. *178-179*
9.3 The percentage of each trade's total output, by value of wages, performed in each quarter of a vessel's construction. *180*
9.4 Cycles in time wages, unemployment, tonnage launched on the Clyde, and earnings of all workers at the Leven Shipyard, 1863-1913. *183*
9.5 Weekly time and piece rates in the shipbuilding trades, 1886-1906. *185*
9.6 Fluctuations in the price of ships and in productivity in shipbuilding, 1883-1913. *188-189*
9.7 Employment and real earnings of all workers, Leven Shipyard, 1852-1913. *191-192*
9.8 Wage rates of selected trades, Leven Shipyard, expressed as percentages of laborers' wage rates, 1880-1910. *195*
9.9 Wage rates of selected trades, Leven Shipyard, expressed as percentages of laborers' wage rates, 1846-1880. *196*
10.1 Naval expenditure on contracts, some major firms. *215*
10.2 Shipbuilding for the Royal Navy, thousands of displacement tons, 1866-1885. *216*
10.3 Shipbuilding for the Royal Navy, thousands of displacement tons, 1885-1913. *217*
10.4 Shipbuilding expenditures for the Royal Navy. *218-219*
B.1 Money earnings of all workers and of joiners and joiners machinemen, Leven Shipyard, 1852-1914. *239-240*
B.2 Real earnings of all workers, Leven Shipyard, 1852-1913. *241*
B.3 The unemployment rate of members of engineering, metal, and shipbuilding unions in the United Kingdom, 1851-1913. *242*
B.4a Wage rates of blacksmiths, carpenters, and joiners machinemen, Leven Shipyard, 1880-1912. *243*
B.4b Wage rates of engineers, blacksmiths hammermen, and laborers, Leven Shipyard, 1880-1912. *244*

B.5a Wages in shipbuilding and engineering, 1866-1914: overall indices (1900 = 100). *245*
B.5b Wages in shipbuilding and engineering, 1866-1914: indices for various districts (weighted average for all districts = 100). *246-247*
B.6 Total labor costs as a percentage of total hull costs, ships built by William Denny and Brothers, 1881-1913. *248*
B.7 Mercantile shipbuilding output of the chief shipbuilding countries, 1892-1914. *249*
B.8 Merchant tonnage built in the United Kingdom, 1860-1913. *250-251*
B.9 Regional shipbuilding output in Britain, 1870-1913. *252-253*
B.10 Assets of the Clydebank Shipyard, 1884-1913. *254*

FIGURES

1 Shipbuilding output, 1870-1913. *26*
2 Assets of the shipbuilding department of the Clydebank Shipyard, 1884-1913. *78*
3 Hands through which the proposed ship passed before construction commenced. *102*
4 The Shipbuilding Department of Harland and Wolff, Belfast, 1912. *113*
5 The Naval Construction Works, Barrow-in-Furness, 1901. *114*
6 The positioning of the cranes at the Vickers Shipyard, Barrow-in-Furness, 1914. *125*
7 Diagram of a tower crane at the Vickers Shipyard, 1914. *126*
8 Money wages and earnings, and tonnage launched on the Clyde, 1863-1913. *181*
9 Money earnings and tonnage launched on the Clyde, and the unemployment rate of all workers in engineering, shipbuilding, and metalworking in the United Kingdom, 1863-1913. *182*
10 Denny and Connell labor productivity (Feinstein price index) and money and real earnings of employees at the Leven Shipyard, 1884-1912. *193*
11 Denny and Connell labor productivity (Pollard-Robertson price index) and money and real earnings of employees at the Leven Shipyard, 1884-1912. *194*
12 Structure of the shipbuilding departments of the Admiralty in 1886. *210*

MAP: Important shipbuilding towns and rivers. *50*

The British Shipbuilding Industry • 1870-1914

Introduction

IN RECENT decades, economic historians have paid increasing attention to the world economic situation between 1870 and 1914. During this period the United States and Germany reached economic maturity and by many standards displaced the United Kingdom to become the world's leading industrial nations. It was also a time of remarkably free movement of capital and persons across the continents. Some scholars also search for explanations of the economic collapse of the interwar years in the conditions of the supposedly "golden era" that preceded the upheaval of 1914-1918.

In particular, a formidable scholarly literature has accumulated on the performance of the British economy in the decades before 1914.[1] Although the issues involved and the explanations evoked are complex, at base the question is whether Britain's rates of growth of output and productivity, especially in manufacturing, compared unfavorably with those in earlier and later periods of British development and with those of Germany and the United States during the same period. Some economists contend that British performance was not inferior, that it only appears to have been so because other participants in the debate have been measuring the wrong variables.[2] If the British record was indeed poor, as a great deal of statistical evidence indicates that it was, then it might have been the result of economic factors, such as a "natural" deceleration or climacteric in the development of a mature economy, or a slow rate of growth of demand because of a low rate of population growth. Such factors might have induced a low rate of economic growth even if entrepreneurs reacted correctly and intelligently to market stimuli. Alternatively, a lackluster economic performance might have stemmed from poor management or poor entrepreneurship, in which case it could perhaps have been significantly improved if British businessmen had been more astute or assiduous in the pursuit of maximum profits.

Industrial organization in Britain between 1870 and 1914 has been one area of controversy. Among the explanations of the relative decline of the British economy in this period put forth by various experts are (*a*) an unwillingness or inability on the part of British businesses to establish and maintain lasting relationships with customers, and (*b*) the fragmented nature of many British industries in which entrepreneurs did not combine effectively with each other to take advantage of potential cost-saving innovations (for example, through economies of scale) or to exert the stronger control over markets that is open to oligopolists and monopolists.

British marketing practices, particularly in the export sector, have been criticized on several counts. According to W. A. Lewis, "Britain equipped herself in the first half of the nineteenth century to sell cotton and railway materials, and she thus got into a rut which unfitted her to sell steel and machinery at the end of the century."[3] Lewis lists three reasons for this deficiency. The first is the tendency of investors to put their funds into established industries instead of searching out new opportunities. This tendency is alleged to have led to an overcommitment to sectors that had already passed their prime, at the expense of newer industries with greater potential for growth. Secondly, the merchant house, the primary institution of foreign trade in the first half of the century, was suited to selling consumer goods in agricultural countries but inappropriate to the expanding European market for sophisticated industrial equipment. As Kindleberger notes, the merchant house is an example of industrial disintegration that resulted in a loss of income by making a portion of the returns external to the manufacturers.[4] If firms had developed their own marketing organizations, salesmen would have been forced to concentrate on their products rather than yield to the temptation to sell foreign goods. Moreover, the merchant house created a barrier that made it difficult for firms to determine the exact needs of buyers and to produce accordingly. Lewis's third reason is the failure of the British to offer intermediate credit terms. While British financiers excelled at three-month bills and twenty-year bonds, the Germans offered six months or allowed payment to ride until the order was renewed. Although dangerous, the German method was better suited to the trade and had the further advantage of tying customers through a bond of debt.[5]

The continued organization of several important British industries into small, self-financed, and fiercely independent firms is sometimes cited as another reason for their slow rate of growth after 1870, especially in comparison with the cartelized industries of Germany, many of which were coordinated by the great banks. David Landes contends that German cartels and finance capitalism worked well in practice, if not in theory, because they led to greater vertical integration, higher rates of investment, and increased rationalization. All of these effects boosted the international competitiveness of German industry.[6] In Britain, cartels and coordination were less common. Combinations, where they did appear, were "in the wrong industries, or if in the right ones, for the wrong reasons; they were often founded by promoters rather than producers, and the initial over-evaluation of capital burdened subsequent performance: the very multiplicity of their adherents complicated their task; and here too the absence of tariff protection exposed the prosperous ones to the incursions of interlopers—success was almost as dangerous as failure."[7]

It has also been charged that British managers neglected to adopt innovations because they were hostile to change. It is claimed that in several important industries the British lagged behind their competitors in the use of new equipment, and that they were equally tardy in adopting so-called no-cost improvements, which are made by rearranging existing plant and machinery. An example frequently cited is the tinplate industry, in which the Americans used the same machines as the British but deployed them more efficiently.[8]

This indictment may be widened to include the alleged British neglect of scientific and technical research and education. According to Derek Aldcroft, British industrialists were a race of "practical tinkerers" who were reluctant to depart from "rule-of-thumb" methods and were "proud of the fact that they carried out little original research or employed few technicians."[9] Although the theoretical benefits of increased education may have been appreciated in Britain, there can be no question that both the state and private individuals were loath to finance secondary and higher educational facilities on the scale that prevailed in many continental countries and in North America. As a result, even if more extensive research had been thought desirable, undertaking it would have been difficult for lack of properly trained personnel.

Several writers have developed sociological theories to explain why the British may have been slower than others to grasp the advantages of technological improvements and increased research and education. One of the most recent and comprehensible explanations is D. C. Coleman's division of entrepreneurs into "gentlemen and players." Between roughly 1850 and the outbreak of the Second World War, Coleman theorizes, British industry was increasingly dominated by second- and third-generation entrepreneurs whose goals were more social than pecuniary and who had been educated in the public schools and ancient universities to affect a gentlemanly disdain of technology which is more generally associated with the older landed classes. At the same time, there was an important admixture of "players," men who had come up through the ranks, whose social aspirations were as strongly felt as those of the more firmly established gentlemen, and whose experience as "practical men" militated against any belief in the efficacy of theory, be it scientific, technological, or managerial. Between them, these two groups had a strong inclination to maintain the status quo.[10]

Sociological explanations have been derided by many economic historians, however, on the grounds that in any advanced industrial society market forces will induce entrepreneurs, on balance, to react "rationally" as profit maximizers. Unfortunately, even within the bounds of conventional Western economic theory, what behavior is rational varies depending on the assumptions adopted by the entrepreneurs. In particular, although it is not always explicitly acknowledged by the participants in the debate, many of the differences in their judgments of the success of British entrepreneurs may be traced to differing opinions as to the suitability of the time horizons that the British employed.

It is clear that a variety of time horizons have legitimate roles in day-to-day business management. Investments in inventory items with a rapid rate of turnover require a different perspective from investments in plant and equipment that are expected to last for many years. Once a time horizon has been selected, it is logical to inquire whether the entrepreneur has acted rationally within the framework he has set for himself—as in most cases he will no doubt have done. It should also be permissible to question the appropriateness of the time horizon itself. Examples are numerous of firms that have suffered because their managers placed so much emphasis on long-term planning that they neglected to take precautions against short-term dangers. Similarly, performance may be impaired if proper attention has not been given to long-term trends. Short-term profit maximization is no guarantee of long-term success; conceptually, the long term is not simply the sum of the shorter terms that it comprises. Options once rejected may no longer be feasible at a later date. In the case of investments with very long periods of return, such as those in education, the consequences of an improper time horizon may take decades to work themselves out.

S. B. Saul contends that in some instances it is impossible to make judgments on time horizons "without an improper use of hindsight." As an example, he asks, "What reason could there be for not investing in cotton mills in 1905 when profits expected and realised up to the war were comparable with any elsewhere?"[11] This is a fair criticism, but historians nevertheless must consider the problem of time horizons in some cases. Entrepreneurs themselves employ diverse time horizons, and this fact must be considered in our attempts to reconstruct and interpret their actions.

Although not phrased in these terms, the question of time horizons is implicit in some of the theories developed in the 1950s by historians of an entrepreneurial or sociological bent. The British, so the argument goes, may have erred in being too rational in their decision making. In a strictly pecuniary sense, they were often correct, in the short run at least, in refusing to adopt innovations because the relative inelasticity of the home and foreign markets, together with the existence of so much equipment that was still usable, made investments in techniques that appeared to have a potentially short period of amortization unattractive. The Germans, with an "irresistible urge to invest," and the Americans, under the influence of a "non-rational super-charger," were more interested in technical efficiency than in minimizing costs. As a result, they tended to innovate for innovation's sake, to invest when they would have been better advised, in strictly economic terms, to stay with what they had. In the peculiar circumstances of the late nineteenth century, with its rapidly expanding markets, particularly in America, this overoptimism paid off. Through what John Sawyer calls an "entrepreneurial accelerator," investors were able to speed up growth by overanticipating the limits of present objective opportunities.[12] As Landes later conceded, however, the distinction between "technological rationality" and "economic rationality" is to some extent a misleading one. German (and American) entrepreneurs were not economically irrational. Rather, "the difference . . . was one of calculation: the German industrialist and financier had a longer time horizon and included in their estimates exogenous variables of continuing technological change that their British competitor held constant."[13]

Some circumstances may force entrepreneurs to adopt a short-term perspective to the virtual exclusion of long-term considerations. The most obvious example arises when a scarcity of information makes long-term prediction and preparation unacceptably risky. A variation of this example is what might be called a steady-state hypothesis, in which entrepreneurs do forecast their long-term prospects as accurately as they are able, but find them so similar to the short-term outlook that they act in much the same way as they would have if they had been attempting to minimize risk by taking a short-term position. This is essentially the situation that Landes and Sawyer described in distinguishing late-nineteenth-century British attitudes from those in Germany or the United States. If the British decided not to invest because they felt their markets to be relatively inelastic, their decision reflects not a difference in time horizons but a more pessimistic assessment of their long-term prospects than the Germans and Americans had arrived at for their own markets. In order to protect themselves, the British were obliged to be less adventurous and more wary of risk in their long-term planning, thereby perhaps giving the impression of an overly rational, purely short-term, emphasis. If the British had been correct in their assessment, profit maximization would have ensued, but if they had been unduly pessimistic their prophecy might have been self-fulfilling: firms that do not invest in new technologies may lose their markets to more progressive competitors. Bearing in mind differences in behavior between firm and firm, industry and industry, this mechanism does provide a possible clue to an aspect of British performance that has puzzled economic historians.

A great deal has been learned about these general problems, but much remains to be done on the level of the individual firm or industry. The lack of research on this level is unfortunate, for, as is well known, macroeconomic data mask a multitude of significant and contradictory patterns. Such important issues as the effectiveness of marketing practices, the rate of technological change, the degree to which the increasing scarcity of natural resources constrained growth, the general economic rationality of entrepreneurs, and many others can be most profitably assessed by a close examination at lower levels of aggregation. Only through the construction of an elaborate mosaic based on the behavior of many firms and industries can scholars obtain an informed and balanced picture of the interactions of the entire economy.

Our description of the shipbuilding industry provides only one of the tiles in the mosaic, but it is an important one. In the three decades before 1914, the total value of new merchant vessels accounted for approximately 1.25 percent of the gross domestic product of the United Kingdom, and wages paid in the industry came to more than 2 percent of total wages from employment. Value added within the industry was somewhat smaller (although it is impossible to estimate with any precision), but the relations between shipbuilders and their many outside suppliers increased the impact of the industry far beyond that which the value of its own output would indicate. Almost 30 percent of total British steel production was incorporated in ships, for example, and the severe cyclical fluctuations in the demand for new vessels inevitably reverberated throughout the economy.

In the chapters that follow, we take a detailed look at the British shipbuilding industry and conclude that in general British shipbuilders acted rationally between 1870 and 1914. Sources of supply were efficiently organized, and markets were sought and maintained. Within the industry a high level of competition among builders helped to lower costs and prices. We find sociological explanations to have little relevance in explaining the rate of technological change, but the concept of a steady-state hypothesis has considerable value. In shipbuilding, as in some other industries, British entrepreneurs were slower than their competitors abroad in adopting new equipment and techniques. This slowness, we believe, was one of the greatest strengths of the industry and contributed to Britain's continued dominance to 1914 by reducing overhead charges and allowing the British to produce vessels at a lower cost than was possible in many of the highly capitalized yards in other countries. Experienced British yard owners realized that the severely cyclical character of the industry could lead to long stretches in which expensive plant and equipment were underutilized. They were by no means uncritically opposed to innovation, and indeed the technical nature of both ships and shipbuilding was dramatically transformed over the period. British shipbuilders did insist, however, that the profitability of innovations be thoroughly proved before they invested. Foreign competitors took a more optimistic view of the growth prospects of the industry in their own countries and of their ability to weather downturns in output. Moreover, they did not have a well-trained workforce of manual laborers as the British did, and this weakness forced them to invest heavily in machines in the hope that these could be substituted for skilled workers. As it turned out, up to 1914 the guarded view of British builders was the more correct and Britain was able to withstand the challenge from competitors in other countries. The British reluctance to invest as heavily in plant and equipment represented not a shorter time horizon or an irrational love of outdated methods, but simply a cautious long-term policy based on an accurate prediction of their prospects, if the long term is limited to the period before the First World War.

British shipbuilders took a similar attitude toward research and technical training, but here the results were more questionable. Certainly progress was made in both fields, but much less than in Germany and the United States. The British approach may be justified by the fact that British ships and the level of skill of British workers were still as advanced in most respects as any in the world in the years before 1914. Investments in research and education, however, often have exceptionally long gestation periods. A failure to invest sufficiently in education at the right time may have particularly bad long-term consequences because it is more difficult to train or retrain older workers and yet they usually cannot be replaced immediately by a more appropriately trained labor force. It is conceivable that the legacy of poorly trained and undereducated workers from before 1914 contributed to the decline of British shipbuilding in the interwar years.

Nevertheless, ours is a success story. From 1886, at least, the rate of growth of output for the shipbuilding industry exceeded that for the economy as a whole. Moreover, shipbuilding was one of the few heavy industries in which Britain maintained an international supremacy after

1870. As late as the years 1900-1913, the tonnage produced in British shipyards averaged almost 60 percent of total world production, despite heavy subsidy payments to many foreign shipyards and other vigorous attempts to reduce Britain's dominance. Clearly, the caution shown by British shipbuilders was justified by the results.

1.

The Evolution of the Steel Steamship

Wooden Vessels before 1870

THERE WERE few industries in the latter half of the nineteenth century in which Britain was more truly the workshop of the world than in shipbuilding. Favored by an island position, sheltered ports, a large share of the world's trade, a relative abundance of capital, and technical skills second to none, Britain needed only an economical supply of the necessary raw materials to become the world's great shipbuilding center.

In the first half of the century ships were generally built at small yards, strung out along the forested coastline of Europe and North America, wherever a seafaring population furnished a local demand. The British lead in shipbuilding at the end of the Napoleonic Wars seemed assured by the rapid growth of British overseas trade, yet by the middle of the century the Americans had caught up in both the quantity and the quality of the ships built. The British-American rivalry in this period well illustrates the technical changes and foreshadows the future history of the shipbuilding industry.

The main carrying trade of the world was still borne by timber-built sailing ships. There had been few changes in the sailing ship employed in coastal or short-distance trade, but with the increase in commerce and growing overseas migration, improved types of sailing vessels, such as the clipper ships, and new forms of organization, such as the regular sailing packets, had been introduced into the long-distance trades. In both these fields, the Americans had taken the lead. The Black Ball Line, which started its regular sailings in 1816, conveyed emigrants and passengers across the Atlantic and was the first of a number of new lines of well-built ships. At the same time, Yankee captains, encroaching on trade along new coasts, needed fast vessels to escape from British patrol vessels and the Spanish Guardacostas, while British builders rested content behind their protective Navigation Acts. The discovery of gold in California in 1848 and the gold rush to Australia in 1851 called for "extreme clipper ships, and the United States turned out the largest and finest wooden sailing ships afloat."[1] These were beautiful vessels, with sharp bow and stern and with a large spread of canvas and masts of 200 feet or more in height. Moreover, before the Civil War the Americans, as well as the North American colonists, could build cheaper ships than the British, although it was widely held that Britain could build the better class of ship as cheaply as any other country in the world.

Regular sailing packets reached their peak between 1825 and 1850. They were built on the New England coast of American softwood, and attained very high speeds by relentless driving over the entire day, but did not yet incorporate any new designs. New designs followed the opening of new routes and the desire to achieve greater speed, a larger cargo-carrying capacity, or better seaworthiness. In time they led to the evolution of the clipper.

The first clippers were the opium clippers, which appeared after the repeal of the East India Company's monopoly of trade with China. They had to be especially fast and maneuverable to escape Chinese patrol boats and pirates and to stand up to the monsoons in Chinese waters. There were various types: the clippers proper, making the journey from India to the China sea; the coasters, smaller vessels delivering goods to the ports; and the receiving ships, heavily armed and bulky, with storerooms and clerks on board. The main period of activity was in the two decades from 1830 to 1850, after which time they were superseded by steamers and their builders turned to the China tea clippers.

The tea clippers were originally an American creation. With their fine lines, sharp bow, broad beam well aft, and tall raking mast, they reached speeds far above any previously obtained. The first China clipper, the *Helena,* 650 g.t., was built by William H. Webb in New York in 1841, but the most important era of the American tea clipper extended from 1846 to 1860. After the repeal of the Navigation Acts, these vessels usually sailed from the Atlantic coast of America around the Horn to California, unloading gold seekers and manufactures, then across the Pacific to China, loading tea, and finally on to London. The annual tea races of the fastest clippers from Foo-Choo-Foo to the Thames aroused immense interest and excitement in London and were very profitable.

While American interests were thriving, British shipping and shipbuilding languished. Output in the United Kingdom did continue to grow as the industry concentrated on the production of longer-lasting and slower hardwood vessels. The repeal of the Navigation Acts, however, had the immediate effect of swamping British and colonial ports with

American vessels. The United States reciprocated only halfheartedly, still retaining its own coastal trade monopoly, which included journeys from the Atlantic to the Pacific coast. The marked depression among British shippers, evident in 1850, increased with the passing of the Passenger Act of 1855, which ensured minimum safety regulations for emigrants on British ships. Foreign ships were said to enjoy an unfair advantage in the emigrant trade by their exemption from the Passenger Act. This remained a constant source of irritation to British shipowners, who also continued to agitate for the freeing of foreign coastal trades by threatening to retaliate against any country that restricted its own.[2]

Before 1849, no British shipbuilder could think of competing with the American fast sailers. The arrival of the *Oriental* in London in 1850 with a load of tea from China created a sensation. Draftsmen were sent from the Admiralty to copy her lines for the benefit of the construction department of the navy. But soon thereafter the first British builders, Hall and Company of Aberdeen, built tea clippers for Jardine, Matheson and Company, including the small *Stornoway* and *Chrysolite*, and the larger *Cairngorm*. In 1856 the *Lord of the Isles*, built by Scott's of Greenock, beat all American ships in the tea race from Foo-Choo-Foo. British-built clippers were heavier and narrower with equally fine lines, but were of hardwood and later of composite build, and therefore exempt from the seepage and rapid deterioration of the American vessels. Although the American clippers were generally cheaper and faster, the advantage in strength and safety already lay with the British ships.

The share of British ships in the British trade declined in the 1850s, but it increased rapidly afterward at the expense of American vessels. In the trade from the United Kingdom to the United States, British tonnage increased from 946,000 in 1860 to 1,853,000 in 1866, as American tonnage dropped from 2,245,000 to 484,000. In percentage terms, the British share of the American trade increased from 18 percent in 1855 to 65 percent in 1865 to 84 percent in 1885 while the American share was declining from 79 percent to 26 percent to 10 percent.[3]

The suddenness of the American decline was caused by the Civil War, in which 40 percent of American seagoing tonnage was lost, partly to raids by Confederate cruisers and partly to the transfer of American vessels to British registry. In the long run, however, the real causes of the decline were the steam engine and the iron hull. American attempts to compete in steamship development with Britain, with her high output of cheap coal and her large and skilled body of engineers, were bound to end in failure. The introduction of the iron hull further tended to favor Britain as a shipbuilding center. The advantage enjoyed by the British iron shipbuilders was so great that British shippers were accused of favoring iron specifically for the purpose of ruining foreign yards, an accusation that seemed to be supported by some British statements.[4] There seems little reason to suppose, however, that any of Lloyd's rules, involving the business fortunes of underwriters and shipowners, sacrificed commercial interests for national reasons. All the rules were based on the then-prevailing opinion of the strength and safety of iron and of different varieties of timber, as indeed was recognized by less prejudiced American observers.[5]

The extraordinarily rapid growth of American shipbuilding in the first half of the nineteenth century had been based on the abundance of timber on the North American continent and on America's favorable position in relation to the new sea routes that were being opened up. When the iron steamer became efficient enough to displace the wooden sailer, however, Britain was bound to enjoy a considerable advantage. The changed position showed itself first after the crisis of 1857, and became more apparent during the Civil War. From that period onward, Britain not only became the leading shipowner and shipbuilder but for some time practically monopolized iron ship and steamship building.

Britain had little to fear from the commercial rivalry of other shipbuilding nations. In France, the virtual serfdom in which labor was held and the liability to conscription in the dockyards outweighed the advantages of labor's cheapness. John Scott, who had established a yard at St. Nazaire based on British capital and supervision and French labor, found no imitators and later gave up his own yard. The French output of merchant ships was never very large: in the 1860s, any one of the large Thames shipbuilding establishments, such as the Thames Iron Works or the Millwall Iron Works, had a capacity as large as all the French yards combined.[6]

Canadian shipbuilding, like the American, was based on cheap supplies of softwood timber. Its decline at the end of the 1850s was reversed by a period of prosperity during the American Civil War. After a further peak in 1874 of building cheap "bluenose" sailers, it collapsed finally in 1885 with the iron ship.[7]

The Baltic, with its local timber, tar, and other naval stores, was the only other important shipbuilding area during the middle years of the century. The Swedes, enjoying the benefit of their own cheap iron for many fittings, the Finns and Russians with their large supply of Russian pitch and English iron imported duty free, and the Germans, could all build ships more cheaply, if not better, than any British yard. The Germans had to import most iron work such as anchors, chains, and capstans, from Britain, but they could use local supplies for the rigging, and they could build fully equipped and rigged sailers at competitive prices.[8]

The introduction of the iron steamer, however, spelled doom for the timber shipbuilding industry everywhere. Not until the turn of the century did British shipbuilders again have to face serious competition from abroad, from the two new industrial giants, Germany and the United States.

Steam Power and Metal Hulls

The two most important changes in shipping in the nineteenth century were the increases in vessel size and power. Both changes were based on technological improvements in the engineering and metallurgical sectors, in which Britain had been in the vanguard for generations. And both changes, which led to gigantic alterations in the complexity and scale of ship construction, forced the transformation of shipbuilding from an industry of small handicraft firms into one of large, highly capitalized concerns employing laborers and semiskilled machine workers as well as craftsmen, many of whom handled complex and power-driven tools.

By 1860 the difficulties of continued timber shipbuilding had become serious enough to threaten the very existence of the British shipbuilding industry. Clearly, sufficient oak to supply the navy and the merchant marine could no longer be grown in Britain: in the 1860s it was calculated that 400,000 acres of timber were needed to build the annual British output.[9] The supply from abroad also gave cause for anxiety. A good deal of British capital had been invested in Canadian timber, which had become competitive after the Baltic duties of 1809. Vested interests in Canadian timber were sufficiently strong to prevent the repeal of timber duties from 1815 to Gladstone's budget of 1860, forcing British builders to pay exorbitant prices for Baltic timber, which they preferred. The Admiralty bought foreign timber with the intention of breaking the monopoly of James Morris, who had cornered the entire British oak supply by 1810. In 1861-1862 the emergency plantings of oak of 1810 were not yet mature, and there were considerable fears at the Admiralty about the future timber supply, especially for sternposts.

Timber is of varying quality. To obviate delays in building, the best was not always used. Pieces of the proper curvature and strength had to be selected from the timber yard in a way that would allow the remainder to be useful for secondary items. The heavy stocks of wood that had to be carried, for seasoning and for selection of the various curves needed for the hull, became crippling for shipyards competing with iron.[10] Not only might one-quarter to one-half of the weight be lost in planing, whereas with iron only a minute fraction was lost in this way, but half of the weight of timber in a wooden ship was wasted, its only use being to hold the other half in position. Even so, a wooden ship had great stresses as a structure. The absolute limit of its length was 300 feet, and it was liable to "hogging" and "sagging" in addition to being unable to withstand the local strain of the screw propeller.[11]

These problems were overcome by the introduction of iron frames and plating. Stocks of materials could be reduced because of the ease of bending the metal to the desired shapes. Furthermore, the ends of iron plates, unlike wood, could be overlapped and riveted to achieve a water-tight structure of great length. The greater strength of iron made it possible to decrease both the weight of the vessel and the thickness of its hull, allowing increases of up to 35 percent in deadweight cargo capacity and of from 20 to 50 percent in hold space in proportion to the exterior dimensions. Because of prejudice and because of iron's higher cost, however, shipowners were slow in adopting iron ships. As early as 1787 John Wilkinson launched an iron barge, and the first iron steamship, the *Aaron Manby*, was built in 1821. Not until around mid-century, though, did iron become the dominant material. Charles Harley has demonstrated that this continued use of wooden hulls for sailing vessels up to 1890 was based not on ignorance or prejudice but on differential rates of technological change and on costs for wood and iron respectively. Although the cost of iron fell faster than that of wood, wooden shipbuilding in the United States and British North America continued because of the immobility of labor and other factors of production in Maine and the Maritime Provinces. As builders in those areas had few alternative opportunities, they were willing for decades to reduce their prices to competitive levels in order to remain in business.[12]

The introduction of steel for oceangoing merchant vessels in the late 1870s brought further advances in hull construction. Steel, which allowed an additional 15 percent reduction in hull weights, was adopted very quickly as its cost decreased. By 1890 virtually all ships built on the Clyde were of steel.[13]

Increases in power were based on the introduction of the steam engine. The first commercially successful steamship, Robert Fulton's *Clermont*, was launched as early as 1807, but the substitution of steam for sail proceeded slowly. Not until 1838 was the first crossing of the Atlantic under continuous steam made, and not until 1870 did the annual steam merchant tonnage constructed in the United Kingdom surpass that of sailing ships. Into the 1880s, sailing vessels remained supreme on long voyages such as those to the Far East, on which practically all cargo space on steamers would have been eliminated because of the vast amounts of coal that would have had to be bunkered.[14]

Under favorable wind conditions, sailing vessels of the mid-nineteenth century could still travel faster than steamers. Furthermore, most of the increases in shipping productivity between 1815 and 1885 took place on routes where sail predominated. As late as 1875-1890, freight rates for sailing vessels declined sharply on voyages from Britain to California. This decline was made possible by decreases in the cost of vessels and by higher productivity resulting from improved designs and reduced manning requirements. Dependence on unreliable winds meant, however, that sailing vessels could not keep to prearranged schedules and that on average a sailing ton could carry less than a steam ton over any given period. Thus the strain placed on British shipping in the 1850s by the discovery of gold in California and Australia and by the Crimean War led to a shift from sail to the faster and more reliable screw steamers on the Baltic and North Atlantic routes.[15]

The opening of the Suez Canal in 1869 greatly shortened the route from Europe to the East, but it was above all decreases in coal consumption that made steam vessels competitive with sail, even on the long voyages to the East. Early steamships used simple engines working at a pressure of 6-7 pounds per square inch and consuming up to 10 pounds of coal per indicated horsepower (I.H.P.) per hour. Although by 1852 steam pressures had been increased to 15-20 pounds and coal consumption reduced to around 5 pounds per I.H.P. per hour, the first major step toward reducing fuel consumption to an economical level was the perfection of the compound engine by John Elder and his partner Charles Randolph. By raising pressures to around 30 pounds and using two cylinders—the first operating at high pressure and a second, larger one using the expanded low-pressure steam that had not been spent in the operation of the first—Elder and Randolph were able by 1859 to reduce coal consumption to under 2.5 pounds per I.H.P. per hour. Improvements in boilers and the development of carbon steel eventually allowed pressures in compound engines to be increased to 90-100 pounds with additional increases in economy, but as late as 1871 only four major shipping firms, notably Alfred Holt and Company of Liverpool, had adopted them. Finally in the 1880s, reductions in the price of high-quality steel and the development of the triple-expansion engine reduced coal consumption to the point where long voyages such as those to the Far East or to California became profitable enough to eliminate most competition from sailing ships. By 1914, after the introduction of the quadruple-expansion engine, boiler pressures had been raised to over 200 pounds and coal consumption reduced to around 1.25 pounds per I.H.P. per hour.[16]

The most original technical improvement following the development of the compound engine was the marine turbine of Sir Charles Parsons. In its simplest form, a turbine consists of a shaft connected to a drum furnished with projecting blades. The drum is enclosed in a cylinder fitted with fixed projecting guidevanes. As steam is admitted into the cylinder, it strikes and expands against the blades, causing the shaft to rotate. Early turbines were particularly effective at higher speeds, allowing savings in fuel of up to 30 percent at speeds over 20 knots. By 1914 two important further modifications had been introduced. The first was the connection of low-pressure turbines to quadruple-expansion reciprocating engines. The exhaust from steam that had first passed through the reciprocating engines was applied to low-pressure turbines to generate still more horsepower. The second modification was the development of gearing that allowed turbines to be used efficiently in low-speed cargo vessels.[17]

As a result of their efficiency, turbines rapidly came into use in high-speed vessels. Although the first nonexperimental turbine ship, the British destroyer *Viper*, was not completed until 1900, some 378 ships, mostly war vessels, aggregating 5,841,000 horsepower, had been completed or were under construction by 1910. In contrast, the oil-powered diesel internal combustion engine, developed at about the same time, was adopted much more slowly, especially in Britain.[18]

Although standardized tramp steamers were more suitable for mechanized mass production, the ever-increasing size and speed requirements of naval vessels and passenger liners probably contributed most to the evolution of shipping and ship-building practices. There were several justifications for increasing the size of vessels. In the case of merchant ships, increasing size usually allowed for greater earning capacity, because the usable areas of a ship grew at a greater ratio than cost factors, especially labor costs. Naval vessels also grew as thicker armor and heavier guns were introduced. Because of their ability to carry greater firepower per displacement ton, large battleships were considered to be cheaper per gun carried, although the cost of each ship was much higher.[19] The huge and heavy hull components of these naval and passenger vessels, more than anything else, forced yard owners to expand their yards and adopt cranes, railways, and other expensive equipment for haulage.

If size was a source of economy, speed was a source of cost, because more than proportional increases in horsepower and provisions for bunkering were necessary for each additional knot attained. As higher speeds could be economically justified only in naval and passenger vessels, new models of engines were first incorporated in these types of ships. Steam turbines, for example, were used first in a naval destroyer in 1900 and in a passenger vessel two years later. Not until 1910, after reduction gearing to slow the revolutions of the shafting had been introduced, were turbines first employed in a cargo ship.[20]

Passenger liners were also responsible for the addition of a wide variety of comfort and safety features. Early passenger steamers had rudimentary cabins and cramped public rooms. Conditions changed after

1870 as lines began to use luxury and speed to lure customers and promote prestige. An entire school of interior designers who specialized in ocean liner decoration arose, and the rooms in the first-class sections of a ship were patterned after the most lavish accommodations available on shore.[21]

By 1859, the paddle-wheeler *Great Eastern* had demonstrated the technical, if not the economic, feasibility of constructing large iron ships. Designed by Isambard K. Brunel and built at Millwall on the Thames by John Scott Russell, she measured 680 feet between the perpendiculars and had a capacity of 18,914 g.t. By comparison, the first iron Cunarder, the *Persia* of 1855, measured only 376 feet and 3,300 g.t. Originally intended for the long run from Britain to Australia, the *Great Eastern* was built on the principle that a gigantic vessel would be able to bunker enough coal to make the voyage profitable. The coal consumption had been seriously underestimated, however, and the ship never traveled to the East. After several unrewarding voyages across the Atlantic, she was used to lay submarine cable and ended as a sort of floating billboard and amusement hall before being sold to wreckers on the Mersey in 1887.[22]

Although forty years were to pass before another ship as large as the *Great Eastern* was built, the trend was toward larger and more powerful vessels in all oceangoing classes. By 1875 steamers in the range of 3,000 to 5,000 tons had already proved profitable. Among tramp steamers, there were few if any oceangoing paddle ships. After mid-century the transition was directly from sailing ships to steamers with reciprocating engines driving screw propellers. One tramp built around 1870 measured 1,192 g.t., and as late as 1892 the first "turret" tramp had a capacity of only 1,890 g.t. From 1899, however, the "representative" cargo steamer of the shipping journal *Fairplay* had a capacity of 7,500 g.t. A typical cargo vessel built in 1912 was said to be 470 feet in length with a beam of 54 feet, a depth of hold of 31 feet, and a capacity of 7,760 tons gross or 4,870 tons net.[23]

Liners changed considerably both in size and in methods of construction and propulsion. The first passenger steamer built for the Atlantic trade, Brunel's *Great Western* of 1838, had a wooden hull and was propelled by paddles and sails. She measured 1,321 g.t. and was 212 feet long with simple steam engines that developed 750 I.H.P. The White Star Line's *Adriatic* (1872) had an iron hull and was propelled by screw and sails. She was 420 feet in length, had a capacity of 3,883 g.t., and was powered by compound inverted engines developing 3,000 I.H.P. The *City of Paris* (1888) was built of steel and propelled by twin screws. Her hull measured 10,500 g.t. and was 525 feet in length. 18,000 I.H.P. were developed by two sets of triple-expansion inverted engines. The 31,938 g.t. *Mauretania* of 1907 was powered by steam turbines, which delivered

70,000 I.H.P. to her quadruple screws. Lastly, the *Aquitania*, the largest British liner built before World War I,[24] was 901 feet in length and measured 47,000 g.t. Her turbines developed 60,000 I.H.P. Throughout the period, speed increased as well. The *Great Western* averaged 8.5 knots; the *Adriatic*, 14.75 knots; the *City of Paris*, 19 knots; and the *Mauretania*, the fastest of the prewar Atlantic liners, 25 knots.

The size of battleships and battle cruisers also increased rapidly.[25] The last of the pre-Dreadnoughts weighed 16,500 displacement tons and had a designed speed of 18.5 knots. Ships of the Royal Sovereign Class of 1913-1914, completed in 1916-1917, displaced 25,750 tons and had a speed of 21 knots. Battle cruisers, which had lighter armor and carried fewer and smaller guns, increased from 17,250 displacement tons and 25.5 knots in the Invincible Class to the 28,500 displacement tons and 28 knots of the *Tiger* of 1914.

Specialized Ships

In the last quarter of the nineteenth century the differentiation of ships and their adaption to specialized trades reached a high level. In the days of sail there were few differences between the vessels in different trades at any time. At the beginning of the century, for example, the large, clumsy, and heavily armed East Indiamen and the faster and more maneuverable West Indian sailers practically exhausted the categories of seagoing vessels under the British flag. By 1900 specialization had gone very far, and vessels could be grouped in at least three different ways: first, according to their employment, into liners and tramps; secondly, according to special use for which they were adapted, into mail and passenger liners, tankers, bulk cargo carriers, and so on; and thirdly, according to their design, into well deckers, trunk steamers, turret vessels, and many other types. As the construction of these different varieties of vessels required different kinds of labor skills and capital equipment, increased specialization of shipping heralded a trend toward specialized shipbuilding as well.

The distinction between liners and tramps was fundamentally not one of the vessel itself, but of its use. "A liner service implies to-day a fleet of ships, under common ownership or management, which provide for a fixed service, at regular intervals, between named ports, and offer themselves as common carriers of any goods or passengers requiring shipment between those ports and ready for transit by their sailing date."[26] A tramp, on the other hand, has been defined as "a vessel employed by her owners in seeking freight wherever it may be found."[27] Tramps would not only sail to any port required by their cargo, but could also be hired on "time charters" of three, six, or twelve months at the complete disposal of the shippers, or could on occasions be placed "on the berth," undertaking to sail for a definite port on a fixed date, like a liner.

The evolution of the ocean "greyhounds," express passenger steamers traveling at much higher than economical speeds, has already been discussed. In addition there grew up a class of "intermediate" liners, "of large dimensions and tonnage, but moderate power and speed, and having seagoing qualities and features of safety and comfort not a whit less remarkable than the latest type of ocean steamer. Large cargo-carrying power and more leisurely and comfortable pace on the ocean are features which distinguish these vessels."[28] The intermediate steamers tried to gain in popularity what they lost by lower speed. They began to cater to the professional and lower-middle classes, raising the standard of second-class cabins and lowering their cost, and reserving for them an increasing proportion of their space. Competition was great, for the transatlantic passenger "rings" made agreements and allotted quotas only on steerage passengers, while rivalry in cabin accommodations continued.[29]

The first deliberate intermediate design, the *Pennsylvania* class, was adopted in 1895 by the Hamburg-Amerika Linie. This was quickly followed by the White Star Line's *Oceanic* of 25,000 g.t. and 21 knots, and the sister ships *Celtic, Cedric, Baltic,* and *Adriatic* of 21,000-25,000 g.t. and 17 knots, built between 1900 and 1906, and by the Cunarders *Caronia* and *Saxonia* of 15-16 knots, built in 1899. Harland and Wolff were the chief builders of these giant intermediate steamers, which all British lines had adopted at the end of the century, turning their backs on the costly speeds achieved by German and other express liners. This obedience to the dictates of economy was welcomed in the columns of the *Economist* but was bitterly resented by the Navy League and other chauvinists who were counting on potential cruisers. The potential savings of lowered speeds were skillfully exploited by shipowners when asking for mail subsidies.[30] When building for profit, the White Star Line designed its *Teutonic* and *Majestic* for 21 knots, 2 knots below contemporary German liners, because the gain of 12 hours on the Atlantic crossing would have raised the necessary power by 30 percent, raised the coal bill by 20 percent, and lowered earning capacity by 40 percent.[31]

From the 1870s onward, specialized liners were also built to carry freight and passengers cheaply and at low speeds. Among these were the Cunarders *Bothnia* and *Scythia* of 1874 and the Inman Line's *Kensington* and *Southwark.* Later, "cabin class" steamers were designed to carry cabin and steerage class passengers only. In the 1880s and 1890s the major lines also built smaller up-to-date ships for special routes, such as to Boston or Canada, instead of pressing into service the obsolete vessels from the main New York route. In the same period the first large liners especially for carrying cargo were built by firms such as the White Star Line and the Atlantic Transport Line, an American company.[32]

Although smaller and less advanced technologically than liners, tramps were a larger segment of the British merchant fleet and of shipbuilding output in the decades before 1914. The Royal Commission on Shipping Rings estimated in 1909 that liners made up one-third of total British tonnage and tramps between 50 and 80 percent. Walter Runciman, Jr., calculated in 1903 that British trade was carried by 1,300 liners, 7,000 steam tramps, and 7,000 sailing vessels. Five years later Sir Norman Hill, secretary of the Liverpool Steam Ship Owners' Association, estimated that the British merchant fleet included 1,200 liners of 4,200,000 net tons (n.t.) and 2,400 tramps of 5,760,000 n.t.[33]

Before 1913 few tramps were more than 450 feet in length or carried more than 8,000 deadweight tons. They were generally full-ended single-screw vessels (unless of a special build) designed to have a low first cost and economy in working, with an I.H.P. or 1/4 or 1/5 per ton of weight driven. By contrast, intermediate liners had 1/2 to 3/4 I.H.P. per ton driven, while the fast turbine-driven "greyhounds" were fitted with 1 to 3 I.H.P. per ton and had little cargo space.[34] These costly liners, like the specialized cargo ships, could be built to order only, and their orders were confined to shipyards that had had previous experience with them. Speculative building in depressions, therefore, was inevitably limited to the general, low-powered, medium-sized cargo tramps with no special fittings, which were turned out on a large scale on the North East Coast.

Among ordinary cargo carriers—cargo liners or tramps—a bewildering variety of types evolved in the last quarter of the nineteenth century. With the different needs of special trades and the desire to keep register tonnage low and to minimize running costs, the solutions suggested were very diverse. The forerunners of the cargo carriers of the 1870s and 1880s were the sailing vessels of the previous decades. Over the flush deck, offering no obstruction to the working of the rigging, a bridge and deckhouse were built to protect the engine space, a forecastle to protect the deck, and a poop to house the anchor, forming the "three island steamers." These superstructures were flimsy and not part of the structure. As screw steamers were built for heavy bulk cargoes such as grain or ore, the propeller shaft casing reduced the cargo space aft, causing the vessel to trim by the head. To preserve an even trim, a raised quarter deck was added.[35]

For heavy bulk cargoes, single deckers were preferred because of their easy loading, but the difficulties of the trim caused the engine room to be moved forward. This proved satisfactory for the laden ship, but when sailing in ballast the propeller was too lightly immersed. A longer poop was then tried, to be filled with lighter cargo, but it naturally raised the tonnage dues. The addition of a second deck caused the bulk cargo to shift and made the vessel unstable.[36]

Meanwhile the raised quarterdeck had become popular on the North East Coast by about 1865. It was safe when loaded with bulk cargoes, there was no intermediate deck to allow the cargo to shift, and because the fore and aft holds were of equal capacity, the vessel had her stern well down when in ballast. As the bridge was brought well forward, there was a space left between the bridge and the forecastle, which the owner could not normally have loaded because of the trim, and which was therefore left open to reduce the register tonnage. Over the space—the "well"—the sea swept unhindered and washed out by the scuppers. Accordingly, these well deckers were viewed with suspicion by the Register Societies and by the Board of Trade. Many of the vessels were detained for unseaworthiness, and the builders were forced to narrow the well by bringing the bridge farther forward, a move that raised the tonnage without giving them any useful covered spaces, except when loading a very light cargo. But the Hartlepool owners and builders, who had evolved the well decker, formed their own underwriting society and organized a campaign to influence the Load Line Committee in their favor. The safety of these vessels was finally established, and between 1885 and 1888 no other types of vessels were launched in West Hartlepool. By the latter date, 350 well deckers, over half a million tons, had been built in West Hartlepool alone.[37]

"Shelter deck" vessels came into use in the mid-1880s and became very popular in the 1890s. They had a continuous, lightly constructed upper deck with openings aft, which was not considered fit to carry cargo and was not included in the tonnage. Their low register tonnage was one of the chief attractions of the shelter deckers. When the vessels put to sea, however, they frequently closed the openings of the shelter deck, filled it with bunker coal, and filled the bunker with cargo.[38]

For the transport of eastern pilgrims and coolies on deck, the openings of the shelter-deck vessels (or shade deckers) were closed, but the material was flimsy enough to exclude the enclosed space from tonnage measurement. This type was known as the "awning deck" vessel. Both types had their load lines fixed early by Lloyd's. For the carrying of ordinary passengers, the upper deck was strengthened still further, and had to be included in the tonnage, to form the "spar decked" vessel. In the late 1870s, vessels for that purpose had been built as three deckers, that is, the top deck was of the same scantlings as the rest of the hull, but they were too deep with too narrow a beam, as breadth was thought to add to the resistance. Many were lost as a result. The spar deckers were lighter on top, and were therefore forced to allow a higher freeboard, which made them safer. They took the place of the three deckers in the 1880s and 1890s.[39]

Ships were also designed specifically for special cargoes. Oil was one of the first commodities that required specially constructed vessels. In 1861 the first cargo of oil in barrels was sent to London in the brig *Elizabeth Watts*, and in 1863 Gibson of the Isle of Man built the *Ramsay*, an iron sailing vessel, to carry oil in tanks from America. The Petroleum Carrying Company was formed in the same year to carry oil to Europe in specially designed tank vessels. Two iron sailers were built for the firm at St. Peters, Newcastle. These vessels were by no means tankers, however; they had separate containers fixed inside their holds to take the oil. In the following 22 years further developments in the carriage of oil across the Atlantic were negligible. The *Charles* was said to have carried oil successfully to Europe in iron tanks from 1869 to 1872, and three steamers were built by Palmer's to carry oil as well as passengers. Difficulties created by the American and Belgian governments and by the Register Societies led to the reconversion of most of these specially adapted vessels to general cargo carriers. Oil continued to be sent across the Atlantic in barrels in sailing vessels, although if loaded in that way a ton of oil filled 80 cubic feet and about half the deadweight carrying capacity was wasted. In 1886 more than a thousand ships were engaged in the trade, exporting nearly two million tons of oil annually from the United States. Together with the makers of wooden barrels and tin cans, and with the conservative Register Societies, the vested interests clinging to that method appeared altogether too tenacious for rapid change.[40] Yet in the mid-1880s the modern tanker was designed and accepted almost overnight to meet the threat of competition from the recently discovered oil fields in the Caucasus.

In 1885-1886 no fewer than 11 vessels were converted or built for carrying oil in bulk. Two of these, the *Gluckauf* and *Vorwarts*, built by Armstrong's for a German owner, incorporated most of the features of the modern tanker. Carrying oil to their skin, they had expansion trunks fitted. Their holds were divided by a longitudinal bulkhead and protected by coffer-dams (ballast tanks) at the extremities. Also, the double bottom was omitted, and the oil tank was filled with water if in ballast. Their design clearly made them unfit for any other trade, and in 1885 the *Gluckauf* was classified as a "petroleum steamer" by the Bureau Veritas.[41]

By 1889, 53 oil tankers of 191,000 deadweight tons (d.w.t.) capacity had been specially built, apart from those converted. In 1893, 47 oil bulk carriers had been classified by Lloyd's Register. In 1892, as the British interest in Middle East oil grew, tankers were allowed through the Suez Canal. Five years later, Lloyd's laid down special rules for tanker construction. Altogether, there were 193 seagoing tankers in existence in 1900 with a deadweight capacity of 637,000 tons, and 441 in 1914, carrying 2,344,000 tons.[42]

In 1875 the first consignment of chilled meat arrived in England from the United States, and by 1880 imports totaled 120,000 tons. The first refrigerated meat to be sent through the tropics arrived from Buenos Aires at Rouen in 1877 in the *Frigorifique*, in a refrigerating plant designed by the Frenchman Charles Tellier. A second French vessel, the *Paraguay*, was similarly fitted in 1878, but Australia and New Zealand adopted the new invention most eagerly. The *Strathleven* arrived in 1880 with her first Australian cargo, kept frozen by a Bell-Coleman plant, and the *Dunedin* carried the first cargo of frozen meat from New Zealand in 1882. Before long, the Orient, P. and O., Houlder, Shaw Savill and Albion, and Tyser lines entered the trade of frozen mutton and beef from Australia, as did the Australian Land Company, the New Zealand Shipping Company, and the Shire Line from New Zealand. In 1883 the first load of frozen meat was shipped from Argentina, and in 1899 the first chilled meat cargo arrived from the Plate.[43] Vessels with refrigeration equipment in their holds became uneconomic for any but the cargo for which they were designed. Lloyd's register adopted rules for the inspection of refrigerating machinery in 1898, and 108 vessels with an aggregate capacity of 10 million carcasses had been certified by 1906.[44]

The first load of bananas preserved with the help of refrigeration arrived from Jamaica in 1896. The following year, Joseph Chamberlain introduced subsidies for the Elder-Dempster Line to ensure a regular transport of bananas from the West Indies in specially designed vessels, a service which began in 1901. In 1897 the Union liner *Scot* brought the first large cargo of fresh fruit from South Africa. The first large cargo of fresh dairy produce from Australia arrived in 1911.[45]

With the building of specialized vessels, and the dredgers, trawlers, cross-channel steamers, ice breakers, and others, the general undifferentiated vessel became much less common after 1880. The advantages of specialized building for a certain task were so great that the risk could be taken of making the ship unfit for any other use. Many of these special vessels were built in one or a small number of yards only, which concentrated on their construction.

WITH THE exception of the turbine engine, most of the important innovations covered in this chapter were well-established by 1880. The compound engine and steel hull allowed increasing size and speed, and the initial breakthrough in fuel consumption made large-scale, long distance steam voyages economically feasible. By comparison, later improvements such as triple- and quadruple-expansion engines were individually minor in both the technical and economic senses. Even the development of specialized ships often merely involved variations on recognized themes of ship design and construction. Taken together, however, these innovations led to the evolution of larger and faster ships of far greater average carrying capacity than earlier wooden sailing ships.

If these increases in the average size and speed of oceangoing ships had come in an era of stagnant or slowly growing world trade, the number of vessels built would have declined and the shipbuilding industry would probably have been reduced to a few large firms. The very rapid increase in world trade between 1870 and 1914 and Britain's comparative advantage in shipbuilding, however, allowed the British to increase their output dramatically. By 1913, despite increased foreign competition, there were more than 50 British yards with an annual output in excess of 10,000 g.t. apiece.

2.

Fluctuations in Shipbuilding Output

The Pattern of Output

BETWEEN 1860 and 1913 the value of world commerce increased by more than 400 percent, and between 1870 and 1910 British foreign trade doubled in value and more than doubled in bulk. Furthermore, almost half of the world's shipping capacity was British. In July 1914, 42 percent of the world's steam tonnage was on the register of the United Kingdom; if ships registered in the overseas dominions and colonies were included, the proportion rose to 45 percent of the world total. These ships not only carried 71 percent of the total seaborne trade of the empire, internal or with other countries, but 30 percent of the trade between foreign countries outside the empire. In all, about 52 percent by value and 50 percent by volume of the seaborne trade of the world was transported in British bottoms.[1]

The growth in world trade and the high proportion of tonnage that was British-owned made it possible for the British shipbuilding industry to expand rapidly after 1870. A number of series giving output over the period are available, but because they are expressed in different tonnage units they are not readily comparable.[2] Nevertheless, they give a good indication of the growth of the industry. Between 1860 and 1864 the average annual output was 307,054 net tons (n.t.). By 1870-1874 the figure had risen to 463,509 n.t., and by 1910-1913, to 1,037,094 n.t. The composition of output also changed. Between 1860 and 1864 an annual average of 196,267 n.t. of sailing ships and 110,787 of steamships was built. In 1870-1874 the average net tonnage of steamships had increased to 358,669. By 1910-1913, only 37,373 n.t. of sailing vessels were built in an average year, but the average output of steamships had increased to 999,721 n.t. From the early 1900s, approximately 20 percent of British output was for foreign buyers.

Their strong international position, rapidly growing output, and ability to specialize in particular classes of vessels all encouraged British shipbuilders to enlarge their yards and workforces and to invest in capital equipment. The highly cyclical nature of the industry, however, weakened them financially and tempered their enthusiasm for expansion. The amount of tonnage produced fluctuated tremendously over short periods (figure 1) and facilities built during a boom were often underemployed for five or more years. In 1884, 40 percent less tonnage was launched in Britain than in the previous year. By 1886, launchings had fallen by more than 60 percent, and not until 1889 was the level of 1883 again reached. On the Clyde, the amount of tonnage launched in 1883 was not surpassed until 1896.[3] The pattern repeated itself down to 1914, with each rapid expansion being followed by an almost equally sharp contraction.

A number of economists have studied shipbuilding cycles to establish the relationship between freight rates, the price of new ships, and shipbuilding output.[4] The theoretical aspects are beyond the scope of this book, but an understanding of their empirical findings is central to any meaningful account of the shipbuilding industry. They have found, not surprisingly, that freight rates were based on the demand for shipping services relative to the existing pool of tonnage. As new tonnage was only a small portion of the total amount available, freight rates affected shipbuilding output much more strongly than the converse, although

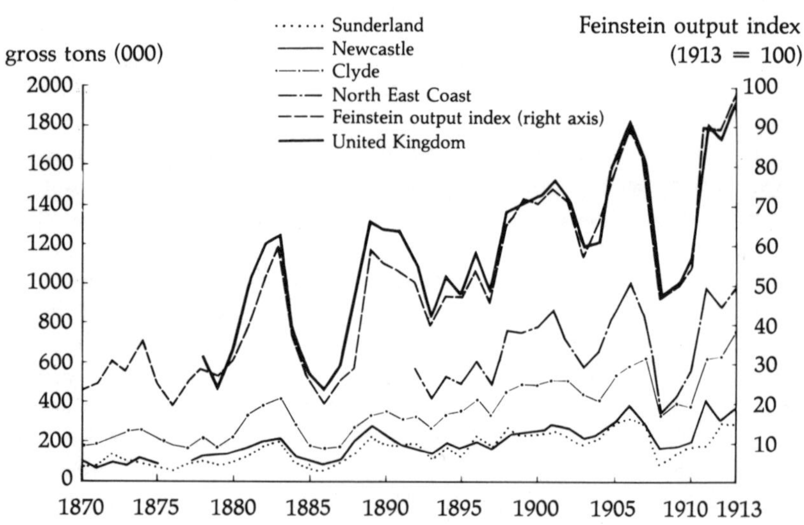

Figure 1. Shipbuilding output, 1870-1913. (Appendix table B.9; Feinstein, *National Income, Expenditure and Output of the United Kingdon*, table 52, pp. T114-115.)

new tonnage did help to drive down rates at the peak of the cycle of freights. Secondly, there were two distinct groups of purchasers: (1) large lines that had enough capital in reserve to be able to buy when ships were cheaper near the bottom of a depression; and (2) a more numerous group of small owners who could afford to buy new tonnage only after they had begun to collect high profits from rising freights. As a result the shipbuilding cycle had two phases, a small increase in output when large lines purchased near the trough of the cycle, and a much bigger increase later, after freights had begun to recover. Finally, because most new shipping was purchased by the small owners, the price of ships was influenced much more strongly by conditions on the demand side—freight rates—than by those on the supply side—the cost of labor and shipbuilding materials.

As described by the economists who have studied the phenomenon, a typical shipbuilding cycle might proceed as follows:

1. As freights begin to rise, old vessels that have been laid up during the preceding slump are reactivated and, after a lag of a quarter or two, orders for new vessels are placed in anticipation of further improvements in demand.

2. If freights continue to rise, more orders are placed. Because of the time necessary to construct a ship, however, freights have been rising for some time before a substantial amount of new tonnage is ready for service.[5]

3. When new tonnage begins to be placed in operation, increases in freights slow down, but, because a large proportion of buyers can afford to purchase a ship only after they have already profited from high rates, some new orders continue to be placed.

4. Freights start to decline, because large amounts of new tonnage usually become available just as the demand for shipping reaches a peak. Older and less profitable ships begin to be laid up.

5. After freight rates and the price of new ships have been falling for an appreciable time, perhaps several quarters, large lines place orders, creating a small temporary increase in shipbuilding output. When these ships are delivered, however, freights fall even further because the demand for shipping has not increased proportionally.

6. When enough ships have been laid up and the demand for shipping services increases, freight rates rise and the cycle begins again.

Shipbuilders were in a precarious position because in order for output to be stabilized it was necessary for the demand for shipping to increase at a constant rate. A decline in the rate of increase would lead to a decrease in shipbuilding output even if the absolute demand for shipping services was still growing.[6] Moreover, because the price of new ships was influenced most strongly by factors on the demand side, shipbuilders could go through periods of "profitless prosperity." This varied considerably with the type of work undertaken by the firm. The output of builders who specialized in liners was relatively stable, but builders of tramps were severely affected by fluctuations in orders. As the peak of the cycle of freights approached, tramp owners, most of whom could not have afforded to purchase earlier and were inferior credit risks in any event, recognized that conditions would soon worsen and were willing to pay more for tonnage that could be delivered immediately than for ships that would not be completed for some months.[7] At this point, therefore, perhaps two or three quarters before the peak in freight rates and even longer before the peak in the shipbuilding cycle, orders for new cargo vessels (which would probably be delivered after freights had begun to decline) could be obtained only at lower prices than had prevailed earlier in the cycle. Much if not all of the construction of these ships, however, would take place before the peak in the shipbuilding cycle, when the costs of steel and labor would still be rising because of increased demand created by orders that had been placed previously. Thus, much of the tonnage was built under conditions of falling prices and rising costs. Near the peak of the shipbuilding cycle, profit margins were considerably reduced and losses might even be incurred. This created a paradoxical situation in which the best years in terms of tonnage constructed were among the worst for shipbuilders' profits.[8] Builders could ill afford not to bid during booms since a large output might bring substantial cumulative profits even if the margin per ton were small. Because the industry was highly competitive in this tramp cargo class, which constituted the majority of its output, builders were forced to gamble on the course of input costs and to accept the prices offered, thereby risking losses.

Even in good periods the output of individual firms experienced large variations in which the tonnage launched was as much as doubled or halved in a single year. Admittedly, tonnage launched is a somewhat misleading measure, because a firm might be active in constructing the frame of a large ship or in fitting her out for a year before or after the launching, but it indicates fairly accurately the strain placed on shipbuilders by fluctuations in production, as different shops, machines, and skilled laborers were needed for each stage in the construction of a vessel.[9]

The severity of fluctuations in output was of great importance in matters of mechanization, labor force organization, industrial organization, and other aspects of shipbuilding, which are discussed in later chapters. The principal task of the shipbuilders was to minimize total overhead expenses while maintaining the ability to build large and complex ships at prices that were competitive on the world market. The highly cyclical nature of the industry amounted to a special form of limitation on the extent of the market, which discouraged the division of labor and mechanization by placing a premium on factors of production that were versatile (unspecialized) or could easily be dispensed with. The experience of German and especially American shipbuilders had demonstrated the dangers of heavy capital expenditures. Despite their more modern equipment, German builders were unable to meet the prices of their British competitors, largely because even with better machinery German workers were less productive than British workers. In the United States, vast overheads crippled builders in all but the best years. British yard owners were able to take advantage of their more highly skilled workforces by investing only in equipment that was absolutely necessary to manipulate the large, heavy, and hard components of modern ships, and by refusing to purchase as many labor-saving machines as German and American builders did. Thus the conservatism of the British was motivated not by a shorter time horizon or an irrational distrust of innovation, but by an awareness (which the less experienced German and American builders did not share) of the hazards of the industry.

The cyclical nature of the industry also created problems in labor management. Wage rates tended to vary with output, rising sharply in good periods and stabilizing or falling during depressions. The number of changes was great as the peak of the shipbuilding cycle neared, with as many as three rate increases in a single year. This added to the difficulty of controlling costs after contracts for vessels had been signed, but builders were reluctant to refuse worker demands because their losses would increase if their yards were closed by strikes.

The recruitment of labor was also complicated by shipbuilding cycles. During slack or ordinary periods it was easy to recruit an adequate workforce, but as cyclical peaks approached the competition for labor drove wages up and made it more difficult to find skilled workers. In depressions workers could be laid off, but builders did not like to let go any more than was necessary, as it might not be possible to rehire them when conditions improved.

One solution—to replace as many workers as possible by machines—would have only increased the problem of lowering overhead expenses during the slack periods. The difficulties were never satisfactorily resolved, although yard owners were able to reduce labor expenses by placing more emphasis on manual training and not providing as much costly theoretical education as German builders did. By such means, British shipbuilders lowered their costs and also helped to create the skilled labor force that allowed them to get by with less machinery and ultimately to reduce overheads.

Finally, many British yard owners attacked the problem by attempting to eliminate the fluctuations themselves. This was done through vertical integration with suppliers and shipping lines. Variations in the price of steel, the primary raw material, were a special problem.[10] As early as the 1850s, Palmer's of Jarrow began to produce its own steel, and by the beginning of the First World War a number of steel and armaments manufacturers had formed connections with shipbuilding firms, thereby assuring a more stable market for the steelmakers and steadier costs for yard owners.[11] Similarly, shipowners and builders found it advantageous to unite: owners could obtain vessels more cheaply during slack building periods, and shipbuilders could make more continuous use of their yard machinery and labor forces.

Shipbuilding in the British Economy

The unusually heavy fluctuations in output contributed to giving the shipbuilding industry a significance within the British economy greater than was warranted by its employment or output alone.[12]

It is difficult to calculate the value of shipbuilding output, because cost components did not change uniformly over the period. The increasing predominance of steamships with their expensive propulsion systems seems to indicate increases in the average cost per ton of shipping constructed between 1870 and 1914, especially because the trend toward higher speeds necessitated larger and more complex engines. Similarly, the increased use of shipboard appliances, safety devices, and, in the case of passenger vessels, improved accommodation, would have tended to raise the cost of ships. On the other hand, hulls of large ships cost less per ton than those of smaller vessels, so increases in average vessel size would have worked to lower the average cost per ton produced. In addition, exact information on naval vessels is difficult to obtain, thus distorting total output calculations. And finally, the figures, which are based on dates of launchings, do not necessarily indicate the amount of work done within any given year. Ships launched in the early or late months might well have been largely constructed in the preceding or following year.

The best estimate of the value of output is probably that from the Census of Production. According to this census, the value of new tonnage produced in British shipyards in 1907 was between £31.71 and £32.21 million, or between £28.20 and £28.70 if warships are excluded. Total output for the private shipbuilding, marine engineering, and repair sectors of the industry was £41,660,000, with a value added in the industry of £17,678,000, or £96 per person employed. In addition to this information, Karel Maywald has published an estimate of the average cost of newly constructed steamers per ton per year between 1850 and 1938. He gives figures for both hull costs and total costs per gross ton, but to use them effectively it is necessary to have an accurate ratio for converting the official output figures, which are expressed in net tons. Charles Feinstein suggests using a coefficient of 1.64, the average ratio of gross to net tonnage calculated from the data for steamers built from 1886 to 1913, when both measures are available.[13] Clearly, to apply a figure derived from such a long period will lead to inaccuracies in annual calculations: for 1913, for example, multiplying the official figure of 1,201,539 n.t. for steam output by 1.64 yields a gross tonnage of 2,020,350, whereas Lloyd's tabulation of both steam and sailing output for that year is only 1,932,153 g.t. To apply the ratio further back in time, when both engine room and bunkerage requirements were drastically different, would be even more dangerous and has not been attempted here. Finally, Maywald's estimates are based on the costs of inputs and, as he points out, tend to fluctuate less severely than would a series based on the actual prices at which the vessels were sold.[14]

Our calculations for the value of merchant ship output between 1886 and 1913 are given in table 2.1 along with a few independent estimates for the years before 1900. Official data on annual naval building expenditures from 1896 are also given. As the official data do not distinguish between work performed in the Royal Naval Dockyards and work contracted out to private firms, however, they cannot be used to arrive at annual estimates for the total value of output of private shipbuilders.

All of the series in table 2.1 show significant cyclical fluctuations. The values of merchant hulls and engines both more than quadrupled between the poor year of 1886 and the good year of 1913. Between 1889, also a good year, and 1913, the value of merchant hulls produced increased by 62.8 percent, that of merchant engines by 57.4 percent, and that of total merchant vessels by 60.5 percent. These increases in the value of output may be attributed to the growth in tonnage produced, increasing input prices, and the greater technical sophistication of ships. As we will argue later (chapter 9, part 2), however, because of increased productivity in shipbuilding, which was passed on to consumers in the form of lower prices, it is likely that the *real* cost per ton actually decreased over the period, especially when account is taken of technical improvements. This would have been particularly true after 1900. Fluctuations in the value of naval vessels produced depended primarily on the government's perceived need for naval tonnage. Although the shares attributable to private builders are indeterminate, it is known that they increased after the turn of the century and were sometimes a countercyclical cushion in periods of reduced merchant ship output.

Employment statistics, though grievously inadequate, are the most complete of the economic indices available. The decennial census statistics furnish some basic data. Although censuses from 1861 to 1911 enumerated shipbuilding as a separate category in the occupational statistics, the data are not of uniform value to the historian. The form of the census was determined in the era of timber and sail shipbuilding, and the detailed subdivisions of trades into rigging, sail-making, and similar occu-

Table 2.1. The value of British shipbuilding output, 1847-1913 (£000,000).

Year	(1) Value of merchant hulls	(2) Value of merchant engines	(3) Total value of merchant ships[a]	(4) Value of output	(5) Naval expenditures
1847	—	—	—	10.7	—
1879	—	—	—	4[b]	—
1880	—	—	—	6[b]	—
1881	—	—	—	8[b]	—
1882	—	—	—	9[b]	—
1883	—	—	—	19[b]	—
1886	6.7	2.1	8.9	9.5[b]	—
1887	7.6	3.2	10.8	—	—
1888	11.1	4.9	15.9	—	—
1889	16.4	6.8	23.3	—	—
1890	16.1	6.0	22.1	—	—
1891	14.5	4.7	19.2	—	—
1892	13.3	3.8	17.1	—	—
1893	9.6	3.1	12.7	—	—
1894	11.3	3.5	14.7	—	—
1895	10.5	3.2	13.7	—	—
1896	12.1	3.6	15.7	—	8.66
1897	10.7	3.1	13.8	—	6.76
1898	15.5	4.6	20.1	42	8.07
1899	18.4	5.2	23.6	—	9.30
1900	19.9	5.3	25.2	—	10.03
1901	19.0	6.0	25.1	52	10.34
1902	17.6	6.0	23.6	—	9.78
1903	13.8	5.3	19.1	—	12.40
1904	16.2	6.8	23.0	—	13.18
1905	19.8	8.8	28.6	—	11.37
1906	22.2	9.3	31.5	—	10.49
1907	20.9	8.2	29.1	41.66	8.85
1908	12.3	3.5	15.8	—	8.52
1909	12.4	4.3	16.7	—	11.08
1910	14.0	4.1	18.1	—	14.75
1911	17.4	7.9	25.4	—	15.15
1912	23.2	8.5	31.8	48	16.13
1913	26.7	10.7	37.4	52	16.88

Source: Columns 1-3 are based on the net tonnage figures given in table B.8, multiplied by cost estimates in Maywald, "The Construction Costs and Value of the British Merchant Fleet, 1850-1938," p. 50. Column 5 is derived from annual *Admiralty Returns*.

a. Because of rounding, the figures in columns 1 and 2 may not sum to those in column 3.

b. Excluding repair.

pations remained until 1891, when its importance was hardly more than formal. On the other hand, the occupations connected with iron and steel shipbuilding and marine engineering were not subdivided at all until 1901; and there is an overwhelming suspicion that other trades connected with fitting marine engines and auxiliary engines, foundry work, and other preparatory activities were not classified at all but rather included in the general engineering or metalworking sections.[15]

The census of 1901 was the first that showed a reasonably satisfactory modern classification. It was still an occupational census only, however; not until 1911 was an industrial census taken as well. In 1911 the numbers of the shipbuilding trades and the shipbuilding industry showed a startling difference: by adding persons employed in the shipbuilding industry but not in shipbuilding trades (plus one-half of an uncertain intermediate group), and subtracting persons in shipbuilding trades employed in other industries, the total employment figure was raised from 163,834 to 195,080, or by 19 percent.

The census figures will, therefore, have to be raised to correspond more closely to the definitions of shipbuilding, firstly, in order to bridge the gap between the shipbuilding "trades" and the "industry"; and secondly, in order to include the trades that were lumped together under general engineering and metal working until 1891. For the census of 1901, an addition of 19 percent—the same addition that rectified the 1911 census—seems reasonable. For earlier years, other considerations will have to be borne in mind. On the one hand, the nonshipbuilding trades in the yards—the draftsmen, clerks, machinemen, fitters, and so on—formed a smaller proportion of the shipyard labor force before 1901, and a smaller one still in earlier years; on the other hand, the range of trades left out was perhaps wider in 1871-1891 than in 1901.

No clear-cut decision can be reached on this point. In table 2.2, 20 percent has been added to the data for 1871-1891 on the assumption that both these influences canceled out over the period, and only 10 percent has been added to the 1861 figure, since fewer yards at that time had a wide range of ironworking occupations.

Apart from G. F. Young's bold, if unreliable, estimate of employment in the ship and repair yards in 1847 (table 2.2), few national statistics are available to interpolate the census data, because contemporary estimates were based on census data. The Census of Production of 1907 forms the only reliable independent estimate.

Our estimates of local employment before 1901, given in chapter 3, may supplement the census figures. They have been multiplied by a coefficient, representing the proportion of the local to the national employment enumerated in the nearest two censuses, a method that ignores the difference in the rates of local growth within each decade. Independent

Table 2.2. The economic weight of the shipbuilding industry.

Year	Employment	Aggregate wage bill[a] (£000)	Aggregate wage bill[b] (£000)	Total investment (£000,000)
1847[c]	(79,600)	(4,968)	—	(16)
1861	60,600	—	2,333	—
1865	71,800	—	3,123	—
1866	75,800	4,230	3,184	—
1871	76,400	4,263	3,411	—
1877	85,500	5,318	4,232	—
1881	90,400	5,492	5,379	—
1886	121,800	7,229	7,064	19
1891	118,000	7,723	8,083	—
1898	169,100	11,700	12,006	—
1901	171,760	12,135	13,655	34.4[d]
1907	184,557	—	14,672	150[e]
1911	195,080	—	17,070	73[f]

a. Based on extrapolations from 1906 average wage figure.

b. Based on average earnings at the Leven Shipyard of William Denny and Brothers (see table B.1).

c. C. F. Young's estimates for 1847 in his evidence to the Select Committee on Navigation Laws, 1847, qs. 6103-16, and 1848, q. 8337. These estimates are in parentheses because they are highly tentative.

d. Based on Schwarz and Halle, *Die Schiffbauindustrie*.

e. Crammond, *British Shipping Industry*, p. 37. This figure is unreliable.

f. Based on Mackinnon, *Social and Industrial History of Scotland*, p. 94.

figures are available for 1865 (North East Coast), 1866 and 1877 (Clyde), 1866 (Belfast), 1892 (North East Coast), 1894 (Clyde), and 1898 (North East Coast).

Despite the obvious imperfections of table 2.2, the steady increase during the period 1861-1911, when employment in shipbuilding more than trebled, is beyond doubt. If the figures are to be trusted, the chief periods of rapid increase were 1861-1865, the years after 1881, and the last years of the century. There was little increase in employment in the boom years of 1871-1873, when presumably the growth in the number of iron workers was counterbalanced by the falling off in that of ship's carpenters and other wood workers. The years of high activity 1888-1890 similarly show no increase in employment over the previous cycle. Apart from technical factors which improved the efficiency of labor, the boom itself was not as buoyant as previous ones.

There are no reliable estimates of the aggregate wage bill of the industry before the Board of Trade enquiry of 1907, which covered 117,000 employed persons. For these, the average wage in 1906 was £70.13.0. We have taken this crude average as the standard, adjusted it by the index number of earnings, and applied it to the years for which estimates of the number employed are available. It appears in table 2.4 together with an alternative estimate based on the average earnings of all workers in the Leven Shipyard of William Denny and Brothers. The alternative estimate assumes 50 working weeks per year. The large discrepancy between the wages bill based on the 1906 figures and that based on the earnings of Denny employees can be explained in large part by the lower relative wages that prevailed in Scotland (including the Dennys' home of Dumbarton) as compared to the rest of Great Britain, a differential that narrowed drastically as the nineteenth century progressed (see chapter 9, part 3).

The estimates for total investment in the shipbuilding industry (table 2.2) are too few to allow of any conclusions, but they may help to fill in the picture of the relative economic importance of the industry.

The direct impact of the shipbuilding industry on the British economy is difficult to gauge. In 1907, the only year for which data are available, value added in the industry amounted to 0.91 percent of gross domestic product (GDP). As is shown in tables 2.3 and 2.4, although the shares of the costs of both hulls and total merchant ship output remained approximately constant as a proportion of GDP between 1886-1890 and 1909-1913, wages in shipbuilding as a proportion of total wages (excluding salaries) more than doubled over the period. In part, this reflects the continued dominance of skilled labor in the industry at a time when mechanization was proceeding faster in other sectors, but it also results

Table 2.3. Value of merchant hulls and of total merchant output as a percentage of Gross Domestic Product, 1886-1913.

Period	Merchant hulls	Total merchant vessels
1886-1890	0.93	1.30
1896-1900	0.96	1.23
1909-1913	0.91	1.26

Source: Data for value of hulls and total merchant output are from table 2.1. Data on Gross Domestic Product are from Feinstein, *National Income, Expenditure and Output of the United Kingdom, 1855-1965*, table 4, p. T12, column 5.

Table 2.4. Aggregate wages in shipbuilding as a percentage of total British wages from employment, 1871-1911.

Year	(%)[a]	(%)[b]
1871	1.1	0.9
1877	1.2	0.9
1881	1.2	1.2
1886	1.6	1.6
1891	1.3	1.4
1898	1.8	1.8
1901	1.7	2.0
1907	—	1.9
1911	—	2.2

Source: The shipbuilding wage bill is from table 2.2. Total British wages (excluding salaries) are from Feinstein, *National Income, Expenditure and Output of the United Kingdom, 1855-1965,* table 21, p. T54, column 1.

a. Based on aggregate wage bill extrapolated from the 1906 average wage figure.

b. Based on average earnings at the Leven Shipyard of William Denny and Brothers.

from the increase in labor costs as a percentage of total hull costs (see appendix table B. 6) as the relative price of steel, the primary material input, decreased.

By its strategic position in the economy, however, shipbuilding exerted an influence far beyond the weight of its own labor or capital employed. The speed of the changes in the prosperity of shipbuilding made up for its small mass and exerted a notable influence on the prosperity of other related activities. Because of its geographical concentration, the multiplier effects of employment changes in shipbuilding, and the social effects of its rapid growth, were strong influences on the local life of a large number of outsiders. The large exports of the industry's products, new and secondhand, were an important strand in the fabric of world trade and formed part of one of the most influential aspects of the British economy of that period, the export of capital.

The relationship between shipbuilding and the steel industry was particularly important. Although it was little understood at the time, nitrogen impurities made Bessemer steel unsuitable for ship plates. The rapid increase in the use of steel in shipbuilding made the open hearth sector of the steel industry in Britain, in contrast to America, far more dynamic from the 1880s than the Bessemer sector, which depended largely on the stagnating export demand for rails. After 1890, roughly half of the open hearth output in Britain was for ship plates and angles.[16] The severe fluctuations in shipbuilding production were rapidly transmitted to this other basic industry and from it to the remainder of the economy.

Foreign Demand and Foreign Competition

After 1890 shipbuilding output increased rapidly in other major industrial nations, but Britain's leading position in the world shipbuilding industry remained unassailable. Britain's lead was even more dominant in shipbuilding than in shipping, and was maintained by the export of new and secondhand vessels. British purchases of foreign-built ships remained negligible, apart from the early 1860s, when a large tonnage of Union sailers was acquired, and from the winter of 1897-1898, during the Engineers' lockout.

The rise in the sales of shipping abroad was greater than the general increase in tonnage launched. In the period 1869-1883 an average of 12 percent of British output was sold abroad; in 1900-1913, it was 24 percent, chiefly merchant vessels.[17] The sales of secondhand vessels rose in proportion. The tonnage of vessels sold from the British Register multiplied 3.8 times between 1875-1884 and 1904-1913, while the tonnage on the British Register rose by only 75 percent between 1879 and 1908.

Throughout the period, the sale of secondhand tonnage was a most effective method of keeping the British merchant marine up-to-date and supreme. As obsolescent vessels were sold abroad, they were replaced by the latest types.[18] Foreign owners might still be able to employ profitably vessels that were no longer competitive in Britain, however, because of their cheaper labor, laxer safety regulations, and, perhaps, legislative protection. Information for 1896-1901, published in the report of the Royal Commission on Steamship Subsidies of 1902, shows clearly that most secondhand ships sold to foreigners were at least 10 years old.[19]

Both the new tonnage built for foreign owners and the old tonnage sold abroad showed fluctuations that were in step with variations in the general state of trade. It can therefore be assumed that foreigners bought British tonnage in years of prosperity, when demand and prices were high, rather than during depressions and periods of low prices, when secondhand tonnage fetched no more than nominal prices. There is, unfortunately, no way of arriving at an estimate of the prices of the vessels sold abroad.

Norway, with the largest shipping tonnage per head of population but with no modern shipyards, was a prominent buyer. Her cheap ships, largely used for conveying timber from the Baltic, were bought from Canadian and American owners and from British owners after the Plimsoll Acts. In the years 1894-1903 inclusive, Norwegian owners bought 242,000 g.t. of secondhand sailing vessels from Britain, more than were bought by any other nation. Germany and Italy were the only other countries to which a large tonnage of used sailing ships was sold in that period. Steam tonnage from the British register went chiefly to Italy (425,000 t.), Spain (406,000 t., 116,000 t. of which was in 1899 alone), Norway (307,000 t.), and Germany, France, Japan, Sweden, and Greece, in that order. The Japanese, French, and Italian purchases were irregular, depending on the level of shipping bounties and the support given to their domestic shipyards.[20] The other countries looked to Britain as the natural source of their tonnage.

The sale of obsolete tonnage abroad and its replacement by new vessels were not an unmixed blessing for British shipping. In many cases tonnage that might have had to be scrapped (or "lost") under the British flag continued to compete on the world freight markets by virtue of the less stringent safety or tonnage regulations of other countries. Any conclusions drawn from the simple statistics of tonnage sales must be modified, however, because in many cases vessels registered under foreign flags continued to be British owned and managed, while in other cases American owners bought up large numbers of foreign vessels, including British, which continued to fly their original flags. This was done partly to take advantage of bounties or other subsidies. This was one of the main objectives behind the absorption of a number of British lines into J. P. Morgan's International Mercantile Marine.[21]

Between 1870 and 1890 all modern fleets were in the habit of ordering their new tonnage from British yards, usually on the Clyde. For example, only a system of pressure and subsidies diverted some German orders to home yards in the 1890s. By that time French orders were also falling off, as the bounty law discriminated against foreign building, but they were supplanted by orders from Japan, Austria, Holland, and Norway. After 1905, when both Germany and Japan had built up their merchant navies and had sufficient building capacity to supply their own replacement demand, the chief British markets were the colonies and dominions and Belgium's new trade with the Congo. The annual tonnage built for Japan decreased from an average of 20,000 t. in 1896-1901 to 8,000 t. in 1904-1914; and that for Germany, from an average of 67,000 t. in 1896-1906 to 25,000 t. in 1907-1914. But these decreases were more than offset by increased sales to other nations.[22]

The world merchant navies were thus dependent to a striking extent on the British shipyards. This applied, in particular, to economically backward European countries that nevertheless had a large foreign trade consisting mainly of the export of food and raw materials and the import of capital goods and luxuries. The examples of Germany and the United States showed that an existing heavy industry allowed the foundation of a modern shipbuilding industry in a very short time.

In the middle decades of the nineteenth century, the development of the German shipbuilding industry was slow, despite the progress made in the engineering and iron-making sectors. Costs were high because of the distance of coal and iron mines from the coast, and because of the German habit of ordering to British specifications and measurements. Even the tonnage measurements were taken from Lloyd's Register. Although Krupp was supplying heavy components by the early 1860s, German vessels themselves were always built in British yards.[23]

After 1870, German shipbuilding capacity became a function of the interest taken by the state in its expansion. General von Stosch, the Minister of Marine, insisted on building German warships in Germany and from German materials only. Kiel and Wilhelmshaven, the imperial yards, were opened in 1869, and in the 1870s plate and armorplate works were set up in Germany. But mercantile shipping found the Clyde yards to be cheaper. In 1879 the German customs law was altered to admit all shipbuilding materials free. The following year the German plate manufacturers changed from puddled iron to steel, but only after 1890 was their output large enough to make their prices competitive. An order that the Hamburg-Amerika Linie placed with a German yard in 1881 was clearly the result of official pressure, and there were no further important home contracts until 1885, when the Nord Deutscher Lloyd ordered six steamers in Germany—a condition of the mail contract that the line had been granted. These were the first large merchant vessels built in Germany, and from that date onward German shipbuilding advanced rapidly.[24]

In the late 1880s German production, though rising, was still largely dependent on official encouragement, on warship work, and on mail subsidies. But after 1890 it advanced on its own merits. The makers of Martin steel improved their deliveries in the 1890s (after 1893 Krupp steel became the standard for the armor of all navies), and the launchings of the German shipyards, especially their own "greyhounds", were not inferior in quality to the British. A growing share of Germany's ships came to be built at home.[25]

In 1899 the German shipbuilding industry employed 27,000 persons in private yards and 17,500 in the imperial yards, of whom 65 percent were metal workers, and only 5 percent wood workers. There were 24 yards, plus 3 river yards, with electrically driven machines, and 9 with pneumatic power stations. The total number of shipyards grew from 8 in 1881 to 35 (141 berths) in 1901. The capacity of the German heavy plate rolling mills increased from 105,000 tons in 1880 to 818,000 tons in 1901, and over the same period production of ship plate increased from 17,000 tons to 207,000 tons.[26]

The German shipyards clearly were not yet in a position to supplant the British, but their rise from a mere 6 percent of the British launchings in the early 1890s to 15 percent in 1899 appeared threatening. Moreover, Germany's output was of the highest quality, her ships among the largest and fastest, and her competition keenly felt on the shipping routes and in the colonial scramble. German yards were better equipped than the British with the latest mechanical tools, and German naval architects were believed to be better trained. In 1901, excluding the imperial yards, 12 yards had a capital of 2 million marks or more, and 11 employed more than a thousand workers. The strong impact on British public opinion of the reports of this sudden expansion combined with the threat of increased naval building in Germany was understandable. The main period of German expansion was over by 1903, however. The German merchant fleet had found its level and only replacements were needed, many of which were still ordered from Britain. German vessels were too expensive on the world market, and the builders had failed to attract a significant number of foreign orders. German mercantile launchings until World War I remained at about one-sixth of the British level.[27]

The rise of the American shipbuilding industry at the turn of the century was even more spectacular than that of the German, though rather less surprising. The United States, with its high level of overseas trade and technical skill and its abundance of cheap steel, could not be expected to remain permanently without a shipbuilding industry.

Iron shipbuilding began in the United States in the 1830s. Throughout the 1840s and 1850s a small iron tonnage was produced, primarily at Wilmington on the Delaware River. Around 1860, Reaney, Son and Archbold invested one million dollars in outfitting an iron shipbuilding yard at Chester, Pennsylvania, also on the Delaware. The industry received a strong impetus from the demand for iron naval vessels during the Civil War, but many of the firms founded at this time were hothouse creations that expired soon after the fighting ended.[28]

Although the American output of iron vessels was reduced during the next three decades, several firms were fairly large even by British standards. Most were located on the banks of the Delaware, close to supplies of coal and iron in eastern Pennsylvania. Unlike their British competitors, most of these yards were founded by engine makers or manufacturers of railway cars; one Wilmington builder, Pusey and Jones, survived the slack period following the Civil War by repairing and manufacturing paper-making machinery. Cramp's firm at Philadelphia was the only Delaware River yard to switch from wooden to iron shipbuilding. John Roach and Son, who had purchased the Chester yard of Reaney, Son and Archbold from receivers for $450,000, had an output of 146,693 g.t. between 1870 and 1883. Roach consciously patterned his yard on the best practice on the Clyde, and the firm's employment reached 1,800 men in 1882. He had a fully integrated business, which mined coal, forged iron, and built marine engines, although he did not own vessels as did C. M. Palmer and Christopher Furness in England. In the late 1870s Charles Cramp was able to secure orders from Russia; by 1886 his yard was capitalized at $4 million. Nevertheless, the industry remained miniscule by British standards. Total annual output averaged only 18,688 g.t. in the 1870s and 37,438 g.t. in the 1880s. Much of this was composed of either naval vessels or small craft. Most of the yards were obliged to build a variety of types of vessels, and one of the most important firms, Harlan and Hollingsworth of Wilmington, was prevented by its cramped situation from building ships of more than 300 feet in length.[29]

There were diverse reasons for the American failure to establish an important shipbuilding industry at this time. On the supply side, labor and materials costs were too high to permit competition with British builders. On the demand side, the United States did not develop a large oceangoing merchant fleet. Government regulations, the weaker seafaring tradition, and higher returns from other types of investments combined to make shipping and shipbuilding unattractive alternatives for American capitalists, whereas it was alleged that in Scotland "the shipping interest is more than a business—it is a sentiment pervading the whole mass of the people," and the farmer's surplus "goes to sea as so many sixty-fourths of a new ship, commanded probably by the surplus son of the family."[30]

In no other country has the modern shipbuilding industry been so directly a child of naval policy as in the United States. In the middle 1880s the overseas adventures of the European powers alarmed the country, whose navy was composed entirely of coastal defense vessels. A new navy, able to sail the high seas, was to be constructed, and it had to be built at home.

The growth of American shipbuilding capacity was slow at first. Young naval architects were sent to study in Glasgow, Greenwich, and Paris. In 1892 the Society of Naval Architects and Marine Engineers was founded, mainly by naval interests. Yet in 1893 most warship designs were still obtained from London. That year marked a turning point: investment opportunities in railways, mining, and oil began to show limitations. In the following few years the American shipbuilding industry expanded rapidly.[31]

Encouraged by the needs of the navy, the Bethlehem Steel Company acquired forging equipment from Whitworth's and armor plate equipment from Schneider-Creuzot. In the same period the extensions of railroad companies into shipping also furnished a market for mercantile vessels. Cramp built his first two large express liners in 1893, and in the following years investment in shipbuilding expanded with amazing rapidity. Yards were laid out to full size at once. In some of them (such as Trigg's warship yard of 1899 or the Eastern Shipbuilding Company yard of 1900) construction of the workshops and of the ships began simultaneously.[32]

Because of the lack of skilled labor in an industry founded overnight, and of the need for speed, expensive equipment was the rule in the American shipyards in 1894-1901. On the Great Lakes the large yards combined to form the successful American Shipbuilding Company, within which each constituent yard specialized in the mass production of one type of lake steamer. On the seacoast the mushroom growth led to an overinvestment in costly equipment on the basis of a temporary demand, and to a rise in the price of steel plates after 1899, which brought some of the yards to near bankruptcy in 1901.[33]

In 1901 there were nine yards employing more than 100 workers. The total capital invested in shipbuilding rose from $20.1 million in 1880 to $77.4 million in 1900 and $122 million in 1905; the value of output rose from $36.8 million in 1880 and $29.8 million in 1891 to $74.6 million in 1900 and $82.8 million in 1905. Workers in the industry numbered 21,300 in 1880 and 50,800 in 1905. The high output of 1900-1903 consisted of some very large and fast steamers, the ordinary cargo vessels for the Great Lakes traffic, and a large proportion of wooden sailers. But output failed to keep pace with the capital invested, and the industry subsequently found itself in serious difficulties for which a state subsidy seemed the only remedy. The British anxiously watched the efforts of the shipbuilding and steel lobbies, but although subsidies seemed certain more than once, they were never enacted. The U.S. Shipbuilding Co. (not to be confused with the Great Lakes combine), which had acquired five of the largest yards in 1902 in an attempt at "rationalization," proved to be merely a field of further speculation and overcapitalization and was put in the hands of a receiver in 1903. The industry was then reorganized under the aegis of Bethlehem and settled down to a lower annual output maintained by the needs of the navy and of the Great Lakes.[34]

The French shipyards, which in the 1860s and 1870s had been the most serious rivals of the British, were never truly competitive in price, and their output stagnated until the turn of the century.[35] The defeat of 1871 and the temporary setback to French heavy industry reduced French capacity, although naval launchings maintained their excellent quality. From 1881 onward French shipbuilding depended fully on state support, and the Subsidy Act of that year was the main factor in the history of French shipyards until 1914.

The payment scales for construction and navigation under the Subsidy Act had the effect of encouraging sailing vessels rather than steamers, but for the first 10 years of the Act some navigation bounties were paid to the owners of foreign-built ships. These subsidies helped to keep the French shipping industry in touch with the world market, and the object of the

Act, which was to encourage the erection of new shipyards in France, was not realized.[36]

In 1891 the Subsidy Act was altered, and neither building nor navigation bounties were granted to foreign-built ships. The importation of foreign-built vessels virtually ceased, and the international competitiveness of the French shipping industry declined still further. The building of sailing vessels, in particular, was encouraged to an absurd degree. At the turn of the century, the great international wave of investment in shipbuilding washed over even France, and with the support of the steel works and the large shipping companies, a number of up-to-date yards were laid down, including the Chantiers de France at Dunkirk, the Ateliers et Chantiers de Provence, and the Chantiers de Construction Maritime at Nantes. Altogether (including the naval yards) France had eight large shipyards in 1901, each with a capital of at least 2 million francs and an employment of at least 1,000 men, and an additional nine medium-sized yards. The total number of men employed in shipbuilding has been estimated at 35,000-37,000 and the total capital invested at 75 million francs. But the equipment was still obsolete and the number of vessels built in each yard small. The overheads were high, and the building of competitively priced merchant vessels was out of the question. Despite further alterations in the bounty act, French shipbuilding remained well below the level of the German or American after 1901.[37]

Other countries with small but modern shipyards included Italy and Japan. In Italy naval shipbuilding capacity grew rapidly with the aid of British capital, as in the Ansaldo Yard (Maudslay, Armstrong), the Pattison Yard (Thornycroft), and the Vickers Terni Armour Plate Works. Mercantile shipbuilding developed after the passage of the Subsidy Law of 1885 and grew more rapidly when the tariffs on imported raw materials were lowered. Subsidies were important in the growth of Japanese shipbuilding, as well, and by 1914 Japanese mercantile output had risen to 86,000 g.t. On the naval side, in 1912 orders were placed for dreadnoughts of 27,500 tons.[38]

The modern shipbuilding capacity of the major maritime countries was laid down in the 1880s and expanded rapidly in the last years of the century. The growth was slower after 1900 and was interrupted by two setbacks after 1903 and 1906. The British share of the world's output declined in the late 1890s, but Britain's predominance was at no point seriously in doubt, despite the government support accorded to most competitors (see tables 2.5 and 2.6). From 1892 the tonnage launched by the British for buyers abroad consistently exceeded the total launchings of either German or American yards. This does not prove, of course, that British builders could not have been more efficient than they were or that their share could not have been larger, but that they did so well in the

Table 2.5. British tonnage built for foreign owners and total tonnage built in German and American yards, 1892-1913 (in thousands of gross tons).

Year	British tonnage built for foreign owners	Total German tonnage built	Total American tonnage built
1892	173	65	63
1893	133	60	27
1894	139	120	67
1895	199	88	85
1896	388	103	184
1897	266	140	87
1898	357	153	173
1899	314	212	224
1900	366	205	334
1901	359	218	433
1902	272	214	379
1903	244	184	382
1904	226	202	239
1905	382	255	303
1906	394	318	441
1907	550	275	475
1908	400	208	305
1909	369	129	210
1910	224	159	331
1911	413	256	172
1912	445	375	284
1913	503	465	276

Source: Lloyd's Register, Annual Returns.

face of foreign discrimination suggests that they operated at a highly competitive level. Quite simply, Britain continued to build the cheapest ships.[39]

In the 1870s Britain was practically identified with steam and iron shipbuilding, and displaced foreign yards wherever the steam or iron ship gained ground. In the boom of 1881-1883 some iron and steam building capacity was erected abroad, particularly in Germany and France, and the Royal Commission on the Depression of Trade (1886) listened to a good deal of evidence on the competitive power of continental shipyards.

Before the commission, little was made of the superior technical education of naval architects in France, which had agitated John Scott Rus-

Table 2.6. The British percentage of the world's shipbuilding, 1892-1914.

Year	All vessels	Seagoing steamers of 3000 g.t. or more
1892	81.7	—
1893	81.4	—
1894	79.1	—
1895	78.1	—
1896	74.0	—
1897	71.5	—
1898	72.2	—
1899	68.8	—
1900	62.6	—
1901	58.2	—
1902	57.0	—
1903	55.5	—
1904	62.8	—
1905	64.5	—
1906	62.6	81.4
1907	57.9	71.5
1908	50.7	63.6
1909	61.9	75.0
1910	58.4	73.5
1911	68.1	78.0
1912	59.9	71.3
1913	58.0	64.7
1914[a]	59.0	64.0

Source: Based on Lloyd's Register, gross tons. Compare Board of Trade, *Industrial Survey of the North-East Coast*, 1932, appendix table 2, p. 276; also statistics on a slightly different basis in Neumann, *Volkswirtschaftliche und soziale Bedeutung der technische Entwicklung in der Schiffbauindustrie*, p. 45, and *TINA*, 1926, opposite p. 2.

a. Based on incomplete information.

sell so greatly in the 1860s; the greatest emphasis was reserved for the lower wage rates and the bounties abroad. J. Price, of Palmer's, quoted German wages that, for longer hours, were about one-half of the British, while German Siemens-Martin steel was as cheap as steel on the North East Coast and cheaper than in Glasgow. J. Scott of Greenock, who had built and managed a yard in France for some years, claimed the same advantages for French yards: their wages were lower, their hours were longer, and they could get cheap steel from South Wales. Lowthian Bell referred to the cheap labor abroad, which caused iron and coal to be shipped to Norway and Denmark to build vessels there, and to the erection of shipyards on the Meuse working on English account.[40]

Robert Knight, secretary of the Boilermakers Union, had no difficulty in showing that the British output was nevertheless cheaper and was still growing at the expense of others. The wage rates quoted were day rates, and the hourly output of the British workman was so much greater that labor costs per ton were lower in Britain. Two commissioners of the union who had been sent to France discovered slack trade, apart from naval building, antiquated equipment, and mercantile orders only when British yards were full. At least one Bordeaux builder had given his opinion that to succeed in France a new yard would have to employ British managers, masters, workmen, and administration throughout. Not only were most shipyard foremen in German, Danish, and Russian yards Britishers (and members of the Boilermakers), but so were many of the men.[41]

The other witnesses agreed that even the bounty law of 1881 had had little effect in France, merely increasing the size of some yards, but not raising their output after 1883.[42] Shipyards abroad remained more expensive than British yards in the following two decades, even though the skill of their naval architects and workmen was not markedly inferior, their basic industries were equally efficient, their equipment was more up-to-date, and their yards were more efficiently laid out.

One cause of the British superiority was the large market that allowed builders to specialize, particularly in the building of cargo tramps. Many yards built one type only, on a standardized method, and could thus frequently build on speculation ahead of demand. For the building of those cheap and simple vessels little mechanical equipment was needed, reducing overhead costs in contrast to some American yards which built tugs on berths equipped for building liners. Orders could thus be completed sooner in Britain, an advantage that American builders stressed. The Americans noted that their specialized yards on the Great Lakes could build as cheaply as the British, while battleships, which not even in Britain could be built on standardized plans, were also no more costly in the United States than on the Clyde. It was primarily in the important class of oceangoing cargo vessels that the British ability to specialize gave them a distinct edge.[43]

Because of the size of the British shipbuilding industry, British steelmakers supplied a much higher proportion of their output to the shipyards than did steelmakers in other countries. Their longer "runs" of certain sizes were cheaper, and they were willing to go to some lengths to retain the shipbuilders' custom. As late as the 1890s even German builders preferred to draw their plates and angles from British makers for these reasons. While in 1910 only a doubtful 4 percent of German steel output went to shipbuilding, 17 percent of British pig iron output went to the shipyards as early as 1878-1884, and 30 percent of British steel output by 1910-1913. In the shipbuilding districts the proportion was higher, and the Clyde yards consumed two-thirds of the output of the Scottish steel works.[44]

At the turn of the century, supplies of steel were cheaper to British yards than to the German or American yards, and this difference was accentuated by American dumping practices. After 1898, U.S. steelmakers sold plates to Scotland at $8-$15 per ton cheaper than at home. Altogether, the cost of raw materials for shipbuilding was estimated at 30-40 percent more in America than in Britain, but American builders were reluctant to import foreign plates as the ships might be used in the coastal trade where all material had to be of American origin. After the turn of the century, German steelmakers also began to dump semifinished products in Britain, although not until 1910 could the Germans actually supply their plates, angles, and bars to the British ports at prices below those of domestic makers. In all, 43 percent of the steel castings and 10 percent of the steel plates used in Britain in 1912-1913 were imported, mostly from Germany.[45] Britain's yards were favored, therefore, not only by the location of her iron and steel supplies, but also by the cartels, dumping, and protective legislation of her two chief rivals.

Labor costs were highest in the United States. Piece rates there were on a par with the British, but time rates were 50 percent higher, so that the U.S. wages bill exceeded the British by 25-33 percent. German wages, on the other hand, were lower by one-third, and the men worked 60 hours even in 1910. Yet even with his more modern equipment the German iron worker's output was below that of the British boilermaker.[46] Those British shipyards building the more expensive classes of vessels—Harland and Wolff; Swan, Hunter and Wigham Richardson; Clydebank; Vickers—were as well equipped as any abroad. But it is more than likely that expensive equipment could not compensate for the lower level of skills and more irregular output in Germany and America, especially in the construction of mercantile tramps. Thus, despite Britain's inferior capital equipment, the output per man was still highest in Britain at the end of the century.

Germany appeared to have the lowest costs outside Britain. According to Francis Elgar, her cargo carriers and ordinary passenger steamers were cheaper than the British; according to John I. Thornycroft, it was her largest ships and the small types of vessels that were cheaper. Tugs and trawlers, in particular, were said to be built at lower cost in Germany and Holland. Yet foreign orders continued to come to Britain. Even if the long traditions and connections, the financial ties, and the shorter delivery dates are taken into consideration, the German prices must also be presumed to have been in general somewhat above the British.[47]

Of the two sister ships built for the Atlantic Transport Line in 1901, the liner built in Belfast cost £1,419,000 and her sister ship in America, £1,649,000. Of four other sister ships, the two built at Belfast cost £535,000 and £486,000, while the two built in the United States cost £720,000 and £729,000. The *Statist* put the price of liners in 1902 in America at 30 percent above the British, and in 1906 American prices were reported to be 37-47 percent above the British, 50 percent in the case of freighters. A French official report put the cost of steamers at 300 francs per ton in Britain and 420 francs in France; and in 1891 the cost of a warship was estimated at £30.5.0. per ton in Britain, £46.9.0. in France, and £87.5.0. in Russia, with the price of engines varying in proportion. The British shipbuilding industry continued to operate at costs below the costs of its rivals.[48]

THIS COMPETITIVE advantage of British shipyards originated in the period 1860-1880, when no other major country, save perhaps America, whose resources were occupied elsewhere, had the iron, steel, and engineering capacity to build modern steamers on a large scale. The advantage continued to 1914 largely because of three factors: access to a large market, which provided the opportunity for mass production, specialization, the full use of fixed capital, and external economies; cheap supplies of raw materials and components; and the skill and experience of management and men.

Britain's large output of tonnage led in turn to a different type of specialization, on a geographical basis. As the size of ships increased, so did the external economies resulting from a concentration of builders in a few centers of production, generally close to supplies of cheap iron and steel and coal, where they could share valuable capital equipment, such as giant cranes and docking facilities, and draw upon larger pools of skilled labor. In the process, the industry contracted from the spatial diversity of the days of wooden shipbuilding, when firms built primarily for local buyers, to a small number of ports whose builders served the national and international markets.

3.

Regional Shipbuilding

ALTHOUGH THE number of ports involved in shipbuilding in Britain contracted when wooden vessels became obsolete, the industry was still spread over several geographic regions, as no single river or harbor had enough suitable frontage to handle all of the necessary shipyards. The final distribution of the industry was by no means random. Accessibility to cheap supplies of labor, capital, raw materials, and land, a ready market for ships, the existence of subsidiary industries, and the availability of repair work were all important influences on costs and profitability. As a result, in the second half of the nineteenth century shipbuilding came to be increasingly concentrated in a small number of centers that were exceptionally well endowed with these prerequisites. This high degree of concentration, often in small communities, allowed yard owners to exercise a high degree of control over their workforces and to dominate the social and political as well as the economic lives of many of the shipbuilding towns.

Location and Concentration of the Industry

In the first half of the nineteenth century, when timber sailing vessels of around 200 tons burden were the standard output, there were shipyards in all major and minor ports dotting the coastline of Great Britain and Ireland. By the 1870s, the Clyde, the Tyne, the Wear, and the Tees accounted for nearly all the ordinary tonnage launched in Britain, while trawlers and other small and specialty craft were built on the Mersey, the Humber, and the Scottish East Coast. At least three other important building areas developed later—Belfast, Barrow, and the Hartlepools. The differences in location and concentration between 1841 and 1913 are striking. In 1841 the industry was widely scattered with its center in the

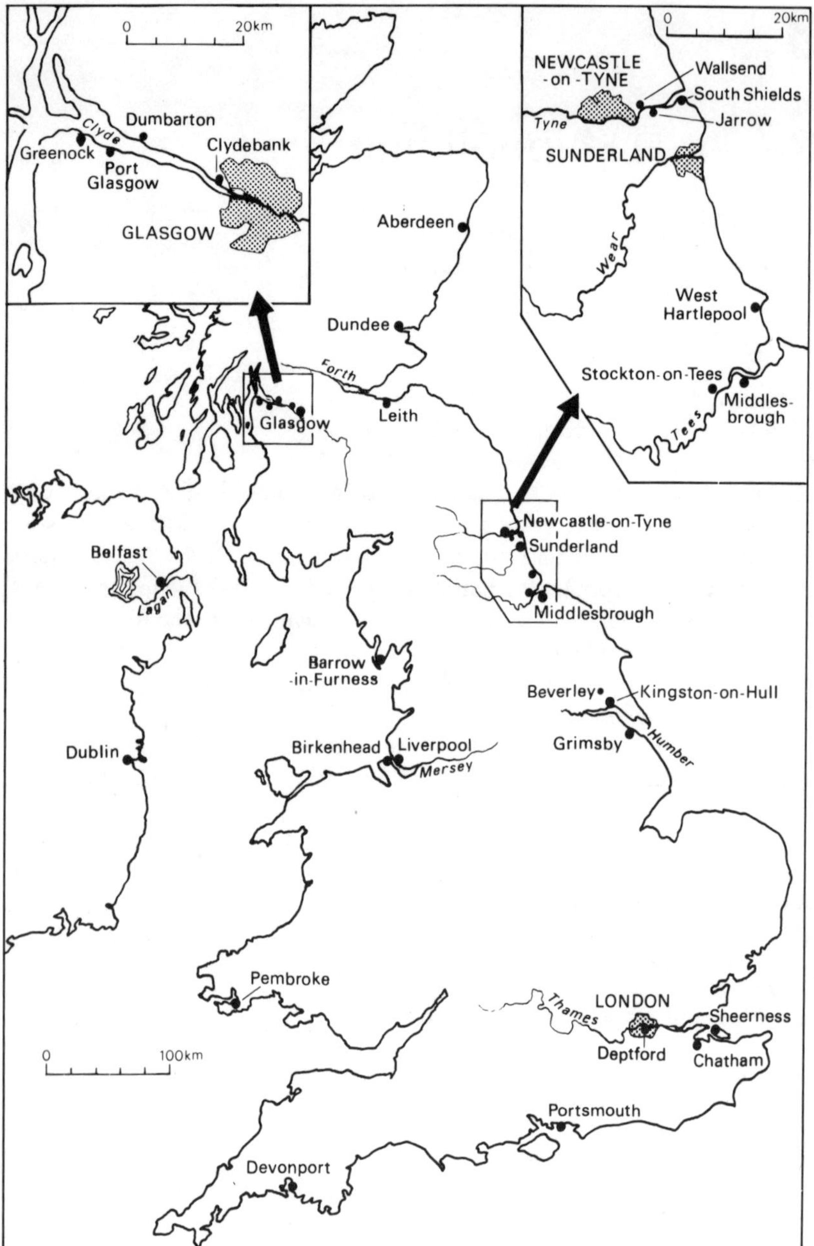

Important Shipbuilding Towns and Rivers

south, on the Thames; in 1913 it was practically confined to three major areas—the Clyde, the North East Coast, and Belfast (see table 3.1)—and half a dozen minor ports. The explanation for this change can be found largely in the availability of cheap and docile labor, ready supplies of raw materials and other inputs, and easy access to markets.

A large pool of skilled labor, such as existed on the Clyde, the Mersey, or the Thames, was an attraction when a new yard came to be laid out or an old one expanded. The supply of labor was more elastic in such areas and rush orders could be more easily met by extra hands. Also, the "amphibian" labor—plumbers, painters, joiners, electricians—required at some stages of the building could easily be found among the building trades in the large cities. On the other hand, in the new areas, in which large-scale shipbuilding was not firmly established, labor was easier to control once the initial expenditure of attracting skilled men had been

Table 3.1. Locations of important shipbuilders.

Shipbuilder[a]	Town	River
Alexander Stephen and Sons[b]	Linthouse (Glasgow)	Clyde
Armstrong-Whitworth[c]	Newcastle	Tyne
Barclay, Curle	Whiteinch (Glasgow)	Clyde
Caird and Company	Greenock	Clyde
Cammell Laird[d]	Birkenhead	Mersey
Charles Connell and Company	Scotstoun (Glasgow)	Clyde
Fairfield Shipbuilding and Engineering Company[e]	Govan (Glasgow)	Clyde
Furness Withy and Company[f]	West Hartlepool	Tees
Harland and Wolff	Belfast	Lagan
Irvine's Shipbuilding Company	West Hartlepool	Tees
J.L. Thompson and Company	Sunderland	Wear
John Brown and Company[g]	Clydebank	Clyde
Northumberland Shipbuilding Company[h]	Howdon	Tyne
Palmer's Shipbuilding Company	Jarrow	Tyne
R. and W. Hawthorn, Leslie and Company[i]	Newcastle	Tyne
Readhead and Sons	South Shields	Tyne
Richardson, Duck and Company	Stockton	Tees
Ropner and Sons	Stockton	Tees
Russell and Company	Port Glasgow	Clyde
Scott and Company	Greenock	Clyde
Short Brothers	Sunderland	Wear
Sir James Laing	Sunderland	Wear

(cont.)

Table 3.1. (*cont.*)

Shipbuilder[a]	Town	River
Sir Raylton Dixon	Middlesbrough	Tees
Swan, Hunter and Wigham Richardson	Wallsend and Newcastle	Tyne
Vickers, Son and Maxim[j]	Barrow-in-Furness	—
William Beardmore[k]	Dalmuir	Clyde
William Denny and Brothers	Dumbarton	Clyde
William Doxford and Sons	Sunderland	Wear
William Gray and Company[l]	West Hartlepool	Tees
William Hamilton	Port Glasgow	Clyde
Workman, Clark and Company	Belfast	Lagan

a. These 31 firms, forming about one-quarter of the total number of firms, produced approximately two-thirds of the total British output in 1870 and between three-quarters and four-fifths after 1910.

b. At Kelvinhaugh (Glasgow) to 1870.

c. Armstrong-Mitchell to 1897.

d. Laird Brothers to 1897.

e. John Elder and Company to 1886.

f. E. Withy to 1890.

g. J. and G. Thomson to 1897; Clydebank Company to 1899.

h. Edwards Shipbuilding Company to 1898.

i. A. Leslie and Company to 1886.

j. Barrow Shipbuilding Company to 1889; Naval Construction and Armament Company to 1897.

k. R. Napier to 1895.

l. Denton, Gray and Company to 1872.

met. Mobility to the new districts was high, even over distances as great as from the Clyde to Belfast or Barrow.

It has commonly been assumed that high labor costs were primarily responsible for the decline of the industry in many of the older centers. It seems plausible, at first sight, that the powerful and restrictive shipwrights' unions in London, Liverpool, Bristol, and Dublin made competition with the builders on the northern rivers impossible. A brief enquiry, however, shows this explanation to be inadequate.

In the early decades of the nineteenth century, the cost of labor was nowhere more than one-third of the total cost of a vessel, and usually less than one-quarter. Wages seldom varied by more than 20 percent between the highest and lowest wage areas, and generally by less. This is especially true when the quality of work in the high wage areas of the Thames and the Mersey, and the large amount of repairs there, are borne in mind (see table 3.2). Differences in labor costs between the highest and lowest regions were, therefore, unlikely to be more than 7 percent of total cost, and were probably more often about 5 percent, far less than the differences in the prices charged for finished vessels. Moreover, there were also strong unions in Sunderland and on the Tyne, and the labor factor could on no account have lowered total costs by more than 1-2 percent as compared with the Mersey or the Thames. Yet the industry declined in the latter areas while it flourished and expanded on the North East Coast.

On the other hand, while the labor factor cannot be held solely responsible for the relative decline of London and Liverpool, it may have been more influential in determining the move from the larger ports on the Clyde, the Tyne, and the Wear to the smaller ports nearby and to new areas such as Belfast, Barrow, and the Hartlepools where wages were lower.[1] But low wages were not the only attraction. In those isolated communities of the smaller shipbuilding boroughs a builder could hope to keep labor under very strict control, even outside the confines of his own yard. In some of them—for example, in the smaller burghs along the Clyde or in the Tyne ports—local trade, prosperity, and social life were intimately bound up with the fate of the shipyards. Others—including Jarrow, West Hartlepool, Barrow-in-Furness, or Wallsend-on-Tyne—would not have existed at all but for the firms for which shipbuilding remained the main activity.

In those "company towns," the shipbuilders and their immediate assistants were inevitably in control of the local government, and thereby

Table 3.2. Wage indices of eight maritime districts, shipbuilding and engineering (all districts, 1883 = 100).

District	1866-1872	1873-1882	1883-1889	1890-1899	1900-1906	1907-1914	Mean 1867-1914
Mersey	96	105[a]	102	113	119	121	110
Tyne	88	102	103	111	116	117	107
Wear	93[a]	101	105	109	114	116	106
Hull	90[a]	99	98	104	110	110	102
Clyde	90[a]	91	95[a]	104	111	112	99
Belfast	78[a]	87	91	102	109	112	97
Barrow	80[a]	90	90	100	104	105	95
Dundee	71[a]	83[a]	87	95	99	100	89

Source: Appendix table B.5b.
a. Incomplete data.

controlled not only the economic life, but also the education, housing, and health services of the inhabitants. Jarrow, for example, was the town whose occupation, livelihood, housing, and other amenities depended entirely on Palmer's Shipbuilding Company. "So completely, in fact, is the town identified with the works, that it might more appropriately be called 'Palmer's town,' " wrote one admirer in 1904. "There is probably no other town whose interests are more closely identified with the success of a single firm," Lord Aberconway wrote proudly in 1927, only a few years before the empty yards and desolate streets of Jarrow became a symbol of national distress.[2]

The history of Barrow, at least until 1897 when Vickers took over, was similarly determined by a single group of industrialists who controlled the varied industries and works of the town through a number of nominally independent companies. Here too, the town council, the educational system, the press, and the workmen's institutes were in the hands of a single syndicate, and corruption and profiteering of building contractors, hand in glove with local authorities, were common.[3] As a local paper wrote in 1881,

> The land is in the hands of the existing powers who govern all. They employ the labour, they let the majority of workmen's dwellings, they act as magistrates, and also as town councillors, so that a man's work, his house, his workmen's clubs at his works are simply rent collecting agencies, for if a man meets with an accident and is entitled to 9s. weekly, he finds when laid on a bed of sickness that his rent, 6s. weekly, is deducted from his club money, leaving him the munificent sum of 3s. weekly to maintain his family.[4]

The municipality became a dependency of the works. Malcolm Dillon, general manager and director of Palmer's, stated in 1905, "We pay one sixth of the rates of Jarrow. Every time the Council or the Education Authority, or the Board of Guardians spends £1, we pay 3s. 7d. of it. I mention this to show how largely interested we are in local expenditure, and how necessary it is that we should have a greater control over it."[5] In Barrow in 1913, one-quarter of the whole population, or two of every three employed males, worked for Vickers. In 1914, Caird's was estimated to represent 20 percent of the economic worth of Greenock, a town of 80,000 inhabitants. The Lord Mayor of Newcastle, in seconding the annual report of Armstrong's in 1912, informed the shareholders that 120,000 persons, or more than one-third of the population of Newcastle, were dependent on the Elswick works for their livelihood. Very similar was the influence of William Gray in West Hartlepool, whose "rise and progress [has been] co-extensive with the growth of this shipbuilding and marine engineering concern,"[6] and much the same conditions prevailed in the Wallsend of Swan, Hunter, the Clydebank of John Brown and Co. and J. and G. Thomson, and other shipbuilding towns.

When the Vickers Company bought the Barrow shipyard in 1897 and expanded its production in the following years, the firm had to lay out a new housing estate on Walney Island, known as Vickerstown, to attract and keep workmen.[7] This was a model estate—across the Walney Channel from the shipyard—laid out in pleasant contrast to the dreary and dingy slums, cottages, and "Scotch flats" erected on Barrow Island itself by earlier local magnates.

Vickers was not the first firm to feel the need of providing houses for its men, or to see the advantages of tied houses during labor disputes. Palmer's houses, built by the Factory Building Society, had just that effect: it was the opinion of J. Burnett that had Andrew Gourlay, later the leader of the nine-hour movement, been given a house instead of being sacked for his union activities, he would have kept quiet like others, "a fact which Mr. Palmer has not failed to note."[8] Almost every important shipbuilder of whom there is a record built accommodations for his workers. There was a "Dennystown" in Dumbarton; the firm of A. Stephen erected houses for 120 families after its move to Linthouse; and from the Wear it was reported that Castletown, built by T. R. Oswald to house his employees, became deserted after he went bankrupt and moved to Southampton in 1872.[9]

The shipyard also controlled the choice of professions. "Before the war of 1914-18, every second boy of the middle classes in such a town as Greenock passed automatically from school to serve an apprenticeship in Scott's or Caird's."[10] Frequently, also, other local industries were in some way connected with the shipyards, like the rope works of Harland and Wolff in Belfast, or the jute works of the Barrow syndicate. In those towns, where the roofs and the stagings and cranes of the yards dominated the skyline, the shipbuilders dominated the lives, the hopes and aspirations, and the economic welfare of the men who filled the dark rows of their tied houses.

It would be difficult to assess how far the cost of labor was responsible for the retreat from the established centers with their independent trade unions, their alternative employment, and their more democratic local government. It is certain that the marked tendency for the industry to leave the large cities such as Glasgow, Newcastle, and Sunderland for the open banks or the small towns of their rivers was, at least in part, due to a hope of greater control over their labor forces. For the more general concentration along the Clyde and the North East Coast, however, other factors were important as well.

The two decisive influences were the origins of the capital employed in the industry and the supply of raw materials. Essentially, the capital for the relatively large fixed equipment came from within the firms themselves and from the shipping industry, with which they were closely connected. Thus shipping provided not only the market, but also the credits on which the builders worked. Wherever possible, in a very real sense, the yards followed the ships. In addition, the repair requirements of a busy port furnished further valuable employment for the builders.

For the successful expansion of shipbuilding it was necessary that local shipping be of a kind to provide sufficient orders and sufficient capital. The character of Aberdeen, Dundee, Leith, Hull, and Grimsby as fishing ports reduced the building on the Dee, the Tay, the Firth of Forth, and the Humber to trawlers, drifters, and whalers. And it was not entirely fortuitous that coasters, colliers, and tramps were built on the North East Coast, and liners on the Clyde and its offshoot, Belfast.

In the timber days, the proximity to the markets was alone important, and shipbuilding output was closely proportionate to the trade of the ports. To some extent, the influence of the market was cumulative. Because the Thames was known as the river on which the highest-rated ships were built, further orders flowed there from owners who were not in a position to form an independent opinion but rather followed the consensus of the market. Similarly, because Belfast had never been known as a major building area, for a long time its orders came almost exclusively from personal friends and close associates of the builders. It was very hard to establish a reputation for a new river, as witness the early difficulties of the Barrow Shipyard. It was infinitely easier to open on the Clyde or the Wear, in full view of potential buyers, as another "Clyde" or "Wear shipbuilder."

Yet the largest of the old centers declined with the introduction of iron and steam, and the traffic of their ports could do no more than delay the decline for a decade or so. Clearly, modern shipbuilding required a further condition besides the shipping market and its capital funding—easy access to coal, iron and steel works, and engineering capacity.

It was not only the cost of transport that made the supply from a distant area less favorable, though that alone would account for the relative advantages of the Clyde and the North East Coast. The certainty of delivery, the chance to prod a laggard supplier, to know at once of any windfall bargains or improved processes, and ultimately to branch out, if necessary, into coal mining or iron smelting and steel manufacture, gave the northern firms their advantage. The links of the shipbuilders with iron and steel makers were only less close than those with shipowners. In the amalgamation period of the 1890s, when price bargaining in a free market gave way to economic strategy, the vertical combination—with its steel works, coal mines, and armor plate works—was in direct line with the early local connections.

Granted the volume of overseas trade, the capital formation, and the engineering ability that formed the background of British strength in world shipbuilding in those years, it was the cheap supplies of iron and steel plates and bars, of machinery for the yards and of coal to drive the steam engines that were ultimately the bases of the location of the industry along the Scottish and northern rivers. Compared with the major factors of the proximity to shipping centers, to iron and coal, and, in a special sense, to the supply of labor, other possible causes for the location of the industry must be rated as of only secondary importance.

Port facilities were a necessary precondition for the establishment of any shipyard, but in the nineteenth century these became a function of the development of the shipping trade. As such, the dock, harbor, and dredging schemes in Britain in that era formed a worthy example of man's mastery over nature. Without the dredging and the continuous improvement of the Clyde up to Glasgow, and of the improvements at the mouth of the Tyne, those two rivers could not have carried the monster ships that were launched into their waters. The shipyards in Belfast, Barrow, Birkenhead, the Hartlepools, and other areas were founded entirely on artificial docks, harbors, or channels. Shipbuilders also benefited from the dry docks and fitting-out basins maintained by all major harbor boards for the repair and overhaul of the vessels trading there, and from the mass of auxiliary industries, fitters, chandlers, and other suppliers. Lucrative repair work, almost the only work done during trade recessions, was a further asset in large ports. River banks were more sought after as sites for shipyards than were harbors directly on the seaboard, for they were more protected and usually had a cheaper frontage. Newly launched vessels ride high in the water and have a shallower draft than fitted and laden ships, so the yards could be laid out on land further up stream than might be competed for by the ports. The growth in the maximum size of vessels, however, sometimes caused a costly move to deeper channels, such as Armstrong's transfer from Elswick to its new site below Newcastle, and Thornycroft's from Chiswick to Southampton.

A number of authorities have cited the cost of land (including ground rents, rates, and the extra cost of river frontage) as a major factor in driving the industry from the largest cities and ports, not only in Britain, but also from New York, Marseilles, and Hamburg. Land costs certainly played a part in Yarrow's decision to relocate from Poplar to Scotstoun on the Clyde.[11] No useful figures are available on the cost of land in Britain, but in Germany ground rents accounted for about one percent of the cost of production. Nevertheless, what was in fact an absolute shortage of usable frontage along the major shipbuilding rivers seems to have been of great importance. Established firms were generally reluctant to abandon the external returns that could come only from being located in an important center of the industry. Large and expanding firms, such as

Harland and Wolff at Belfast or John Brown at Clydebank, could not easily obtain new acreage adjacent to their yards after the 1870s. As they were reluctant to move to out-of-the-way rivers or to literally carve out level plots of frontage along major rivers as Armstrong's had done at Walker-on-Tyne, they were obliged to put up with cramped sites on which berths and shops were haphazardly laid out according to the needs of earlier decades (see chapter 6).

The influence of tradition cannot have been very great in view of the total eclipse of the famous yards on the Thames, the Mersey, and in the Bristol, Dublin, and East Anglian regions. It is difficult, however, to cite other causes for the absence of large shipyards in South Wales, for instance, where ship repair had been carried on extensively since the middle of the century. There remains a suspicion that resort to the explanation of tradition in an industry as flexible and adaptable as shipbuilding in our period is a veiled confession of ignorance.

No single cause can account for the strongly marked migration of the industry. On balance, the strong centripetal forces of the market, or its supply of capital, or of the proximity of centers of engineering, coal mining, and iron making—these were diffused by the centrifugal desire to get the maximum advantage out of a labor force divorced from the relative laxness and freedom that prevailed in large cities. The growing output on the Clyde, the Tyne, and the Wear was launched increasingly beyond the boundaries of Glasgow, Newcastle, and Sunderland. The small trading and fishing ports, on the other hand, located far from supplies of coal and iron, were at best left with the building of tugs and trawlers on the basis of local demand and cheap labor.[12] The medium-sized trading ports with hinterlands of coal mines and iron works came to build the major share of British output. The "company town" seems to have indeed suited the industry best. It was at first necessary in order to attract labor by cheap houses and subsidiary industries, and it later proved suitable for controlling labor through tied houses, political domination, and influence over the magistracy. Thus grew up, about the year 1870 when the modern iron and steel shipbuilding industry took root, the company towns of Dumbarton, Greenock, Port Glasgow, Jarrow, South Shields, Wallsend-on-Tyne, Howden, Barrow, Birkenhead, and whole quarters of large cities including Belfast, Glasgow, and Newcastle. These towns would bear the full shock of all the crises and slack periods in the industry, and would blight the country as "depressed areas" in the twentieth century.

Output and the Ports

The various shipbuilding districts were distinguished not only by their levels of costs and wages, but also by their output and by their methods of work. The sequence of erection and other shipyard conventions differed from area to area, and craftsmen or constructors who removed to new areas had to learn many techniques afresh.[13] Builders generally spoke of other areas as though they were foreign countries. The British shipbuilding industry before 1900 was not one industry, divided into a number of districts: it was a group of local industries producing substantially the same or similar products. Only in the years immediately before 1914 did the methods of work become more standardized as technical training, improved means of communication, and patented equipment brought the areas closer together.

The specialization of output in the different areas may perhaps best be illustrated by table 3.3, from which a wealth of information can be gleaned. The largest ships, mercantile and naval, were built at Belfast, Barrow, and on the Tyne and the Clyde.[14] But while the Belfast average was not substantially lowered by the building of many smaller ships, on the Tyne the ordinary cargo tramps, and on the Clyde and in Barrow

Table 3.3. Average tonnage of vessels built in the principal districts (private yards only).

	Merchant vessels for the British flag only		All vessels over 1000 tons
District	Average 1867-1872 (net tons)	Average 1907-1913 (net tons)	Average 1907-1913 (gross tons)
Belfast	1,180	3,960	7,340
Wear	530	2,030	3,230
Tyne	470	1,620	3,030
Tees	610	1,620	2,750
Barrow	—	1,510	2,120
Clyde	640	1,260	1,590
Mersey	670	470	1,350
Dundee	540	440	1,250
Forth	240	290	880
Humber	270	160	460
Thames	160	70	320[a]
Aberdeen	440	80	260
U.K. (including minor ports)	330	710	2,690

Source: Pollard, "Economic History of British Shipbuilding," tables A.25 and A.26, pp. 577-578.

a. Average, 1903-1908.

many smaller trawlers, dredgers, tugs, and so forth swamped the smaller number of monster ships. The North East Coast showed the typical sizes of tramp steamers. The Humber and the Scottish East Coast had considerably lower averages since their output consisted mainly of fishing vessels, coasters, whalers, and other smaller types. The Thames's low mercantile average, made up of pleasure steamers and barges, was raised by the inclusion of warships in the last column, as was the tonnage on the Mersey, where the average of a more varied mercantile output was enhanced by the inclusion of Laird's naval launchings. The decline of average vessel size in a few areas stands out significantly: Aberdeen, reduced from building the finest clippers of their day to building trawlers and whalers; Dundee and the Humber, reduced from oceangoing vessels to small coastal craft; and Liverpool and London, in the course of losing their shipbuilding altogether. To an extent, however, these lower figures are deceptive in that yachts and other small craft are included only in the reports for the minor ports.

Comparing the actual output year by year in the various districts (see figure 1) shows that the discrepancies between the unofficial and the official statistics are considerable. The latter, collected by the Board of Trade, are more reliable, but do not include any warships of the considerable tonnage built for foreign flags. Moreover, the Board of Trade statistics are in net register tons, which are meaningless for gauging industrial activity, indicating at best the demand for shipping. The trade figures are in gross tons, with warship displacement tonnage added. They are more satisfactory in showing the material and labor that went into building, although a displacement ton represents an uncertain proportion more of both than does a gross ton. On the other hand, the trade figures were supplied by the builders themselves and are not quite reliable on that account. An enquiry made in 1883 showed that the Glasgow press had overstated the output on the Clyde by 34,000 g.t. in three years.[15] Not until 1892, when Lloyd's Register published the first of their invaluable series of annual statistics of shipbuilding, did the "unofficial" figures become at all reliable. With the exception of the vessels under 100 g.t., which are omitted from Lloyd's tables, the trade statistics and Lloyd's are comparable in scope. Considerable variations between the two sets after 1892 confirm the suspicion of unreliability of the trade figures before that date. Nevertheless, the discrepancies are not serious enough to warrant the exclusion of the trade figures before 1892, when they formed the only complete record of the quantity and types of output of the various shipbuilding districts.

In many cases, the specializations of the districts were, in part at least, determined by the effects of local conditions on their development. The Clyde, for example, is not a river favored by nature for shipping development. Its narrow, tortuous course prevented navigation by fair-sized vessels until successive works had dredged and widened the river, secured its banks, and deepened its draught in the first half of the nineteenth century. Since then, further improvements have kept the river abreast of shipbuilding requirements (except, perhaps, for the latest supertankers), and the largest liners and battleships have always been launched along its course. These, along with a multitude of dredgers, tugs, and pilot boats needed by river traffic, allowed the Clyde to remain throughout our period the most important shipbuilding river in the world. About one-third of new tonnage built in Britain was launched on the Clyde between 1870 and 1914. Altogether, the Clyde had the most varied output of all rivers. Among its products were large and small warships; mail liners; fast channel steamers; coastal, harbor, and fishing craft; steel sailing vessels; and after the depression years of 1884-1887, even ordinary and special types of cargo tramps. In addition, many foreign orders were placed there throughout the decades before 1914. Warship orders and foreign orders, and later also those from the largest mail steamship companies, prevented drops of output on the Clyde in depression years as disastrous as those of the Wear or the Tees.[16]

The easy availability of engineering skill and ready access to coal and iron on the Clyde were not the only reasons behind the district's early success in building iron and steam vessels. The early use of steamers on cross-channel trips to Ireland and the coasting trade to Liverpool furnished some badly needed capital from enterprising Scots shipowners who were encouraged to invest in new vessels during their erection. During the era of the timber sailing vessels, however, building on the Clyde was viewed with suspicion. In 1839, Samuel Cunard wrote to Robert Napier, "You have no idea of the prejudice of some of our English builders. I have had several offers from Liverpool and this place; and when I replied that I have contracted in Scotland they invariably say, 'you will neither have substantial work nor completed in time.' The Admiralty agree with me in opinion that the boats will be as good as if built in this country."[17]

After 1866, when iron and steamship building became dominant, it was in the Clyde yards that many of the highest-class vessels were built. Transatlantic shipping lines were mostly founded in Glasgow or Liverpool by Scotsmen or others with a due regard for the capabilities of the Clyde yards. The Cunard Company, for example, had its earliest ships built there exclusively. Thus some of the largest yards, some of the best equipment, and some of the greatest engineering talent were found on the Clyde in the 1870s, including John Elder and A. C. Kirk at Fairfield, William Denny and his family at Dumbarton, Robert Napier at Govan, and Rankine and Thomson at the University. In 1873, 93.5 percent of the

Clyde output was iron steamers, and in 1874, 74.5 percent. Wooden vessels made up only 1.1 and 0.5 percent of output in those two years. In 1876, more iron ships were built on the Clyde than in the rest of the world put together.[18] Steamers too, were built early on the Clyde: over the period 1867-1873, 66.8 percent of the Clyde tonnage of mercantile vessels for the home register were steamers, while the average for the whole country was only 60.2 percent. The short revivals of the steel sailing ship in the depressions after 1883 and 1896 were based on the Clyde as well.

Shipbuilders frequently divided the Clyde building district into the Lower Clyde—Greenock, Port Glasgow, and Dumbarton—and the Upper Clyde—Glasgow and its suburbs. The output of both reaches was similar, and there were some giant yards in both. The number of firms on the river remained fairly constant: about 35 firms between 1857 and 1877, about 45 between 1880 and 1905, and 39 in 1913, not including boat, yacht, and barge builders.[19] The number of men engaged in shipbuilding on the Clyde is less easy to establish. The census figures of shipbuilding trades were underestimates before 1901. Private estimates frequently included engineering and therefore are equally unusable. The following estimates are submitted with the proviso of a possible error of $\pm$ 10 percent: 1852, 11,000 men; 1866, 17,500; 1877, 21,000; 1894, 24,000; 1901, 28,600; 1911, 54,769.

For most years, the Clyde enjoyed the local advantages of cheap supplies of iron and steel and cheap labor. At times, particularly after the formation of the Anglo-Scottish Steel Cartel in 1906, local builders actually received their steel at prices 5/- to £1 above those charged at Belfast by Scottish makers, and in the Scottish wage area—which included Barrow and Belfast—Clyde wages were actually the highest.[20] But by and large costs favored the Clyde builders.

On the North East Coast as a whole more ships have been launched than even on the Clyde since the 1880s. But its various districts—the Tyne, the Wear, the Tees, and, in later years, the Hartlepools, Whitby, and Blyth—have not always had parallel paths of development.

In the timber days, the Wear was the chief shipbuilding district on the North East Coast. The vessels launched there were intended for the coastal or Baltic trades only, but they were built competently and cheaply. The Wear was the district par excellence of small builders, skilled shipwrights who tried independence with limited capital or with loans from the timber merchants. From the 17 yards on the Wear in 1804 and 20 in 1814, the number rose to 34 in 1833, 80-90 in 1840, and 70 in 1852, but declined again to about a dozen in the depressions of the middle 1840s and 1850s. The number of Wear shipyards was estimated at 71 in 1857 and at 100 by 1860, but the collapse of 1865-1866 finally ended the hope of survival of small timber yards. In 1884 there were 17 yards on the river, and from 1888 to 1914 there were 13 or 14. The vessels built there before the iron period were small, "a ship of 400 tons being constructed of prodigious and gigantic size," and, indeed, the river was not suited to any deep navigation.[21]

The Tyne was equally inaccessible to larger vessels until the various improvements, undertaken by the River Commission from 1850 onward, deepened the river, made its banks secure, and provided harbors and docks along its shores. Before 1850 the Tyne was a less important shipbuilding river than the Wear, but the iron, engineering, and armaments works along its course, and the needs of its coal exporters, rapidly created a large iron shipbuilding capacity there. In 1842 the first iron vessel of the North East Coast was launched on the Tyne. Armstrong's works were opened at Elswick in 1847, and in 1852 Charles Mark Palmer built his first iron steam collier. The Tyne, together with the Tees, was among the earliest British districts to change over completely to the construction of steamships. After 1863 few sailers were built there, but on the Wear, the Clyde, and at Belfast, Dundee, and Aberdeen, the change was not completed until 1878. From the 8 small builders, employing 239 shipwrights, in Newcastle in 1825, the number expanded to 20 large yards in 1890, and 15-16 in the last 20 years before 1913. With the phenomenal growth of population along the banks below Newcastle—at South Shields, North Shields, Tynemouth, Jarrow, Wallsend, and Hebburn—the Tyne became the cradle of some of the largest ships, the only river on the North East Coast building ships other than cargo vessels. Among them were the warships of Palmer and Armstrong, the giant liners of Swan, Hunter and Wigham Richardson, and the tugs, trawlers, and drifters of the Smith Dock Company. At the same time dock facilities and repair yards were improved to take the increased traffic of the enlarged port.[22]

The Tees Conservancy Board, formed in 1852, was, like the Tyne Board, commissioned to improve the shipping of coal to the London markets in competition with the railways. This enabled a flourishing shipbuilding industry to develop in Stockton, Middlesbrough, and in the Hartlepools. Most of the eight or nine large shipyards in the region had been founded in the age of iron or steel and specialized in cargo tramps. At the same time, the old building ports of Scarborough and Whitby declined into insignificance.[23]

Though they paid higher wages than the Scottish areas, the North East Coast builders generally enjoyed the benefit of cheaper freight for coal and iron. William Denny also attributed the success of shipbuilding on the North East Coast to the close contact among small owners, the general public, and family builders in the supply of capital (only the Tyne had a number of joint-stock companies); the cheapness of iron; the longer hours worked; the greater proportion of piece work; and the strong competition among a large number of builders, all producing ships of substantially the same class.[24]

From 1890, wages and conditions over the whole of the district were almost identical. The strong competition, as well as the large number of repeat orders that were common in the region with "almost a monopoly" in the building of cargo tramps, led to a high degree of efficiency, ingenuity, and specialization on the North East Coast.[25] On the other hand, this same dependence upon tramps meant that output fluctuated more on the North East Coast than in most other areas.

A segment of the labor force for the new shipbuilding yards—and towns—in the district was furnished by immigrant Scotsmen from the underpaid East Coast areas in search of higher wages.[26] For the district as a whole, shipbuilding employment increased as follows: 1865, 12,000; 1892, 29,000; 1898, 29,750; 1901, 42,802; 1911, 46,906.

The Thames shipbuilding industry suffered a total eclipse after 1865, sinking from the position of the leading producer to that of a repairing area by 1912. Before 1865 the Thames yards produced a large tonnage of ships of the highest quality. The East Indiamen, Blackwell frigates, men-of-war in the sailing era, besides the early marine steam engines and iron ships—all were built on the Thames. In 1863 the annual output of the Thames was estimated at 117,000 g.t., or nearly a quarter of total British output.[27]

After the collapse of 1866, however, the higher cost of wages, rents and rates, and freight for coal and iron supplies, combined with the congestion on the Thames, made competition with the North impossible on a commercial basis. But Thames yards and works continued to turn out marine engines—mainly for warships—naval vessels for home and abroad, river boats, barges, and other special craft. By the 1890s even the building of warships and their engines was threatened by the "fair wages clause," which forced contractors of the government to pay the current wages of the district. Humphrys, Penn, and Maudslay, the three large Thames marine engineering firms, were shown to pay rates below the London level; the Thames Ironworks found it hard to secure any contracts after the building planned in the Naval Defence Act of 1889 had been completed; and Yarrow and Thornycroft moved from the Thames to Scotstoun and Southampton respectively. After a vain attempt to carry on with the help of small contracts from the London County Council and with battleships built at a loss which was met out of the earnings of other departments, the Thames Ironworks, the last important builders in London, closed in 1912.[28]

In the timber era, Mersey shipyards were second in importance to those on the Thames, but the peak of building there had been reached much earlier, in the late eighteenth century. Common shipbuilding ground was leased at 6d. a ton to the builders, who employed apprentices only. In independent contracts, journeymen took on the large amount of repair work found in an important port. During the Napoleonic wars, Liverpool was considered the cheapest repair port in the country, and building flourished there again after 1819.[29]

The City Corporation, however, was dominated by shipping interests who were frequently in league with the shipwrights against the builders. The builders were forced to use the corporation docks, and by short leases of 6 or 12 months they were kept continuously on the move as the corporation, with its expanding needs for dock space, encroached on their yards. As a result no heavy machinery could be acquired, and the existence of the industry was threatened when iron shipbuilding became important. A petition presented by the shipbuilders to the corporation in 1850, asking for a special area to be set aside for the builders, was shelved, and the industry was doomed.[30]

There was a short revival on the Mersey, as on the Thames, in the early 1850s and 1860s before the final decline. In 1853, 46,000 g.t., or one-fifth of the total British output, was launched on the Mersey. In 1863 there were still 10 reputable Liverpool firms in existence, but soon thereafter 4 of these had closed down and 3 others turned to boat building, repair work, and engineering respectively. The Liverpool Shipbuilding Company (Messrs. Quiggin) was then founded and enjoyed a brief prosperity: in 1865 the firm was credited with the launching of no fewer than five steamers on a single day. They built their last ship in 1875, however, as did Vernon's. T. Royden, the largest yard, closed in 1893, and R. and J. Evans followed in 1895. Only one minor yard was left by 1900, when Grayson's, a repair firm, started to build small vessels. By 1911 Grayson's had acquired two other small firms, but, apart from their yard at Garston, only repair work has since been carried on at Liverpool.[31]

The decline of building in Liverpool is not as easy to explain as that on the Thames, for raw materials were easily and cheaply obtainable. The causes are partly to be found in the uncompromising selfishness of the shipping interests, the absence of good railway links to the yards, the high cost of unskilled labor in view of the competition from the docks, and the high wages and strong bargaining position of the skilled men vis-à-vis weak and ill-equipped masters.[32]

Across the Mersey, in Birkenhead, shipbuilding was synonymous with the firm of Laird Brothers. Established in 1824 on waste land belonging to the corporation, Laird's were among the first to take up iron shipbuilding, and they remained important government contractors throughout our period. A tumultuous battle—in which two factions of land and dock speculators led by William Jackson and by the Lairds were locked in bitter struggle, while the development of docks and railways was thwarted by the jealous Liverpool Corporation—ended with the costly failure and financial scandal of the docks scheme in the crash of 1847. The population, which had risen from 2,500 in 1831 to 20,000 in 1843 and to over 40,000 by January 1847, shrank to 20,000 again in December 1847 when the Irish navvies left the town.[33]

Recovery came quickly in the early 1850s. Brassey persuaded the Liverpool Corporation to take over the bankrupt docks. Laird's expanded into new dock areas, and increased trade gradually brought prosperity to shipbuilders, dock owners, and land speculators alike.[34] Laird's weathered the storm of 1866, and after 1900 built a modern extension at Tranmere Bay which came into operation in time for the naval armament drive before 1914. After the amalgamation with Charles Cammell and Company, the Merseyside firm specialized almost entirely in the building of warships.

At first sight, few ports seem less suited than Belfast for large-scale shipbuilding. The shipyards were located 12 miles from the sea, on a river narrower than the Thames. Coal had to be imported from Ardrossan, steel from Great Britain or the United States, and timber from Canada and Norway. Yet its cheap labor and rent, the comparative cheapness of ocean freights for bulk goods, the assistance of the Harbor Board, the rapid adoption of new methods and up-to-date equipment, and its position opposite Liverpool made "the Lagan . . . the shipyard of the Mersey," and one of the most important shipbuilding areas in the United Kingdom.[35]

There had been some shipbuilding on the Lagan since at least 1791 when William Ritchie imported Scottish craftsmen to build his first ships. Ritchie and McLaine launched their first (wooden) steamer in 1824, and the Lagan Foundry built the first iron ship there in 1838. Modern shipbuilding did not begin, however, until the Belfast Port Authority cut the Victoria Channel, built a patent slip, and enclosed a shipyard on four acres of ground thrown up by the excavations. The yard was first leased to Hickson, the owner of a local iron works, who tried unsuccessfully to build iron ships on it. His young manager, Edward Harland, bought him out in 1857, ruthlessly depressed wages by the importation of workers from the Clyde and Tyne, and put Belfast's largest shipbuilding yard on a firm financial basis. The close relationship with the Harbor Board, in which the firm was the dominating partner, continued down to 1900. Each port and dock extension was partly financed by Harland and Wolff, and was laid out largely with an eye to their benefit. "We have never built a ship," Pirrie informed the Clyde Navigation Trust in 1922, "that the Harbour Commissioners of Belfast had not built a graving dock ready to accommodate that vessel the day we launched her." On the refusal of the commissioners to lease more land to the firm in 1899, Harland and Wolff rapidly acquired building and repair facilities elsewhere. Dublin was found to be unsatisfactory, but branches were opened in Southampton, Liverpool, Glasgow (repair), and Govan (building), while Pirrie urged the government in 1900 to transfer their dockyards to the shipbuilding rivers, so that the costs of harbor and river improvements could be borne by the Admiralty but benefit the owners, builders, and repairers as well. After 1880 there were two other shipbuilding firms on the Lagan, of which the larger, Workman, Clark and Company, also specialized in large mail and passenger liners.[36]

The local advantage of Belfast was partly one of cheap labor, especially the native unskilled labor. Skilled men came chiefly from the Clyde, whence they brought their unions and their accepted wage levels, the latter of which could be lowered slightly because of the lower cost of housing. After the disputes of the middle 1890s, wages and hours in Belfast were regulated by agreements on the Clyde, though Harland and Wolff remained outside the Employers' Federation before 1914. The gradual increase in the number of English workers, and the Scottish origins of so many of the Belfast shipyard laborers, are strikingly shown by the increasing dominance of Protestant Episcopalian and Presbyterian workers in the yards, to a much greater extent than in other industries in Ulster. As unskilled shipyard labor, in Great Britain as well as in Ireland, was largely Irish, skilled men would no doubt show a still higher proportion of non-Catholics.[37]

What shipbuilding there had been in the rest of Ireland sank to insignificance during the nineteenth century. Dublin, in particular—a port with an active trade during the Napoleonic wars when it had four builders and over 300 shipwrights—saw its shipbuilding decline in the following 25 years. By 1838 only 100 shipwrights remained. The industry in other ports such as Cork, Waterford, and Drogheda, whither some of the Dublin capital had fled in search of cheap unorganized labor, also languished with the decline of wooden shipbuilding. In the five years 1875-1879, all the Irish shipyards together, except Belfast, launched only 2,500 tons of shipping, of which only 250 tons were steam powered. The decline of Dublin has been blamed on the unions, but more impartial evidence suggests that neither wages nor other union demands were exorbitant, and that, in fact, it was the powerful shipping and dock-owning interests in Dublin that strangled the industry, just as they had done in Liverpool. Minor modern shipyards were laid down at the end of the century in Londonderry and in Dublin, but the other Irish ports, except Belfast, were limited to ship repair.[38]

The shipyard in Barrow was entirely a creation of the iron and steel age, as were the town and port of Barrow-in-Furness. A hamlet of 130 inhabitants in 1830, Barrow became the terminus of the Furness Railway, which was started in 1844 and was financed by a group of local mine and royalty owners led by the dukes of Devonshire and Buccleuch. On the basis of the rich hematite ores of its hinterland, and of the railways and docks built to exploit them, Barrow grew rapidly to a town of 700 in 1851, 10,000 in 1864, 18,000 in 1877, and 55,000 in 1891. The small band of financiers, consisting of the two noble families and their associates W. H. Schneider, James Ramsden, James Fisher, and two or three others, owned among them the railway, the docks, the land, and the iron mines. In the years from 1859 to 1874 they also founded the Barrow Hematite Steelworks, the gas works, the iron shipyard, two steamship companies, the Barrow Rolling Mills, a wire drawing mill, a linseed oil factory, the steam corn mill, most of the housing and slum property in the town to accommodate the influx of workers, and a jute works to attract the workers' wives. They ran the local government, the magistracy, and the *Barrow Times.*[39]

The grandiose scheme of developing a port and industry of the size of Liverpool failed. The hematite ores gave out, the main railway line bypassed Barrow, but the shipyard carried on. Staffed with technicians and craftsmen from the Clyde and with unskilled labor from Ireland, supplied with iron and steel plates from the allied steel works, with coal from associated collieries in Yorkshire and South Durham, and with armor plates after 1897 by the Vickers works themselves, the shipyard survived two reorganizations in 1888 and 1897. With the decline of most other local industries, it became the main local source of employment after 1900.[40]

Besides the shipbuilding districts just considered, there were a number of minor ports that usually specialized in various types of smaller vessels between 1870 and 1914. For example, Aberdeen, Dundee, Grangemouth, Grimsby, and Beverly built trawlers; Southampton and Cowes built fast small warships and yachts; while many more areas, ranging from Faversham in Kent to Bideford and Dartmouth, built smaller river and fishing vessels. There was boat building in nearly every port, but ordinary cargo and passenger vessels were built only by larger firms, and only in the major centers of the industry.

Although large cities such as Glasgow and Newcastle had important yards, shipbuilders in the new regions tended to settle in relatively small towns nearby, where labor control was easier and wages lower because fewer alternative jobs were open to residents. Here, builders were able to overshadow other employers and impose their stamp upon the society and politics of the towns. Changes in the location and concentration of the shipbuilding industry, however, were also due in large part to the changing nature of vessels. Many of the old shipbuilding rivers were no longer well adapted to building as ships increased in size. Rivers such as the Clyde, Tyne, and Lagan, which local authorities were willing to widen and dredge, retained their suitability, but at Liverpool, where the City Corporation favored shipping interests in their allocation of riverside land, shipbuilding was driven out entirely. As iron and steel hulls and steam power were adopted, the availability of raw materials and engineering skills also grew in importance. The Clyde and the North East Coast, which had good deposits of coal and iron ore and a tradition of engine making, were well placed to meet this demand. Although Belfast had neither, her cheap supplies of labor and the low cost of bringing materials by sea from nearby Scotland more than compensated for this lack.

4.

The Firms

MANY OF the forces that occasioned the northward migration of the shipbuilding industry also worked to bring changes in the organization and output of the firms. The trend toward larger and more complicated vessels meant corresponding increases in the size of shipyards and the number and complexity of the tools employed. Although additional capital continued in most cases to be raised through the reinvestment of profits, the traditional partnership form of organization became less and less appropriate as the average capital value of individual firms increased. Especially after 1900, many shipbuilding firms began to use registration as private limited liability companies as a refuge against problems arising from the death or withdrawal of a partner and rises in death duties. At the same time, many builders also specialized in the construction of a limited range of vessels, thereby curtailing their necessary expenditures on land and equipment and, in some instances, helping to smooth out fluctuations in output.

THE SHIPYARDS IN TIMBER DAYS

Before the mid-1860s, most shipyards, both in Britain and abroad, were surprisingly small and financially weak. This fact is often obscured by the disproportionate prominence given to the handful of very large establishments on the Thames, the Clyde, and the Tyne. In other countries where there were no giant firms, the small unit was unhesitatingly accepted as representative.

Wooden shipbuilding was carried on in practically every port of the kingdom, though not everywhere on a permanent basis. Carpenters, or local mariners and fishermen in association with one or two migrant shipwrights, might decide to build a small vessel or two at any place where water and timber were available, and then revert to their former occupations. In other ports, where shipbuilding was carried on more continuously, wooden shipyards were generally no larger than the workshops of other crafts. Very little capital sufficed, and shipwrights passed from the status of journeyman to that of master and back again fairly easily.[1]

The plant required was limited to a large crosscut saw, bolt cutters, borers, and a derrick. The other tools, such as the adze, caulking tools, and saw, were the property of the men, who therefore shared in part in ownership of the means of production. Yards could be laid out quite cheaply. In 1812 R. Morton of Dublin laid out a large shipyard and a house for himself at a cost of £5,000. In 1864 Barry still estimated the capital needed for wooden shipyards at between £1,000 and £5,000. William Stephen's business, which employed more than 30 men in 1813, was considered substantial, since it was assumed that a master and about 20 journeymen and apprentices could build four to six ships of an aggregate of 500-600 g.t. per annum.[2]

During trade disputes the men occasionally banded together without apparent difficulty to build ships cooperatively. The Dundee shipwrights, after a dispute, formed the New Shipwright Building Company in 1825. Although they failed to get the £200- £300 capital which they had hoped for from the benefit fund and had to operate in a period of depression and falling prices, they kept going for more than six years.[3] A. Stephen recalls in his diaries a similar venture, the Tay Shipbuilding Company formed by 15 workmen and in existence at least from 1847 to 1851.

There were some larger yards among the wooden shipbuilders. With more capital and larger labor forces, they carried their own stocks of timber, a service for which the smaller yards had to rely on timber merchants. Their yard buildings usually included draftsmen's offices, a molding loft—a floor on which the outlines of the vessels were traced at full size and "faired"—workshops for smiths and foundrymen making knees and castings, a copper workshop making nails and sheathing, and sheds for sailmakers, mastmakers, and boatmakers.[4] Steam was used to drive the saw and bend the timbers. Carrying was still done largely on the backs of the men, although some firms had erected derricks for lifting heavier pieces such as masts and engines.

Shipbuilding is essentially an assembling industry, building up its products from components supplied by outside firms, rather than working up one raw material into a variety of products. In wooden shipbuilding the assembling was done in the open air, on the building slip, which consequently became the most important section of the yard. Sloping down to the water in which vessels were to be launched, the building slips were, in the wooden days, nothing more than parallel wooden slipways on as solid a foundation as could be had. Some yards had building berths dug below the water level, protected by heavy gates, which were opened at high water for the launch. Dry docks for fitting out and repair relied on the tides for flooding and emptying. Some repair yards, however, had patent slip dry docks in which ships were rested on a carriage below the water level and pulled up inclined rails.[5]

To make their large capacity pay, the more important wooden yards needed long years of tradition, of accumulating a stock of seasoned timber and a clientele. Iron shipbuilding yards, on the other hand, although from the first they were built on a larger scale, could be erected almost overnight. This fact and the larger capital sum required made them more tempting objects of speculation. Barry estimated the minimum capital at £5,000-£25,000, while the largest Thames works needed at least £50,000-£250,000. Hall estimated their cost in America at $60,000, or $200,000-$1,000,000 for very large yards. Yet the equipment was still rudimentary: a few slotting, shearing, drilling, and punching machines; steam hammers of up to 30 pounds pressure; and a steam engine to drive them. In Chatham in 1861 the whole ironworking machinery amounted to only £3,000-£4,000.[6] But there had to be more buildings in an iron yard, and much of the preparatory work was done, not on the slips, but in the workshops. This in turn required improved means of transport within the yards. J. L. Thompson has described the development of shipbuilding practices on the Wear:

> With the coming of iron shipbuilding, the terms Shipbuilder, Shipyard and Shipowner, took on a new meaning. The Shipbuilder from being a working shipwright with a capital of a few hundred pounds, building a ship as a speculation and employing very few hands, became a principal in a big industry, requiring considerable capital, and employing a large number of hands. The Shipyard changed from any site on the river front, selected at will by the shipbuilder, with only enough space to allow for the building of a ship of small dimensions, with accommodation for a screw-pit and a pitch-kettle, to a large fixed area with the machinery, plant and equipment of a factory employing a variety of trades.[7]

Business Organization after 1870

The British shipbuilding industry remained until 1914 a family industry, into which little outside capital or outside influence was allowed to penetrate. Firms were owned by successive generations of fathers, sons, and brothers. Even when they were turned into limited liability companies, they generally remained private, and the shares were held by only a handful of people. Not until the twentieth century did genuine joint stock enterprise make much headway in the industry.

The industry was also characterized by a second factor: after about 1870 few entirely new yards were laid out in the United Kingdom. It was far more common to buy up an old derelict or bankrupt shipyard, or to improve and extend existing facilities, than to start entirely from scratch. Taken together, these two characteristics contributed greatly to the stability of the location of the industry. The initial capital outlay, after 1870, was large relative to the general size of productive enterprises, yet was not of a kind to tempt blind investors by either quick or high returns. Capital and enterprise, therefore, were usually local, from related industries including shipping and engineering, and they had to grow slowly, building on previous generations and ploughing back profits. Moreover, a shipbuilder produced very costly units of capital goods; he supplied a cautious crowd of conservative experts, the shipowners, and not a fickle general public. Methods of high-pressure sales and advertisement are likely to be irrelevant where performance is widely known and publicized. In consequence the yards, which were more permanent than the firms, grew slowly, sometimes changing their ownership but seldom attracting the attention of outsiders.

Some firms were handed down in direct succession within one family for generations. Such were the Scotts of Greenock, whose firm was founded in 1711, the Laings, founded in 1793, the Stephens, going back to 1750, and several others. Sometimes the line of transmission was less regular, as in the case of George B. Hunter, whose firm later formed one of the pillars of Swan, Hunter and Wigham Richardson. The son of a shipbuilder, Hunter served an apprenticeship with his cousin, worked at the Napier yard in Govan for a brief period, and managed his cousin's yard before his family subscribed enough capital to enable him to buy a building-and-repair yard in partnership with S. P. Austin. It was also common to find various members of the same family connected with several shipyards at the same time. A convenient arrangement before the days of interlocking directorates, it was practiced by the Dennys, the Napiers, the Piles, the Hendersons, and the Scotts in the middle of the century. Before the period of limited liability, close family ties were indispensable in spreading risks in an industry of widely fluctuating output in which the death rate of firms was very high.[8]

Some firms retained this close family organization right down to the First World War, even after their incorporation as limited companies. Among the 50 or so important shipbuilding firms in 1914, 13 still had boards comprised of the members of a single family.[9] Twenty-five years earlier, the list would have been very much longer.

By 1914 there were many shipbuilders whose families had been in the industry for one or more generations and who were gentlemen in the sense of owning landed estates or taking an active interest in politics.[10] There were also self-made men who had risen to become partners or directors. There were, however, almost no gentlemen in Coleman's sense (see introduction), since most shipbuilders preferred to educate sons who were going into the business in much the same way as their contemporaries from less-favored backgrounds. For all of them practical training or an apprenticeship was indispensable and appeared far more important than financial ability before 1890. The more fortunate started in the yards of their fathers, their families, or their friends; others had to pay heavily to become "premium apprentices." In both cases their future superior position was borne in mind, and after some brief spells in each of the chief branches of work—wood work, iron work, marine engineering—their main instruction was imparted in the drawing office. Some firms with high reputations, almost invariably due to their work for the Admiralty, were especially sought after as training grounds for premium apprentices. Robert and David Napier, Palmer's, Armstrong's, and Stephen's all trained the founders or heads of several other firms. Some builders, though, such as Harland and Wolff, attempted to educate all their specialists at their own works, from the managing director down to mechanics.[11]

Before 1890 the large majority of British builders and designers had practical training only. The few who attended institutions of higher learning treated their courses in mathematics or engineering as secondary to their main practical training. The luxury of academic training was confined almost entirely to those whose future was mapped out for them in their fathers' yards; others would not have found it useful in seeking employment in British yards. Among the British institutions of higher learning that attracted future shipbuilders and marine engineers, Scottish universities—particularly Edinburgh and Glasgow—and the English Nonconformist academies predominated. Wigham Richardson (b. 1837) and C. A. Parsons were notable among the very few cases of men for whom study seemed the primary and practical training the secondary consideration.

Thus, even in the most competitive period of this highly competitive industry, self-made men were the exception. It was still possible at the beginning of the nineteenth century for mechanics and other craftsmen to establish their own firms, which, in time, could grow to representative size. Later on in the century it might be possible to rise from common apprentice to owner of a firm with the help of capital supplied by members of one's family, as did Alfred Yarrow, G. W. Wolff, and C. S. Swan. However, only a few shipbuilders rose from apprentice to draftsman and to manager, and so reached a position in which they could acquire sufficient capital and connections to set up a new yard. Toward the end of the century, when public joint-stock enterprise invaded related industries and finally shipbuilding, the direction of affairs began to slip into the hands of nonspecialist businessmen who made up for their lack of experience in naval architecture by their ability to organize large masses of capital. Some, like C. M. Palmer, William Gray, and Christopher Furness, entered the industry from the related fields of shipping, coal mining, and iron manufacturing; others, such as Charles McLaren, a lawyer who became chairman of Palmer's in 1897, treated their firms merely as fields of investment and hired specialist aides.

Joint-stock enterprise in shipbuilding began early, immediately after the 1855 Act was passed. There were two distinct phases of company formation in shipbuilding. First came the rapid but ephemeral growth of 1856-1865, with its center of gravity on the older rivers, the Thames and the Mersey. This stage was a response to temporary boom conditions and to the speculative fever of the early years of joint-stock legislation. Then followed a period of slower growth, centering on the northern ports, after 1866. The latter phase did not become significant, however, until after 1886.[12]

In the first period, the capital was commonly blind and the investment had to compete in attractiveness with other industries. Only a small fraction of the capital subscribed had to be paid up at once, and those early joint-stock companies were, in consequence, usually overcapitalized, their yards laid out on too large a scale, their prospectuses too optimistic, and their death rate very high. Their rate of formation increased in those 10 years after the Act, reaching a peak in 1864-1865; 4 effective companies were formed in 1856-1859, 5 in 1860-1862, and 17 in 1863-1865. Of those 26 companies, only 16 survived after 3 years, and only 8 after 10 years.[13] The period 1855-1865 also saw the opening of many private yards. Some of these firms were as short-lived and apparently as speculative as the ephemeral joint-stock companies, but many were the foundations of successful businesses, including Randolph and Elder, Wigham Richardson, Robert Duncan, Harland and Wolff, Charles Connell, John Readhead, and Sir Raylton Dixon.

Of the 92 limited shipbuilding firms attempted in England between 1866 and 1883, 28, or more than 30 percent, were abortive or small. Of the 64 effective registrations, 38 were attempted during the prosperous periods of 1872-1874 and 1881-1883. Only half of the 56 effective public companies survived for more than 10 years, and 25 percent were either sold or dissolved within 3 years of their registration.[14] Information on Scotland is less complete, but to judge from Marwick's research, few builders incorporated before 1900.[15]

Status as a limited company was valuable not only because it protected shareholders in case of bankruptcy but because it facilitated mergers and sales. Incorporation also provided legal continuity in ways that a partnership did not. Increases in death duties before the First World War were a great incentive to incorporate, and for many family firms registration as a private limited company was the solution. In an economic sense, private companies were much like partnerships, in that public subscriptions of capital were not sought and disclosure of accounts was not required. The number of shareholders was often small and the members related or personally acquainted. Before 1914, less than half of the Clyde shipbuilders had incorporated. The total capital of the limited firms exceeded £7 million, of which £2 million was in private companies, but most of the shares in limited firms were privately held and not quoted. Even many of the largest firms were reluctant to incorporate. William Denny and Brothers did not become a private limited company until 1918 when, under the pressure of increased taxation and death duties, it merged with its sister company, the marine engineering firm of Denny and Company. Alexander Stephen and Sons had incorporated around the turn of the century following the death of the principal partner. It did not become a public company until after 1945 when, with evident distaste, the family felt obliged to seek outside capital because of "high levels of death duty and other taxation," which war had again made necessary.[16]

For those companies which did incorporate as public companies, the changeover was slow and did not begin until the depression of the 1870s when it was usually the result of financial troubles or the desire to expand. As capital requirements increased, self-financing became increasingly difficult. The depression of the middle 1880s changed the structure of a number of firms. Apart from the bankruptcy or reorganization of a number of well-known yards, the slow metamorphosis into public companies began. Information collected from the company files at Somerset House shows that this change was usually completed by stages, each new financial embarrassment or physical expansion leading to a higher stage: from partnership to private company; to a wider, though still restricted, professional or local ownership of shares; to the issue of one group of securities—preference or debentures—to the free market; and finally to the public issue of ordinary shares. The process can be traced by the share values adopted. Shares of £25 and up to £1,000 each usually signified private or family ownership; £5 or £10 shares, a wider ownership; £1 shares, a public company. The proportion of shares of smaller value was clearly increasing.[17]

Throughout the period, the growth of individual firms remained a dominant driving force behind company reorganization. Details of this process are severely deficient, since most firms did not publish any information of value to the economic historian. Occasional glimpses, at irregular intervals, must be used to illustrate the growth that went on continuously. This reconstruction from the material available is even more difficult because employment fluctuated very widely. For instance, the number of men employed at the height of a boom might be four times the number a year or two later in a depression. In any case, the available data are usually suspiciously round numbers. Figures for the capital involved before limited liability days are almost entirely lacking, and are certainly insufficient to form a connected story.

The evidence seems to show that up until 1893 growth meant a roughly proportionate increase in labor, capital, and acreage—a growth in breadth. Only in the 1890s and after the turn of the century was there a pronounced growth in the value of equipment employed per unit of labor. It will be obvious that this conclusion is arrived at as much by the study of the development of yard machinery (see chapter 6) as by the fragmentary evidence discussed here.

If the labor employed and the acreage of the yards are at all reliable indicators, growth was continuous throughout the period from 1870 to 1914, as is to be expected given that the average output of most firms and the size of ships were steadily increasing. The growth in the capital employed can be followed only in the latter half of the period, when a sufficient number of firms published balance sheets. The figures collected refer, as far as is ascertainable, to issued (not nominal) capital, including debentures.[18] Their meaning is somewhat ambiguous. In most cases the increases in issued capital—especially in the case of armaments firms—represent amalgamations with independent firms rather than improvements in the technical equipment, which were financed mostly out of reserves. Nevertheless, as far as they go, the figures show a steady rise in capital resources, interrupted by occasional writings down of capital.

A comparison of the amount of capital to the labor employed would be interesting, but cannot usefully be attempted for lack of material.[19] Annual valuations of the Clydebank Shipyard, however, are available from 1884 onward. When allowance is made for depreciation, these show an almost steady increase in all categories of fixed capital in the shipbuilding department, including a more than sixfold increase in the value of the machinery employed in the three decades before 1913. (See figure 2 and appendix table B. 10.) In general, this thesis of an appreciable increase in fixed capital is supported by the fragmentary evidence given in table 4.1. The amount of work in progress, that is, under construction, would vary widely with the state of trade, while the fixed capital would remain fairly constant, so that the final figure is to some extent a function of the state of prosperity.

The figures that follow must be taken as highly tentative. Although it

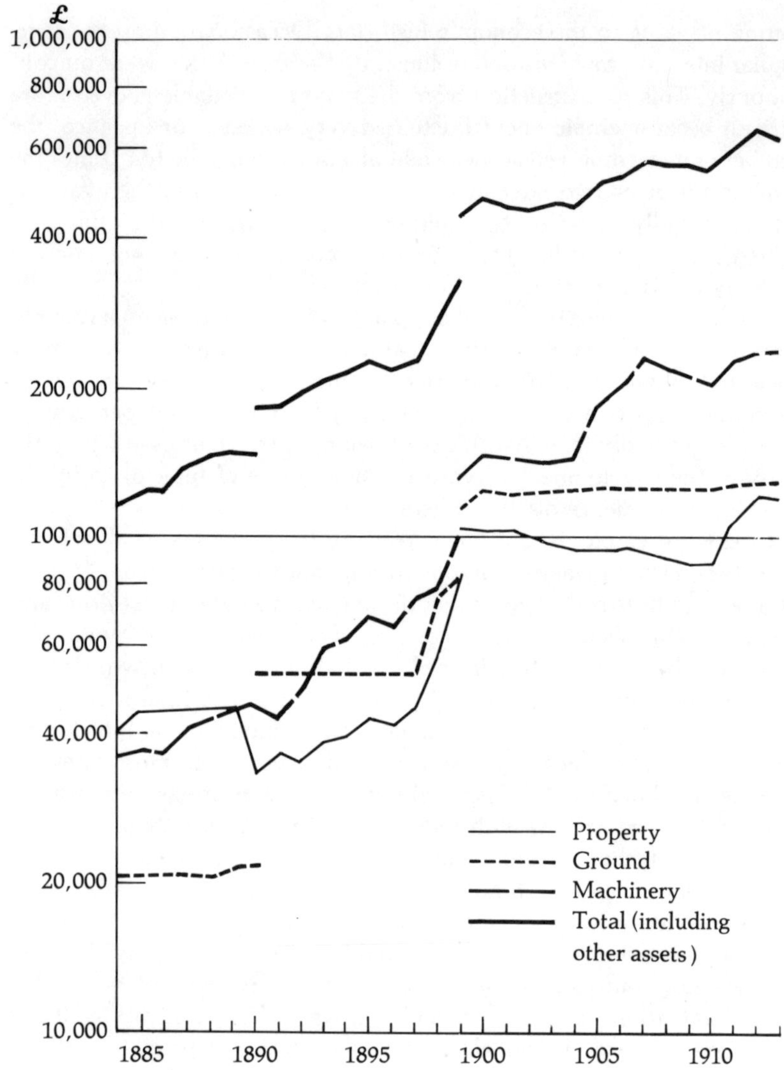

Figure 2. Assets of the shipbuilding department of the Clydebank Shipyard, 1884-1913. (Appendix table B.10.) Note: The breaks in 1890 and 1899 are due to revaluations made in those years.

would be dangerous to generalize from the limited sample of firms whose records we have been able to use, it is obvious that a variety of accounting methods were employed within the industry, with corresponding variations in the accuracy of the data generated. Some concerns, such as J. and G. Thomson, made regular valuations of their plant and other property from as early as the 1880s. Even these must be used with care,

Table 4.1. Fixed and variable capital of some shipbuilding firms (in £000).

Firm and year	Fixed capital[a]	Stocks and tools	Work in progress (W.I.P.)	W.I.P. as a % of fixed capital
Henry Withy, 1891	70	30	31	44
Richardson, Westgarth and Company, 1902	682	58	166	24
Swan, Hunter and Wigham Richardson, 1903	518		131	25+
Sir J. Laing, average 1911/1912	185	39	41	22
Swan, Hunter and Wigham Richardson, 1912	1067	184	155	15
Palmers, average 1912/1913	906	402[b]		22[b]
William Doxford, average 1908/1910	441	115[b]		13[b]
William Doxford, average 1911/1914	441	168[b]		19[b]
S. P. Austin, average 1908/1910	158	13[b]		5[b]
S. P. Austin, average 1911/1914	149	21[b]		7[b]

Source: File of the Registrar of Limited Companies, London; the *Statist;* the *Economist.*

a. Land, buildings, goodwill.

b. Assumes that half the figure represents work in progress.

however, as depreciation policies were frequently arbitrary. Rates of depreciation of machinery, for example, were low, and only in the twentieth century was it found that plant became obsolete long before it wore out, necessitating an increase in depreciation allowances. In 1886 it was revealed that Palmer's assigned "fixed charges" for machinery of 6.75 percent of the total cost of output when the works were totally employed, and 8 percent when they were only 40 percent employed. Armstrong, Whitworth pursued a similar policy after the turn of the century.[20] Some important yards had even more irregular accounting practices. A survey of the records of William Denny and Brothers turned up no regular valuations or depreciation procedures. For decades, it was simply assumed that the firm was worth £250,000. After the collapse of the La Platense Flotilla Company in the early 1890s, the Denny partners wrote off an investment of over £259,000—more than the nominal capital value of the partnership. Nevertheless, the firm was able to continue trading because the loss consisted of bank loans that had not actually matured when they were written off, and the partners subscribed the necessary funds from their own fortunes as the debt installments came due.

Similarly, profit data must be regarded cautiously because of erratic depreciation policies. Some firms such as Denny's did not regularly compute either sales or profits, and it is unlikely that the partners ever had an exact picture of the firm's financial position. These variations in practice can possibly be attributed to differences in the size, complexity, and legal status of the builders. Large, publicly owned concerns like John Brown or Armstrong, Whitworth, which had several diverse divisions, needed to employ stricter accounting practices in order to properly control their organizations and to satisfy their shareholders. Smaller firms such as William Denny and Brothers, however, in which the few partners were actively engaged in the business, could get by with looser bookkeeping because the owners were in a better position to obtain at least an impressionistic picture of the firm's liquidity.

The sums involved in British shipbuilding, especially the fixed capital, seem very large for individual or close family ownership. (See tables 4.2 and 4.3.) Before the 1890s, however, when electric, hydraulic, and pneumatic power transmission and a whole range of new machine tools made the equipment of the works very costly, tools and stocks formed a much smaller proportion of the work in progress. Stores, such as iron, plates, timber, and coal, did not amount to more than 20 percent of the value of annual output in 1868.[21]

In 1875 an American estimate put $160,000 as the minimum cost of a yard on the Clyde. This sum would have been grossly insufficient 20 years later. Few individuals could buy yards outright—A. C. Kirk and J. J. Hamilton bought R. Napier's firm in 1878 for £270,000, and C. H. Wilson bought Earle's, valued at £402,000, for half that sum, but this was not the normal form of a change of ownership.[22] Where a transfer to a joint-stock company was involved, little cash usually changed hands. The vendors, who were proprietors of a private company, generally received shares in payment, and the other shares went to the outside public

Table 4.2. Capitalization of selected British shipbuilding firms, 1900-1901.

Capitalization[a] (£)	Number of firms
0-200,000	5
200,001-300,000	7
300,001-500,000	9
500,001-1,000,000	5
Over 1,000,000	5

Source: Schwarz and Halle, *Die Schiffbauindustrie*, pp. 174-176.

a. Average capitalization of 31 firms = £778,014; average capitalization less figures for 4 firms engaged in armaments manufacturing = £411,409.

Table 4.3. Capital of selected shipbuilding companies, 1900-1912.

Company	1900-01 (£)	1912 (£)
Swan, Hunter and Wigham Richardson	—[a]	1,773,577
William Doxford and Sons	?	450,000
Cammell, Laird and Company	—[b]	2,372,895
Vickers	4,800,000	5,200,000
Armstrong, Whitworth	4,210,000	4,210,000
William Beardmore and Company	1,500,000	2,000,000
J. I. Thornycroft	260,000	346,500
John Brown and Company	2,500,000	3,423,000
S. P. Austin and Company	135,000	135,000
D. and W. Henderson	600,000	525,000
Sir James Laing	?	138,786
Hawthorn, Leslie and Company	—[c]	462,700
Smith's Dock Company	?	600,000
Palmer's Shipbuilding Company	498,765	698,564
Fairfield Shipbuilding and Engineering Company	500,000	500,000

Source: 1900-1901, Schwarz and Halle, *Die Schiffbauindustrie*, pp. 174-176; 1912, Kirkaldy, *British Shipping*, appendix VII, p. 591.

a. In 1900-01, Wigham Richardson and Company had a total capital of £350,000.

b. In 1900-01, Laird Brothers had a total capital of £450,000.

c. Schwarz and Halle give two estimates of the capital of R. and W. Hawthorn, Leslie and Company in 1900-01: £557,300 and £457,700.

in return for the cash necessary for the expansion of the firms. In the few public companies, ownership differed little from the public companies in other trades;[23] the semipublic companies, on the other hand, had mostly local capital, and generally their shareholders were shipbuilders and shipowners themselves. Other firms, such as the Dundee Shipbuilding Company, founded in 1894, included many of the foremen and staff among the shareholders, a development that later copartnership advocates such as Furness tried to foster in other works. Finally, family capital frequently started successful businesses, including Harland and Wolff and Yarrow's.[24]

These diverse origins may account for the large concentrations of capital and labor comparatively early, without the intervention of outside finance. There is no doubt that firms were large for their time. The iron shipbuilding yards employed an average of 574 workers and 109 steam horsepower in 1871, and the figures were highest in the newer areas into which the industry was moving. On the Clyde, for example, there were

852 workers and 176 h.p. per yard.[25] There were a few large wooden shipbuilding yards, but the majority of them did not even come under the Factory Acts at that time. The capital necessary for a fair-sized shipbuilding yard was £250,000 to £500,000 in the period 1866-1900, and £300,000 to £1 million in 1900-1914.

To what extent did these increased investments pay off? Net profit figures, both in absolute terms and as a percentage of invested capital, are available for John Brown and Company and for the Elswick and Walker Shipyards and the Engine Works of Sir W. G. Armstrong, Whitworth and Co., although in all cases only for a short period after the turn of the century (table 4.4). The figures are not entirely straightforward because Armstrong, Whitworth in particular varied its rates of depreciation from year to year depending on decisions by the directors.[26] Furthermore, neither John Brown nor Armstrong, Whitworth was a typical builder. The former specialized in large liners and warships and the latter in warships at its Elswick yard. The Walker yard of Armstrong, Whitworth, which undertook a greater proportion of commercial work, was more representative of the industry as a whole. Nevertheless, the figures illustrate some of the features that would be expected, most notably the severe fluctuations in profits, with losses incurred in poor years. Over the entire period 1900-1910, profits averaged a respectable, but by no means spectacular, 8.10 percent for both shipbuilding and marine engine building at John Brown and Company. As the actual sums lost by the Armstrong, Whitworth yards are unavailable, no comparable calculations may be made, but it should be noted that in 6 of the 11 years rates of return were higher at the Elswick Shipyard, which specialized in warships, than at the more commercially oriented Walker yard. Rates of return at the Armstrong, Whitworth engine works were consistently high, never dropping below 16.69 percent over the period.

Our findings seem to contradict Lewis's contention that at the end of the nineteenth century the British economy was in a rut and overcommitted to sectors with low growth potential. Although descended from an older industry, the iron and steel steamship-building industry was for all practical purposes itself a new field in which both the product and the techniques of production differed almost totally from those of earlier decades. Despite cyclical fluctuations, the rate of growth of output, as measured by the tonnage and value of the vessels launched, was substantial between 1870 and 1914, and the capitalization of the firms necessarily increased rapidly as vessels became more complex. There is nothing to indicate that the industry was starved for capital while outside investors or the builders themselves sent their funds abroad or placed them in consols or in industries with a lower rate of technological progress. If anything, the poor record of many shipbuilding firms that were publicly

Table 4.4. Net profits, capital values, and rates of return for the shipbuilding and marine engineering operations of John Brown and Company, and for the Elswick and Walker Shipyards and the engine works of Sir W. G. Armstrong, Whitworth and Company, 1900-1912.

	John Brown			Armstrong, Whitworth								
				Elswick Shipyard			Walker Shipyard			Engine works		
Year[a]	Net profit[b] (£)	Capital value[c] (£)	Rate of return[c] (%)	Net profit (£)	Capital value (£)	Rate of return (%)	Net profit (£)	Capital value (£)	Rate of return (%)	Net profit (£)	Capital value (£)	Rate of return (%)
1900	71,332	763,329	9.34	—	—	—	—	—	—	—	—	—
1901	29,783	763,340	3.90	—	—	—	—	—	—	—	—	—
1902	42,703	754,325	5.68	14,472	124,546	11.62	43,510	103,409	42.07	49,907	180,925	27.58
1903	85,052	754,207	11.36	38,404	119,993	32.00	38,112	104,645	36.42	42,560	175,065	24.31
1904	14,126	753,684	1.87	122,497	120,169	101.94	4,826	106,434	4.53	62,360	170,271	36.62
1905	72,827	761,654	9.61	88,173	119,534	73.76	896	105,734	0.85	61,990	167,622	36.98
1906	137,011	803,942	17.04	4,449[d]	127,380[d]	3.49[d]	29,311[d]	104,640[d]	28.01[d]	42,529[d]	164,160[d]	25.91[d]
1907	81,230	822,107	9.88	88,878	184,206[e]	48.25	39,393	174,238[e]	22.61	32,945	197,386[e]	16.69
1908	(5,569)	855,814	(0.65)	2,230	279,732	0.79	31,870	181,912	17.52	39,706	207,893	19.09
1909	154,287	845,529	18.25	LOSS	444,874	LOSS	LOSS	181,152	LOSS	36,114	210,869	17.12
1910	23,501	829,156	2.83	64,591	336,619	19.18	LOSS	159,786	LOSS	38,161	146,640	26.02
1911	—	—	—	54,039	331,801	16.28	LOSS	155,350	LOSS	41,597	144,864	28.70
1912	—	—	—	85,290	328,253	25.98	20,252	147,016	13.77	28,740	147,037	19.54

Source: John Brown & Co., File UCS 38/1; R. J. Irving, "New Industries for Old," pp. 157, 159.

a. Figures for John Brown are for years ending in November. For Armstrong, Whitworth, years 1902-1905 end in June; 1906-1912, in December.

b. Figures in parentheses denote a loss.

c. Capital exclusive of tools and utensils.

d. Half year, July-December, 1906.

e. The value of the capital stock of Armstrong, Whitworth was reassessed in 1907.

floated indicates that outsiders responded too eagerly to the opportunities the industry seemed to provide.

Firm Specialization

Except among armaments manufacturers, whose interests were broader and whose capital requirements were far greater, incorporation as a limited liability company was rarely sought by well-run shipbuilding concerns as a means of raising capital from diverse sources. Most firms that incorporated did so for an essentially conservative reason: to use the legal protection of a private limited company to maintain control in the hands of a small group of owners while continuing to manage their yards along old, established lines. To the extent that the existing management was inefficient, this barrier to the introduction of new men with new ideas might well have hampered the growth of the shipbuilding industry in Britain.

The specialization of output by class of vessel, on the other hand, was in two respects a move towards the more efficient use of capital equipment. First, by limiting the variety of their output, builders saved on plant and machinery by investing only in what was necessary to produce their particular classes of vessels (see chapter 6). Secondly, some yards, including those specializing in liners, warships, and, on occasion, popular designs of tramps, were able to some extent to even out fluctuations in output, and thus to use machinery more intensively. Thus specialization of output, which was possible only because of Britain's large volume of production, helped to lower costs and prices and to further her competitive advantage in shipbuilding.[27]

Only a few firms were called upon or offered to build warships. Among the yards building the larger ironclads were Laird Brothers of Birkenhead, the Barrow Yard, the Thames Ironworks, Earle's of Hull, the Clyde yards of Napier's, Fairfield, Clydebank, and Scott's and Armstrong-Mitchell and Palmer's on the Tyne. Smaller vessels were built by Yarrow, Thornycroft, J. S. White, and an increasing number of other commercial yards as World War I approached.

Other firms specialized in the large passenger liners: Harland and Wolff, and Workman, Clark of Belfast; A. Stephen and Sons, Charles Connell, Fairfield, and D. and W. Henderson on the Clyde; Swan, Hunter and Wigham Richardson on the Tyne. The many undifferentiated cargo tramps were built mostly on the Wear and the Tees, by firms such as William Gray, Furness-Withy, Irvine's, Ropner and Sons, and R. Thompson and Sons. Some firms were particularly associated with special kinds of tramp steamers, including the turret-deck steamers of Doxford's and the oil tankers of Laing's. Palmer's, in the first years after

1852, and S. P. Austin of Sunderland concentrated on steam colliers.[28] The Grimsby Shipbuilding Company, Cook Welton and Gemmel Ltd. of Beverley, and Mackie and Thomson of Govan were firms specializing in fishing vessels; William Simons built dredgers; Alexander Stephen built Arctic whalers; Murdoch and Murray built Amazon rubber trade steamers; and Denny's as well as Alfred Yarrow and J. I. Thornycroft built fast river and cross-channel ships. Russell and Company, for a short period up to 1894, specialized in mass-produced steel sailing ships, but after that date they reverted to tramp steamers. Raydon, Potter of Liverpool, Russell and Duncan on the Clyde, Mordaunt at Southampton, and Richardson, Duck and Company at Stockton were firms that had specialized in cheap sailing vessels in the short-lived sailing ship boom in the 1880s.[29]

It should not be thought, however, that firms kept exclusively to one type of vessel. In slack years, especially, any order would be executed. There was also a noticeable drive to rise to the two most coveted classes, capital warships and large subsidized mail liners. The prestige accruing to the firms building those vessels stood them in good stead when tendering for foreign contracts or for orders in times of recession.

From the comparison of the output of some important firms over a number of years, several broad conclusions emerge. The output of builders of cargo tramps fluctuated most clearly with the general trade cycle, after a lag of about one year. Some builders, though, who produced specialized types of tramps that enjoyed a widespread, if temporary, popularity, were immune from setbacks as long as the boom of their products lasted: Doxford's turret ships tided the firm over the intervening depressions with an increasing output from 1894 to a peak in 1906. The fluctuations of the output of warship builders were highly irregular. Not only did the naval programs and the tenders accepted from any particular firm vary according to no known formula, but in addition warships took up to three or four years to complete, so that the series of data showing launchings do not reflect at all reliably the incidence of orders given or work carried out during the year. The output of Palmer's, Vickers's, Laird's may show a slight tendency to vary with the trade cycle—not all their output was naval—but their main fluctuations appear arbitrary.

Firms whose output consisted largely or mainly of mail liners show the least violent fluctuations. Their customers, the liner companies, made their plans for long periods ahead and had sufficient resources to place their orders when building costs were lowest—in depressions—thus acting as a steadying, anticyclical influence. The output of firms such as Harland and Wolff, Workman, Clark and Co., and Swan, Hunter and Wigham Richardson kept remarkably steady.[30]

Despite the tendency toward specialization and the control that armaments manufacturers exerted over warship construction, the firm structure of the industry remained highly fragmented down to 1914. In 1913 there were 102 shipbuilding firms in Britain, with outputs ranging from 120 g.t. to 107,636 g.t. Most of the firms were quite large by the standards of earlier decades, of course. In 1883 there were 17 firms with outputs of more than 20,000 g.t.; by 1913 the number had risen to 39, and 55 firms had outputs of more than 10,000 g.t.[31] However, production did become more concentrated over the period. In 1883 the 10 largest producers accounted for 27.6 percent of total industry output. By 1904 the contribution of the 10 leaders had increased to around 40 percent, where it stabilized until the beginning of the war.[32]

The increase in industry concentration was caused in part by the increasing size of ships, which made it possible for the launching of only one or two giant vessels to place a firm near the top of the list. Before 1900 a larger number of ships had to be produced to achieve a comparable aggregate output. In 1883, for example, the Palmer Shipbuilding Company launched 36 vessels totaling 61,113 g.t., an average of 1,698 g.t. per vessel. Thirty years later John Brown and Company launched only five vessels to gain an output of 82,722 g.t., an average of 16,544 g.t. per vessel. Dependence on large vessels also meant that the industry was less concentrated than annual figures would indicate. The mix of builders in the first 10 places was continually changing because of fluctuations in the output of individual firms. For instance, only 6 firms were among the first 10 producers in both 1912 and 1913.[33]

As A RESULT of the increasing size and complexity of vessels, shipyards were transformed from small plots of ground with few buildings and little machinery into large and complicated assembly plants requiring much greater capitalization than had been needed in the days of timber sailing ships. To meet the challenge, most successful yards expanded steadily through the reinvestment of profits. A few firms, among them the large armaments manufacturers, also incorporated as public limited companies in order to solicit capital from outside investors. Many yards, however, were owned by a single family or a few partners. These were reluctant to incorporate for fear of losing control. When they finally registered, as many of the more substantial firms had by 1914, it was as private limited companies, a type of registration that allowed them to maintain control while gaining the stability that incorporation allows.

Many firms also reorganized their patterns of production. The output of British yards after 1870 consisted of a great variety of ships, and few yards could build them all. As soon as ships began to differ not only in size but also in general features and to become specialized for certain specific duties, shipbuilders, too, began to specialize. This specialization, possible because of the size of the industry in Britain, remained one of the causes of the country's relative cheapness and efficiency in constructing ships, because it led to savings in capital equipment and, in some cases, helped to smooth out fluctuations in production.

5.

Industrial Organization

THE DEVELOPMENT of iron and steel steamships forced yard owners greatly to increase and diversify both capital and labor inputs. The new vessels could not be built without equipment for moving and shaping huge metal components, and the talents of a wide range of skilled craftsmen were essential. The severity of the shipbuilding cycle, the threat of foreign competition, and other hazards, however, led builders to seek efficient ways of organizing their businesses in order to minimize risks and protect their substantial and ever-increasing investments. Two of the methods they hit upon have already been discussed: specialization of output, which allowed reductions in capital investment; and the trend toward limited liability, which made it easier for builders to generate capital for expansion and also offered some protection against taxation.

The growing size and complexity of vessels and the increasing amounts of capital necessary to run an up-to-date yard also affected the relationships between shipbuilders and their suppliers and customers. The response of the builders toward supply problems varied depending on the size of the builder, the extent to which it specialized in particular types of ships, the distance from raw materials and from specialist firms making fittings and components, and other factors. In some instances, builders found it best to integrate vertically with suppliers; in others, to remain independent. The aims, however, were always the same: to purchase components at the lowest prices possible and to secure steady supplies, especially as the peak of a boom approached. Connections with shipping lines were similarly advantageous, for they generated new orders, made specialization of output more feasible, and in certain cases even made it possible for builders to dampen the impact of fluctuations in demand by gaining control over the timing of orders.

The Supply of Components

Marine engineering became the most important of the subsidiary industries when steam replaced sail. Most of the larger firms added a marine engineering department to their yards or obtained control of an outside works. In this way they were able to enjoy complete independence in design, ensure a reliable degree of quality and date of delivery, and retain the profit margin on the engines and boilers. Before the supremacy of the steamer was assured, however, the majority of marine engineering firms had been independent, and many—such as Maudslay on the Thames and Robert Napier on the Clyde—were general engineers as well. Later, semi-permanent links were formed, as an intermediate stage, between certain shipbuilders and marine engineers. Harland and Wolff at first had their engines built at Maudslay's; and the association between John Wood, a timber shipbuilder, and Robert Napier, the engineer, lasted for more than 20 years, until 1841 when Napier began to build iron ships himself.[1] After 1870 this independence tended to disappear. In the years before 1914 there were 83 firms producing marine engines of over 500 h.p. per annum. Of these only 40 were even nominally independent,[2] and some, including the Central Marine Engineering Company and Furness-Westgarth, were in fact controlled by builders. Among the 25 leading producers, only a handful were independent, supplying for the most part tramp shipbuilders on the Wear, the Tees, and the Clyde. Among these firms were the North-Eastern Marine Engineering Company of Sunderland; Blair and Company of Stockton; and David Rowan and Company and Rankin and Blackmore, both on the Clyde. The Parsons Marine Steam Turbine Company, although not strictly comparable since it was founded solely to exploit Parsons's patent, must also be included among the largest independent firms.

Next to marine engineering, ship repairing was the most important adjunct of the shipbuilding industry. It was carried on in most of the important yards,[3] though of course there are no "output" data for repairs. In addition, there were specialist repair firms in ports such as London, Liverpool, and Bristol, which had lost their shipbuilding trade. The advantage of being on the spot enabled the repair firms in those ports to withstand northern competition.[4] Although repair was carried on in the timber days by small and often irresponsible firms (known as "pitch-pot men"), by 1914 many of the concerns had a capital foundation as imposing as that of the builders. Pirrie, for one, disliked building and repair work in the same yard, because of difficulties over wage rates. Repair firms in the smaller ports, on the other hand, found shipbuilding to be a necessity to provide employment in the frequent slack periods.[5]

The outstanding example of a comprehensive and self-sufficient combine was Palmer's on the Tyne. The firm was founded in 1852 to provide ships to take coal from the North East to London, thus preserving a market that the railway had recently made accessible to the Midlands and South Wales coal fields. The venture was an instant success and was the largest vertical combine in Britain when it was turned into a joint stock company in 1865. It possessed its own iron mines at Saltburn and Whitby and an artificial harbor at Port Mulgrave, which C. M. Palmer had built to load ironstone on his colliers on their empty return run from London. The ore was turned into plates and angles by the firm's own blast furnaces and rolling mills at Jarrow, which supplied the shipbuilding yard. Foundries for iron and brass castings supplied the engineering works, and all the accessories such as rigging and carpentry work were also made on the spot. The steam colliers built at Jarrow were then employed in carrying Palmer's coal to London. The firm also owned its own gas works and a bridge-building works at Howden, controlled a number of shipping lines, and after the introduction of the Siemens-Martin process sent its ships as far as Spain and North Africa to bring the hematite ore directly to the company's wharf. At nearly every stage in the process there was a surplus from the Jarrow works which was sold in its semifinished state to outside firms engaged in shipbuilding and other iron-consuming trades. Other yards found it convenient to buy pit iron, plates and bars, steel, and finished sections from Jarrow. This comprehensiveness was exceptional up to 1890, and so was the scale of the enterprise.[6]

Generally speaking, plates, bars, and angles formed the main raw materials of British shipbuilding. In the shipyards they were planed and sheared, bent or rolled to shape, and punched and riveted together on the slips. The manufacture of armor plates was similarly done by specialist firms, mostly steel and armaments works.[7] Gradually, as the benefits of specialization became clearer, and as the necessary machinery—which the shipyards might have to leave idle at intervals—became more costly, the steel works were expected to do an increasing share of the finishing processes:

> The flanging of boiler joints and ends, combustion chambers and furnaces, the rough and finished turning of shafting, the centre-boring of hollow shafting, the slotting or saw-cutting of crank-shaft webs and pins, the preparation of the several parts of crank-shafts of the built type; all these, and numberless other tasks, have now to be undertaken by works which formerly were only expected to supply material in the rough.[8]

A long and growing list of items was generally supplied to the yards by outside firms. Auxiliary machinery had its specialist builders. Windlasses, winches, steam steering gear, pumps, electric generating plant, refrigeration machinery, condensing plant, valves, fans, ventilating systems, electrical equipment, and various types of small fittings were all purchased from outside suppliers. Scientific instruments were provided by different firms, frequently on a patent basis; and as interior furnishings became more sumptuous and elaborate toward the end of the Victorian period, they too were usually supplied by outside firms.[9] As shipbuilding became more clearly an assembling industry, a crop of smaller, specialist concerns took over some of its activities, bringing with them the economic advantages of greater knowledge and skill, a wider market, and more mobile capital for the trade as a whole. Few engineering works at the end of the century prided themselves, as had their predecessors in the 1840s and 1850s, on the universal range of their output. The production of most components was, in fact, limited—either by patent or by the narrow extent of the market—to one or two firms. Furnishings were made by other specialists, and even consulting naval architects set up independently to supply the smaller shipbuilding firms as well as shipowners with plans and calculations.[10]

This specialization went furthest in the main shipbuilding and repairing centers, where large local markets encouraged investment in subsidiary firms—on the Clyde, the Tyne, the Wear, the Tees, the Mersey, and the Thames. Firms that were isolated geographically from the trade tended to make more of their components on their own premises. The outstanding examples are Harland and Wolff at Belfast and Vickers and its predecessors at Barrow-in-Furness.[11]

Where a new patent was involved, the makers either retained a monopoly of manufacture or licensed the rights to a small number of firms, distributed as evenly as possible over the shipbuilding districts of Britain and the world. Where further development appeared more important than immediate production, the patent holders merely continued with experimental work and left production entirely to licensees.[12] Occasionally one of the shipbuilding firms expanded some special line sufficiently to supply other builders as well as its own yard. This branch might continue to work as part of the main firm—as did the brass and copper works of Laing's, which supplied other firms, the government, and Laing's own yard—or be launched as a separate company, like Dr. Denny's Anti-Fouling Paint Company Ltd., formed in 1884.[13]

The high degree of supplier specialization achieved in Britain was almost unique. Only Germany approached the same level: all German yards made their own engines and boilers, used the plates and angles of the steel makers as raw materials, and had the other fittings made by outside firms, frequently by British specialists. American yards, which mostly developed from engineering works, and French yards, which were descended from timber shipbuilding yards, had to make their fittings themselves or apply to British suppliers. Moreover, French yards had their components made in dispersed localities and only assembled them on the slips, thereby appreciably increasing their costs of construction.[14]

C. P. Kindleberger has criticized British builders and suppliers (it is not clear for what period) for "dealing with each other at arm's length," thus impeding "technological change because of the possibility that part of the benefits of that change would have been external to separate firms."[15] It seems clear, however, that before 1914 the state of vertical disintegration was among the strengths of the British shipbuilding industry. It increased the rate of technological change by promoting competition among a large number of supplier firms, which were forced to innovate both in their methods of production and in the goods they marketed; and specialization among the component manufacturers led to increased returns to scale. The large number of firms and the high degree of competition in shipbuilding itself created conditions in which only the biggest concerns utilize the entire output of suppliers, and then only for marine engines and perhaps steel. No one of the builders could absorb enough capstan engines, dynamos, or other of the myriad components to be able to produce them on a scale large enough to compete with independent firms that specialized in their manufacture. It is possible, of course, that some builders could have produced components and then distributed them to the other firms and to concerns in other industries that needed similar equipment—as Laing's in fact did with brass and copper goods. It does not follow, however, that if this had been the general practice it would have resulted either in lower component costs for the majority of shipbuilders or in an increased rate of technological change.

Shipping and Shipbuilding

Close financial ties between shipbuilders and shipping lines could be profitable to both parties. Builders could attract more orders, and shipowners could purchase vessels at lower prices. Perhaps the most famous ties, because of their extent, were the Furness holdings, but the list is almost endless. In 1910 Lord Furness, the greatest individual shipowner in the world, was chairman or a director of four shipbuilding and marine engineering firms as well as controlling collieries, ironworks, railways, and insurance companies. Harland and Wolff was formed in 1858 with money put up by G. C. Schwabe, Wolff's uncle and a partner in the Bibby Line. Of the first 21 vessels built by the company, 19 were for Bibby. In later years, Viscount Pirrie, the chairman of Harland and Wolff, was chairman or a director of at least 10 shipping lines and an associate of J. P. Morgan in the International Mercantile Marine combine. At his death in 1908, William T. Lithgow of Russell and Company held shares in 81 shipping lines, many of them single-ship companies. The Scott brothers of Greenock were large stockholders in the China Navigation Company Ltd.; and the African Steam Ship Company was at different times associated with two shipbuilding concerns. Formed in 1851 by Macgregor Laird of Birkenhead, the line came under the control of Lord Pirrie in 1909.[16]

The close connections between shipping lines and shipbuilders were one of the sources of strength of the British shipbuilding industry before 1914 because they guaranteed repeat orders and saved on the capital required by allowing greater scope for standardized designs. They could also be used to counteract the business cycle by generating orders during recessions. Many of the vessels in which W. T. Lithgow held shares, for example, were built speculatively in slack periods by Russell and Company.[17] Little is known, however, of how such relationships between shippers and shipbuilders were conducted, of the financial arrangements that existed, or of the benefits that individual shipbuilders reaped.

From the unpublished records of William Denny and Brothers, it is possible to form an impression of how very important connections with shippers could be to one large shipbuilding concern. Between 1844 and 1914, Denny partners owned shares at one time or another in at least 19 shipping lines and on occasion operated vessels themselves. A balance sheet from 1859 shows that at that time over £87,000, or 64 percent, of the firm's assets of £136,634 were devoted to shipping interests. At his death in 1895, Peter Denny, for many years the principal partner, owned stock in 12 shipping lines.[18]

Although there were a few exceptions, the shipping investments of the Denny partners were in general unprofitable in the sense that the shares declined in value after they were purchased. Peter Denny, for example, paid £216,071 for shares which by 1895 were worth only £97,835.-15.0.[19] The lines in which the Denny partners were interested, however, accounted for a very large proportion of the shipbuilding firm's sales and profits between 1880 and 1913. Over the period, 60.8 percent of their cumulative sales and 68.5 percent of their profits were derived from transactions with shipping lines in which the Dennys had invested.[20] Whatever losses may have been incurred when the prices of shares in the lines decreased were more than recouped in profits from ships built for those lines.

Connections with shipping lines also allowed an extraordinary degree of specialization, which in turn made it possible for builders to save on capital and overhead costs. Most ships, and in particular large passenger liners, were built according to specifications laid down by their buyers.

Although several ships of the same design were occasionally ordered and yard owners were sometimes able to use similar vessels that they had built earlier as guides, there was seldom an opportunity to build many ships identical in size, shape, means of propulsion, or fittings. As a consequence, while some specialization of tasks and standardization of components were possible, there was in general little scope for the introduction of mass production techniques. It was imperative to design both slips and shops to allow for a maximum of flexibility in the shaping and assembling of components.

Close relationships, formal or informal, between shippers and shipbuilders led to repeat orders and the construction of nearly identical vessels. Again the best information exists for Denny's, whose partners had invested in several firms—including the Irrawaddy and La Platense Flotilla Companies, the Rivers Steam Navigation Company, and the India General Navigation and Railway Company—which used small river craft. Over the years, Denny's sold literally hundreds of vessels to these firms, vessels that were similar in design and in some cases could even be mass produced. Many were barges or small flat-bottomed passenger ships, which were simpler in design than the large liners that Denny's built for other lines. There were fewer options for fittings and the components could be standardized. It is indicative of the simplicity and degree of standardization achieved that many of these vessels were sent abroad in pieces and assembled thousands of miles from Dumbarton.

This specialization paid off handsomely for Denny's. The vessels they built for the river transport firms in which they had invested yielded a rate of profit of 17.99 percent, whereas the rate on the larger vessels they built for the oceangoing lines in which they held an equity was only 2.71 percent; and it appears that the construction of smaller ships for lines in which the Denny partners were not interested also yielded a higher rate of profit. In all, vessels built for the five river transport lines in which the partners held an interest contributed 45.6 percent of Denny's profits from 1880 through 1913 but only 14.0 percent of their sales. The specialization in flat-bottomed river boats, which had begun almost as an afterthought in 1876, paid striking dividends for the firm of William Denny and Brothers.[21]

The formation of shipping rings and combines was also intended in part to take advantage of the benefits that could accrue from specialization in shipbuilding. The first recorded shipping "conference" or "ring" was established in 1875 to regulate traffic to Calcutta, and by 1904 nearly all outward shipments from the United Kingdom were cartelized. Rings generally included liner companies but not owners of tramps, thereby encouraging the building of large vessels by raising freight rates for the liners within the rings above competitive levels in good years. In poor years, rings were again able to keep freights high for their membership, but tramps were hit more severely. The effect on shipbuilding, therefore, was a tendency to replace tramps with liners and to widen the amplitude of fluctuations in output.[22]

The most comprehensive shipping combination of all, the International Mercantile Marine, was formed in 1902. It was backed by J. P. Morgan's financial organization with the object of uniting American steel capacity and railroads with shipping in order to control the traffic eastward, to export steel cheaply, to find a market for the steel industry in shipbuilding, to press for American subsidies, and to prevent the building of the Panama Canal.[23] The I.M.M. was not a complete merger. The member companies retained their identities and did, in fact, revert to independent ownership upon its failure. Despite its ambitions, the combine never controlled more than 40 percent of the North Atlantic traffic and was unable to achieve its other goals. American subsidies and preference were not forthcoming, and as the stock was watered and tonnage had been obtained at exorbitant prices, no dividends could be expected for years. Gradually the constituent lines resumed their independent management and control.

Another combine closely allied with the I.M.M. was formed by the Royal Mail Steam Packet Company, in which Lord Pirrie, head of Harland and Wolff and one of the architects of the I.M.M., held large interests. The combine embraced nearly all important lines trading with the West Indies, South America, and Africa. Among its more important constituent members were the Union Castle mail line (itself an amalgamation of 1900), the Shire Line, the Pacific Steam Navigation Company, Lamport and Holt, and the eight important lines controlled by Elder-Dempster and Company Ltd.[24]

As early as 1911, 3.2 million g.t. out of a total British seagoing tonnage of 11.4 million g.t. were in the hands of 18 companies, most of which were independent only in name. By 1920, after the amalgamation process of the war years, five groups (four of which were linked by common interests) controlled an even larger tonnage.[25]

Those shipbuilders who were connected with the combines naturally did very well. Because of the vast quantity of equipment required to build the largest liners, it was imperative that firms that engaged in their construction maintain a constant stream of orders. Otherwise, there was no way of gaining proper value from their capital investments. This led Harland and Wolff into connections with, among others, the White Star Line, the International Mercantile Marine, and the Royal Mail shipping combine. Eventually, the fixed commitments of Harland and Wolff to these lines became so great that the firm was forced to expand and branch out in order to meet all its responsibilities. In 1907, it exchanged shares and directorships with John Brown and Company, apparently because the extended building program of the I.M.M. proved to be beyond the capacity of the Belfast yard.[26] The need of the Belfast firm to establish repair facilities for the associated lines in the chief ports, and a quarrel with the Belfast Harbor Commission, led also to extensions outside Belfast between 1909 and 1913. Ship repair yards were bought in Southampton, London, and Liverpool. In 1912, Harland and Wolff bought three adjacent Clyde shipyards and engine works, which were then laid out as a single large warship yard, and the firm acquired other properties in subsequent years as it sought to meet its obligations and aspirations.

The Furness interests were equally prosperous. By the end of 1912, Lord Furness owned or controlled 307 steamers of over 1.4 million tons in addition to other holdings in iron and steel and other related industries. In 1883 Furness had obtained control of the shipyard of Henry Withy of Hartlepool. With a firm basis in the demand for his many fleets and with the support of his interests in iron and steel, Furness added to his combine Irvine's Shipbuilding and Dry Dock Company of Hartlepool, the Northumberland Shipbuilding Company of Howden, and a dominant interest in Palmer's of Jarrow. At the same time he acquired marine engineering facilities by associating with Westgarth, English and Company in 1896. In 1900 he combined the new firm, Furness, Westgarth, with the Hartlepool Engine Works and William Allen and Company to form Richardson, Westgarth and Company. His acquisition of the Palmer interests also gave Furness some holdings in Robert Stephenson, marine engineers. Despite its large capacity, however, the northeastern shipbuilding combine was unable to supply all of Furness's needs, and some ships (40 in 1907 alone) were ordered from other firms. The depression after 1908 obliged Furness to "rationalize" his works by closing Withy's old yard and by concentrating the building of warships in Jarrow and the building of cargo vessels in Howden and at Irvine's. Throughout, Furness invested only capital obtained from his own firms, and by 1913 Furness-Withy controlled subsidiaries with a nominal capital of £4.5 million.

The Amalgamation Movement

In his study of the behavior of large American corporations in the first half of the twentieth century, Alfred D. Chandler, Jr., noted several justifications for vertical integration. In some cases, manufacturers integrated forward into marketing in order to reduce dependence on wholesalers or other middlemen and gain control over outlets. In other cases, there was backward integration which was "essentially defensive" in that it "resulted from a desire to have an assured stock of supplies at a reasonable price on hand and available whenever the manufacturing plants needed them."[27] Despite the continued heavy reliance on independent suppliers of many components and furnishings, both types of integration were common in British shipbuilding before 1914 with the important distinction that, while sometimes it was the shipbuilders who were integrating forward or backward in order to capture control over markets or suppliers, in other instances suppliers purchased shipbuilding concerns as part of their own campaigns to secure markets.

The relationships between shipbuilding, on the one hand, and components suppliers and shipping lines, on the other, have already been noted. Around the turn of the twentieth century, close connections between builders and armaments manufacturers also grew up and led through mergers and interlocking directorates to a corporate web embracing many of the most important shipyards in Britain. These combinations of the pre-1914 era were small in comparison to the giant mergers of the postwar period, but not insignificant.

The armaments mergers included, directly or indirectly, the great majority of the private warship building capacity of the United Kingdom, together with its ancillary industries. The early Admiralty contractors who controlled vertical combines, notably C. M. Palmer and Robert Napier, treated their associated coal mines, foundries, and engine works as useful aids in the profitable building of iron steamers. The drive to the later amalgamations, on the contrary, came from the producers of armaments, and of armor plates in particular, to whom the acquisition of shipyards meant the control of the market for their main products. The greatly increased naval estimates of 1889 and the following years in Britain, and a similar growth in naval expenditure abroad, encouraged this vertical and horizontal integration.[28]

Vickers acquired the Barrow shipyard in 1897, John Brown bought up J. and G. Thomson's yard at Clydebank in 1899, William Beardmore absorbed Robert Napier's shipyard in 1900 and developed a larger site at Dalmuir in the following years, and in 1903 Charles Cammell amalgamated with Laird Brothers of Birkenhead, acquiring in the process the large new site at Tranmere Bay. Armstrong-Whitworth, the most important armaments firm, bought Mitchell's yard on the Tyne in 1882 and branched out into armor plate production in 1902.

This was only a beginning. Each of these important armaments firms also acquired coal mines, iron and steel works, small arms and ammunition works, motor and engineering factories, and other plant auxiliary or complementary to their main output. They also developed or obtained control of armament and warship building capacity abroad: Vickers in Italy (Terni, Odero, Orlando), Russia (Tsarytsin, St. Petersburg), Canada (Montreal), Sweden, and Brazil; Armstrong in Italy (Ansaldo, Pozzuoli) and Canada; John Brown in Canada (Cape Breton Isle) and Russia (Nikolayeff); Thomas Firth in Russia (Riga) and the United States; Yarrow in Canada; and Palmer in Spain. Further links, forged among them and including other warship builders by 1914, make it permissible to speak of a single international warship and armaments combine.

The combine was linked by three primary types of association: (*a*) the complete merger or control involving the exchange of shares and directorates; (*b*) the common control of a subsidiary firm; and (*c*) the interlocking directorate. The most important complete mergers were Vickers and Beardmore (1902). Interlocking directorates, it should be stressed, frequently hid a stronger link that can no longer be traced but that may have been a foretoken of even closer combinations between the firms involved.

Expansion into both ordnance and shipbuilding seemed natural for the armor manufacturers. Sir W. G. Armstrong and Company had begun as artillery producers. In 1867 agreement was reached with Dr. Charles Mitchell to build ships at his Low Walker shipyard, which were then taken to Armstrong's at Elswick to be fitted out with armaments. In 1882 the concerns were amalgamated as Sir W. G. Armstrong, Mitchell and Company, and two years later shipbuilding was begun at Elswick as well. The Manchester engineering firm of Sir Joseph Whitworth was added in 1897; an armorplate works was erected there in 1900. Steel for nonordnance purposes had always been built at Elswick, and when a 10-acre steel works for ordnance was built in the early 1880s a journalist described the complex as "an arsenal in itself."[29]

Vickers began to produce armor and ordnance to counteract the effects of the recession of the mid-1880s. In 1884 Albert Vickers became chairman of the new Maxim Gun Company which merged with the Nordenfelt Gun and Ammunition Company in 1888. (From 1870, Armstrong's had produced the rival Gatling gun.) Finally, in 1888, Vickers acquired Maxim-Nordenfelt and the Naval Armaments and Construction Company of Barrow, another Nordenfelt concern, at a total cost of £1,778,000 to form Vickers, Sons and Maxim.[30]

John Brown and Company began to roll steel plates at their Atlas Works in Sheffield as early as 1859. In the 1890s, after compound armor had been superseded by steel plates rolled from single cast-steel ingots, John Brown's modified the Harvey process for hardening plates and licensed its patent to both Cammell's and Vickers. The Clydebank shipyard of J. and G. Thomson was eventually acquired in 1899 for £923,255. Three years later ordinary shares in Thomas Firth and Sons, Sheffield ordnance manufacturers, were exchanged for preference shares in John Brown's and a system of interlocking directorates was established.[31]

The relationships between shipbuilders and armaments producers became even more complicated after 1900, when William Beardmore and Company had purchased the old Govan shipyard of Robert Napier. Although Beardmore's later built the more modern yard at Dalmuir, the firm was plagued by financial difficulties. In 1902 Vickers obtained half of the ordinary shares of Beardmore's in exchange for 362,000 ordinary shares in Vickers. Vickers's ostensible purpose was to keep Beardmore's from entering the ordnance field, but Beardmore's owed Vickers a considerable sum of money, which Vickers wished to secure. Moreover, the firms' armor manufacturing facilities were complementary: Vickers made light armor and Beardmore's, heavy. In 1911 Vickers convinced Armstrong-Whitworth also to purchase an interest in Beardmore's. Vickers gave as their reason a fear that Beardmore was on the verge of regaining control of the firm, but in reality Beardmore's was £750,000 in debt and needed more funds to carry on business.[32]

By 1914 Lord Aberconway was on the board of John Brown, Thomas Firth, and Palmer's and his son on that of Bolckow-Vaughn, the Cleveland steel manufacturers. Another director of John Brown and Bolckow-Vaughn was Colonel Davies. Admiral Sir H. Douglas was a director of Palmer's, Hadfield's and Richardson, Westgarth. B. A. Firth was to be found on the boards of Thomas Firth, John Brown, the Projectile Company, and the Coventry Ordnance Works.[33] J. Sampson combined the directorships of John Brown, Fairfield, Thomas Firth, the Projectile Company, Dick, Kerr and Company, and Harland and Wolff.

An indication of the success of the forward integration of armaments manufacturers may be gained by comparing the performance of the various divisions of Armstrong-Whitworth. Although the rates of return of the Elswick and Walker Shipyards were variable and occasional losses were incurred (see table 4.4), the rates of return of the firm's Openshaw Works and Elswick Steelworks, which supplied armor plates and other steel components for ships, were both more consistent and, in many cases, higher. The ordnance works showed greater fluctuations than the steel and armor divisions, but even here the rate of return never fell below 7.35 percent.[34] Thus, while variations in the quality of management from division to division and other factors may also have been of importance, Armstrong-Whitworth's shipbuilding yards may be viewed as in a sense having subsidized the armor and ordnance divisions by taking reduced profits in order to help secure orders for the firm's other operations. As all returns were internalized, the net result was probably advantageous.

Both Harland and Wolff and the Furness group were directly connected both with the amalgamation movement in shipping and with the armaments interests. In addition to his own mercantile shipping holdings,

Furness shared ownership in the engineering firm of Burmester and Wain with Harland and Wolff; and his control of Palmer's involved Furness in warship production. There were other links. H. F. Swan, a director of two of the Furness steel works, the Cargo Fleet Company and the Weardale Steel Company, was also the chairman of the Wallsend Slipway and Engineering Company—the repair firm for Armstrong-Mitchell—and a director of Armstrong-Mitchell. In 1901 the Wilson Line, a part of the Furness-Pirrie group, bought up Earle's Shipbuilding Yard of Hull, a once-important warship yard that had gone bankrupt. C. S. Swan, the founder of the firm carrying his name, had been set up in business by H. F. Swan, his brother, and Charles Mitchell, both of whom were on the board of Armstrong-Mitchell after the amalgamation of 1882. C. S. Swan combined with G. B. Hunter, and in 1903 C. S. Swan and Hunter (shipbuilders) merged with Wigham Richardson (marine engineers) and the Tyne Pontoon and Dry Dock Company (repairers) to form the Wallsend firm of Swan, Hunter and Wigham Richardson, which in addition controlled the Wallsend Slipway and Engineering Company. This combination, like the Furness group, claimed the advantages of a single management (the properties were contiguous), large capital resources, the prestige of a large firm, large-scale buying of raw materials, specialization, and the ability to coordinate shipbuilding and marine engineering.[35] The buildings and equipment were on the largest scale, compared to Harland and Wolff or John Brown, though the firm had no links with major shipping lines.

This Tyne combine gradually won its place on the Admiralty list and also secured large liner orders between 1903 and 1913. Early in 1913, Swan, Hunter and Wigham Richardson amalgamated with Barclay, Curle and Company, the large Clyde shipbuilding and repair firm. Together, these two firms had a capacity of 220,000-230,000 g.t. per annum, besides their huge marine engineering and repair facilities. Again, it was the opportunity for a larger capacity, economies in the management and purchase of raw materials, and the possibility of further specialization[36] that had prompted the amalgamation. Barclay, Curle had already begun to expand its facilities, and the combine acquired in addition large interests in the Burmester and Wain (Diesel) Company. Further repair facilities were added the following year. Although Barclay, Curle had had no Admiralty contracts of any importance before 1913, the firm's rapid extension was completed in time for early warship orders in the war.

Combinations between shipbuilders and armaments manufacturers were generally "offensive" in character, as the makers of arms and armor plating integrated forward into shipbuilding in order to secure sales. Connections between builders and steelmakers, on the other hand, were classic examples of backward integration for "defensive" purposes. By 1913 shipbuilders used almost 30 percent of total British steel production. The proportion was much higher in areas where production was ostensibly controlled by members of the Scottish and North East Coast steelmakers' associations, who were able to manipulate output and prices.[37] Superficially it appears strange that builders did not make stronger attempts to gain control over the supply of their principal raw material, but with few exceptions shipbuilders and steelmakers remained separate until after the beginning of the First World War, perhaps because of the ineffectiveness of the cartels. The Barrow Shipbuilding Company was originally a part of the complex of steel, shipping, and shipbuilding concerns of the Cavendish family, and Palmer's had produced its own iron and steel at Jarrow from the late 1850s. In 1900, however, only 11 of the 45 chief steelmaking firms had interlocking directorates with shipbuilders. These links were much more common on the North East Coast than in Glasgow.[38]

After 1918 backward integration into steelmaking gathered momentum. In order to guarantee regular supplies at reasonable prices, Harland and Wolff and John Brown acquired a share of the steelworks and collieries of David Colville and Sons. Swan, Hunter and Wigham Richardson bought the Glasgow Iron and Steel Company of Wishaw; Workman, Clark became allied with the Lanarkshire Steel Company; and a consortium of Scottish builders under the leadership of Alexander Stephen and Sons bought the Steel Company of Scotland.[39]

The amalgamation movement in shipbuilding before 1914 was most intimately connected with warship building and armaments for home and foreign governments. In fact, little warship building capacity remained outside the "combine." Some impetus came also from the shipping industry, itself strongly oligopolistic, and in both the Furness and Harland and Wolff groups the drive to expansion undoubtedly came from the shipping interests rather than from the shipyards. Finally, in the Tyneside combine of Swan, Hunter and Wigham Richardson, perhaps the only distinctly shipbuilding combination, the hope for Admiralty orders and for orders from the big lines was the most potent influence.

Nevertheless, according to Hermann Levy, as late as the 1920s both British shipbuilding and armaments manufacturing remained "comparatively" competitive despite vertical and horizontal integration.[40] To some extent, however, the degree of competition was illusory. While there were approximately 100 firms active in shipbuilding in 1914, some of the most important classes of vessels could be built by only a few firms. Thus only a handful of yards—among them Clydebank, Harland and Wolff, and Swan, Hunter and Wigham Richardson—were equipped to handle the largest ocean liners. This was true as well, not only of large naval vessels, but also of river craft, Arctic fishing boats, and other types of ships. Except, perhaps, in the field of moderate-sized cargo vessels, the structure of the industry was oligopolistic as a result of the high degree of specialization of output.

Bidding Practices and Ship Finance

Ultimately, the relationships between builders, suppliers, and customers were determined through complicated negotiations in which the final decisions on measurements, types, equipment, and financial arrangements for the construction of a new vessel were reached (see figure 3). Here, as elsewhere, practice varied. Ships are usually bespoke products. With a few exceptions, the shipowners who ordered their new ships from the builders were not as directly interested in the style of building or structure as in the performance. The specifications laid down by the owners generally included the deadweight capacity, the speed, and the coal consumption; in special cases, such as passenger liners, more detailed needs were specified in addition, such as the number of passengers of each class to be carried. Some shipowners gave further information at the start, including the length, width, and draught, depending on the ports to be visited. On this basis, tenders were invited.[41]

The builders who were invited to tender had to work out further details of the proposed vessels themselves. They had to calculate the dimensions necessary for the given carrying capacity, the immersion resulting from it, and the power for the given speed. Where a ship was one of a series, or the type had been built before on a different scale, these calculations were fairly easy. Where a new type, an untried method of con-

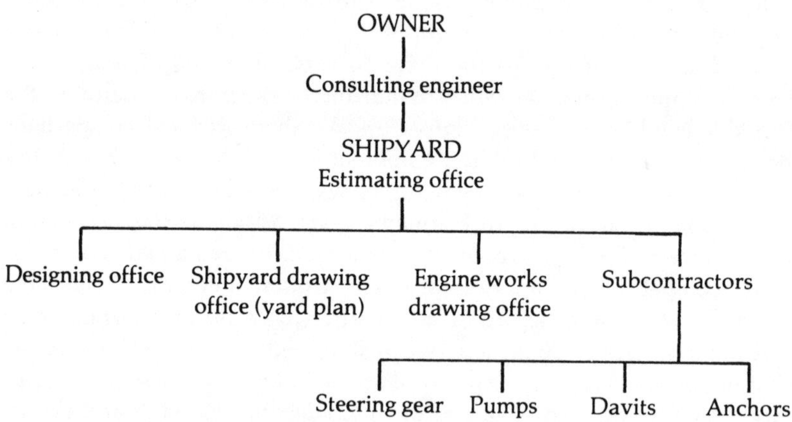

Figure 3. Hands through which the proposed ship passed before construction commenced. (Hardy, *From Slip to Sea*, table A, p. 237.)

struction, or perhaps a new hull material was in question, the details would have to be calculated item by item. The approximate cost was then quoted. The decision to be made by the builder might be a very delicate one. He had to balance the divergent claims of quality and cheapness, speed and safety, reliability and innovation. He had to bear in mind the tenders of other builders, which induced him to lower his standards and prices, but also his future reputation, which induced him to raise them. With these and a number of other considerations in mind, such as the promised date of delivery, the tender would finally be made. It should be stressed, however, that this picture of entrepreneurial initiative and textbook competition was by no means normal, but represents the utmost limit of responsible decision. In most cases orders were repeat ones, the clients and builders had been cooperating for years, and the quality was laid down by reference to the standards of a register society.[42]

When the tender, in its broad outline, had been accepted, the details as to ventilation, light, furnishings, steam, cooking and refrigeration appliances, safety, and so on, were filled in by conferences between builders and owners. This procedure clearly entailed a considerable knowledge of naval architecture on the part of the shipowner, and this the latter did not normally possess. Frequently, he demanded the impossible in the way of capacity or speed with given dimensions, or he showed appalling ignorance of the first principles of hydrostatics, or insisted on too high speeds at the trial run—achieved by very costly engine and boiler power—which he never needed in practice.[43] The builder had to tread warily here: "so anxious are the builders to please their clients in every respect, that they will do all that they are requested to perform, subject, of course, to their own scientifically matured judgement."[44] Other builders were less delicate in their comments: William Denny "could not reconcile himself to the servile task of building ships in accordance with the ignorant prejudice of owners or the antiquated rules of controlling bodies" and suggested to the Worshipful Company of Shipwrights in 1882 that shipping firms should submit conditions and state the special needs of their trade, and leave the specifications entirely in the hands of the shipbuilders.[45] It was here, in the need to supply the shipowner with specialist knowledge in his dealings with the builder, that the consulting naval architect or engineer frequently found his employment.

By 1884 shipowners more commonly had their own permanent designing staff and made their specifications increasingly complete. The large companies had their own drawing offices, and large lines owned an increasing proportion of British shipping. As early as 1873 the Cunard Company was said to prepare its own models and calculate the exact draught and displacement of its steamers—without reference to the builder—before the launch. Similarly, by 1887 the Allan Line designed its own engines for its mail steamers and did not allow its builders any latitude. In later years, when the national interest had been identified with the Cunard Company, the building of the *Mauretania* was supervised continuously by a shipbuilding committee representing the Cunard Line and the Admiralty.[46] Interpenetration could hardly go further.

The clients of the British shipbuilders ranged from mammoth concerns such as the Cunard Line, the White Star Line, or the Hamburg-Amerika Linie (H.A.P.A.G.) to the questionable, puny "one-ship companies" owning a single tramp cargo carrier, and the methods of payment varied accordingly. In the early years of the nineteenth century, when most ships were of the same type, it was common to build ships "on spec" (speculatively) to the general ideas of the builders and hope for a buyer after completion. Of the 31 vessels building in Sunderland on March 11, 1814, 3 had been contracted for, 3 were to be owned by their builders, and 25 were speculative. Of the 69 vessels under construction in Sunderland and Hylton on August 28, 1852, 49 had yet to be purchased. When no buyers could be found, the builder had to run his ships himself—one of the most important causes of close ties between shipbuilding and owning. This practice of building on speculation, though it ceased to be general in the second half of the century, was resorted to widely in times of depression when no other market could be found. Many of the vessels in which W. T. Lithgow held shares, for example, were built speculatively in slack periods by Russell and Company. In other cases, speculative building could form a great temptation for spurious joint stock companies, which bought up these ships, and it was a frequent cause of fraudulent company formation and "overbuilding."[47]

The most common form of contract, however, stipulated payment in installments, according to the progress of the vessel on the stocks. The contract for the first Cunard steamers, dated March 18, 1839, was typical.[48] It called for the building of three vessels of specified minimum dimensions and carrying a minimum number of passengers, to be completed by March 12, April 12, and May 1, 1840, respectively. The purchase price was £32,000 each, or a total of £96,000, of which £60,000 was to be paid during construction— £5,000 at once, £5,000 on April 12, 1839, and on the 12th day of every month until February 1840, a total of 12 payments. The remaining £36,000 was to be paid in three installments of £12,000 on the completion of each vessel, either in bills of 6 months' duration or in three parts in bills of 6, 9, and 12 months' duration, at an interest rate of 5 percent for the time over 6 months. Other clauses stipulated for the right of inspection by the Cunard Company during the building, its right of ownership over the partly completed vessels at each stage according to the installments paid, a 6 months' guarantee of repairs, and a fine of £5,000 on the builders if they failed to complete the ships by the specified dates.

Most contracts in the later nineteenth and early twentieth centuries were similar in form. The shipowners were, in fact, financing the builders by extending credit to them as the building progressed. If the owner was a limited company, as in the case of the Great Eastern Steam Ship Company, its capital was called up as required by the installment payments. If the builders went bankrupt during the building, the owners had reserved their ownership over the partly completed vessel and could find another builder to complete it; if the owners were unable to continue their payments, the builders had to take over the ownership and try to sell the vessels elsewhere. Sometimes, the builders tried to accommodate their clients by taking part of their payments in the form of stock in the owning firm. A case in point is the close relationship between William Denny and Brothers and the Union Steam Ship Company of New Zealand. Denny's received the order for the line's first two vessels, not because it offered the lowest price, but because its terms of credit were the best. Peter Denny subscribed over a quarter of the line's initial capital, and over the years Denny's built 42 ships for the line.[49]

Harland and Wolff's arrangements of cost-plus quotations were drawn up on a different basis. They charged the certified cost plus a fixed percentage that allowed complete freedom for alterations while the vessels were on the stocks—a frequent occurrence when the building period exceeded 12 months—without delicate and controversial adjustments of the price. On the other hand, the system could be introduced only where there was such close cooperation and community of interest between builders and owners as between the Belfast firm and its chief clients, the Bibby and White Star Lines.[50]

A slightly different system, for which a number of advantages were claimed, was used in the contract of the Cunard Company with Swan, Hunter and Wigham Richardson for the building of the *Franconia*.[51] Here the price was divided into four parts:

A. Direct outlay: material and labor. This was originally fixed as a maximum, but deductions might be made if it was not fully used as the work progressed.

B. Establishment charges: overheads. These were a fixed proportion of the original estimate for *A*, no matter what the final figure for *A* turned out to be.

C. Scheduled items: equipment and fittings bought from outside, either by the builder or the owner.

D. Builder's profits: a fixed percentage of the original $A + B + C$ estimate.

It was claimed that both builders and owners would have an incentive to keep the cost of *A* down to a minimum, the owners in order to lower the total price, and the builders in order to get their fixed sums under *B* and *D* on a smaller outlay of *A*. In the contract quoted, £20,000 was paid back to the Cunard Company as savings on the original *A* estimate.

In the case of delay or nonfulfillment of speed, the penalty clause, as it appears in the first Cunard contract, was common practice, but only in exceptional cases, as in the building of the *City of Rome* by the Barrow Company for the Inman Line, did the owners go as far as refusing to accept delivery. After 1871 the shipbuilders on the Clyde inserted a "strike clause" into their contracts, undertaking to complete the vessels by a certain date only if no strike intervened. In 1874 the Admiralty submitted to this, too, and it soon spread to all the other rivers. The penalty clause forced most shipbuilders to draw their supplies of plates and sections from as many different makers as possible, in order to eliminate any danger of default. It was also the usual practice for the shipbuilder to hedge against a rise in the price of his raw material by ordering the total amount of iron and steel required at once after the signing of the contract, for delivery spread over the whole period of building. When the yards were full and orders were booked up to two years ahead at the peak of a boom, however, a builder frequently bought short, hoping for a fall in prices.[52]

In meeting the challenge of building increasingly complex vessels, British shipbuilders in general performed well when judged by the criteria we set out in our introduction. Builders integrated vertically with suppliers and customers in most instances where it could be shown to be beneficial. A very few builders, notably Palmer's, chose to integrate backward, producing almost all their inputs, from coal and iron right through such accessories as rigging and carpentry work. Most added only an engine works, as marine engines were one of the few components that were almost as complex and unique as the hulls. Simpler components, which offered greater economies of scale, were usually left to outside suppliers who served many builders. The resultant competition among suppliers helped to lower costs and to speed up the rate of technological advance.

Forward integration, with shipping lines, was more common. This allowed builders to specialize more because it increased the number of orders for vessels of similar design. In the case of yards building large liners, it also helped to even out fluctuations in production because the great shipping lines such as Cunard and White Star were wealthy enough to place orders in slumps when most other shippers could not afford to do so. Lewis's strictures on marketing are largely irrelevant to shipbuilding. There was no role for the merchant house in the sale of ships. By the nature of the market, there was seldom a barrier between builders and shipowners, for most shipowners wanted their vessels to be designed to custom specifications and approached builders directly to obtain exactly what was needed. British shipbuilders also provided a variety of terms for the financing of their vessels, as seemed appropriate in each instance.

Finally, several important shipyards were themselves absorbed by armaments manufacturers who wanted to find outlets for their armor and ordnance. Many of the armaments firms were connected through exchanges of shares and interlocking directorates. In general, owing to specialization of output, the shipbuilding industry was oligopolistic in organization except in the tramp steamer sector.

6. The Development of the Shipyard

Shipyard Layouts and Shops

THE EVOLUTION of shipyards and shops was affected most strongly by the growing size and complexity of vessels and the resulting increases in differentiation in the building of ships. The area required for shipbuilding increased rapidly between 1870 and 1914. Larger slips and fitting-out basins naturally covered more land. The larger size of ships' components likewise led to increases in area, since railway and crane installations—both of which covered considerable space—were necessary to hold and manipulate the heavy plates and frame sections.

The growth in the number and size of shops was equally significant. As new shipboard appliances came into use, yard owners had to erect additional shops for smiths, electricians, plumbers, decorators, and other tradesmen who had not been important earlier. Hydraulic and electric plants were now needed to provide power. As the size of ships increased, shops were enlarged. Huge molding lofts and forge and foundry facilities were needed to form the girders and castings for the frames. The arrangement of shops within the shipyards was especially important. In order to minimize the expensive and time-consuming transport of cumbersome components, shops had to be placed so that materials would flow sequentially through the plant as they were processed and moved the shortest distance possible for assembling.

Moreover, peculiarities of the shipbuilding industry itself made it desirable to design both yards and shops to allow for a maximum of flexibility in the shaping and assembling of components. Shipbuilding involved both manufacturing and construction activities. Most ships were built according to specifications laid down by their buyers.[1] Although some specialization and standardization were possible, ships were seldom similar and yard facilities had to be designed to provide the greatest adaptability.

The development of shipyard facilities was reflected in changes which were made in a number of important yards over the period. The firms in question were among the largest and most modern, but this makes them no less suitable for comparison, because there were no "representative" shipyards. The layout and facilities of each yard were determined by the amounts and shapes of land available and by the needs of the particular firm. The positioning of berths and wetdocks was dictated by the frontage of the yard and the depth and width of its river. As the size of vessels increased, it was often necessary to dredge river bottoms and cut new basins for fitting out ships after launching. The requirements of local port authorities sometimes conflicted with those of shipbuilders. Expansion of port facilities at Glasgow, for instance, impinged on the yards on the upper Clyde, forcing some to move. The type of business a firm undertook also affected the layout. Yards that built their own marine engines often had engine works connected to or adjacent to their other facilities, except on the Clyde where many builders located their engineering plants in the center of town.[2] A few yards, notably Palmer's and those of the armor plate manufacturers, were attached to steelworks.

Many of the yards of 1914 were enlarged or rearranged versions of yards that had been in use for decades. Originally, they had been designed to build smaller, less complicated ships and then altered as needs and circumstances changed. Rarely was a firm able to start with an open plot of ground on which its facilities could be laid out without regard to existing buildings. As a result many yards were cramped and haphazardly arranged. Even firms that moved to escape crowding sometimes took over yards that had already been abandoned by others. Scarcity of funds usually precluded the extensive changes that would have been necessary to create yards of the most efficient design.[3]

From the 1860s, builders of large vessels were obliged to expand their yards at an unprecedented rate. Near the end of the decade one builder designed a yard capable of handling ships up to 500 feet in length. He felt that it would be adequate for a considerable period, but by 1872 he knew he had been wrong, and by 1880 ships of almost 600 feet were being built. In 1868 the yard of Tod and M'Gregor of Partick, one of the earliest builders of iron steamers, contained only two mechanical workshops —a forge and a small engineering building.[4] Twenty years later, up-to-date yards had six or seven large shops to accommodate the new specialized trades as well as the additional machinery that was used in the older metal and woodworking departments.

The development of the John Brown yard at Clydebank nicely illustrates some of these problems. J. and G. Thomson had moved from

Mavisbank to Clydebank in 1870 in order to build larger ships and because Glasgow Harbor was encroaching on their older facilities. Even before being acquired by John Brown, the Clydebank yard concentrated on large liners and warships; table 6.1 demonstrates the way the firm was forced to expand in order to build the largest class of ships. By 1895 the yard already comprised over 50 acres, but extensive alterations were still necessary when the *Lusitania* and the *Aquitania* were built.

The yard was originally situated almost directly across from the confluence of the Cart and Clyde rivers to allow for the easy launching of large ships, but it proved necessary to add about two acres of water space on the east side of the Cart and six acres on the west side for the launching of the *Lusitania*. When the *Aquitania* was built, John Brown and Company paid £10,000 toward the cost of widening the Clyde from Clydebank to Dalmuir, removing an awkward bend and cutting 13 acres of arable land from Newshot Isle. Although no new water space was added, the Clyde was dredged along the line of the launch from the berth to the mouth of the Cart. Moreover, extensive changes were needed for the yard itself. Two of the berths were combined to allow for the greater width of the *Aquitania*, the fitting-out basin was enlarged and dredged, and buildings at the head of the berths for painting, plumbing, and other trades involved in the final stages of completion were demolished.[5]

As Harland and Wolff expanded in Belfast, the shipyard grew spectacularly, but all changes had to be made within the rigid format of a plot of ground and existing buildings that had not been intended for the purpose, but rather thrown together as the necessity arose. Between its establishment in the 1850s and 1912, the yard of Harland and Wolff grew from 3.5 acres employing 48 men, to 80 acres with an employment of between 15,000 and 16,000 men including those in the engine works. From its founding the concern specialized in ocean liners, and the growth of

Table 6.1. Largest Atlantic steamers built at Clydebank, in successive periods, 1881-1913.

Ship	Date	Overall length (feet)	Gross tonnage
Servia	1881	530	7,392
City of New York	1888	560	10,499
Saxonia	1899	600	14,271
Caronia	1904	672	19,687
Lusitania	1906	785	31,500
Aquitania	1913	902	47,000

Source: *Engineering*, April 18, 1913, p. 515.

ships forced the yard owners to enlarge and rearrange their facilities several times. In 1879 there were six building berths at one end of an island in the River Lagan. As the yard expanded, land was added and the channel between the island and the mainland filled in. By 1885 the number of berths had been reduced to five to allow for the construction of bigger ships. The graving dock was enlarged and an engine works was added. By 1888 the four projected berths at the east end of the yard had been completed and both the shipyard shops and the engine works had been expanded. The greatest changes were made between 1907 and 1912, when the three berths in the northeast corner of the yard were combined to form two giant berths. In the five years before 1912, 165,000 square feet of new shop area were added to the shipyard and 223,000 square feet to the engine works. To minimize the distance over which the heavy boilers would have to be moved, the engine works were placed close to the graving dock and the fitting-out basin.[6]

The yard of the Barrow Shipbuilding Company, which was one of the few in our period to be laid out from scratch, was a most successful facility from a technical, if not an economic, standpoint. Conceived in 1869 as part of the general expansion of industry in the small Lancashire town of Barrow-in-Furness, the firm commissioned Robert Duncan of Port Glasgow to design a model yard at the north end of Old Barrow Island. By taking advantage of the vast open plot available, Duncan was able to lay down a plan of works that was adequate for several decades despite large increases in capacity.[7]

The Barrow yard covered 58 acres with water frontages of 1,050 feet on each side of the island. The plot was divided into two parts: on the west side were the berths and shipyard shops, and on the east, the engine works and fitting-out dock. The buildings were arranged in a quadrangle and in 1880 included a joiners' shop, 300 feet long by 60 feet wide; a sawmill, 400 feet by 60 feet; a machine shed, 300 feet by 180 feet; and a smiths' shop, 200 feet by 120 feet.[8]

The promoters were overambitious, however, and from the beginning the yard faced financial difficulties. The firm proposed to build merchant and naval vessels of the largest size; by 1873 the yard was already one of the biggest and most modern in Britain, capable of simultaneously building 5 ships of over 4,000 tons with room for from 7 to 10 smaller ones. In 1880, although it had 12 building berths with total annual capacity of 40,000 gross tons, little profit had resulted and the expansion was financed almost entirely from the personal fortune of the Cavendish family. In 1881 the *City of Rome,* next to the *Great Eastern* the largest liner then afloat, was launched by the firm but failed to meet contract specifications, and the shipyard was forced to repurchase her from the Inman Line. The recession of the mid-1880s hit the firm severely. Output dropped from over 32,000 g.t. in both 1881 and 1882 to only 1,011 g.t. in 1888. In the latter year the concern was purchased by a consortium interested in producing the Nordenfelt submarine, the name being changed to the Naval Construction and Armaments Company Ltd. Production quickly revived as the more cartelized construction of naval vessels replaced competition in the market for merchant ships. Employment rose from 872 men in 1888 to 5,500 in 1897 when the yard was acquired by Vickers. Between 1897 and 1904 the capacity was more than tripled, and the firm prospered in the naval building boom preceding the outbreak of the First World War.[9]

The early financial troubles were due not to Duncan's plan but to the overoptimism of the founders, who had committed themselves to an installation considerably larger than could be used profitably in the market conditions of the early 1870s. Although modifications and additions were made, the basic design of the yard and shops that Duncan established remained. Almost 30 years later the shops were still regarded as models of efficiency despite the many changes in shipbuilding practice that had taken place in the interim.[10] After the company was reorganized in 1888, a new sawmill and timber yard were added on the south end of the yard and a new brick building for the painters', plumbers', tin smiths', and sheet workers' shops was put up at the northwest corner of the building berths. Duncan's quadrilateral remained the core of the yard, however, with the wood- and metal-working shops as close as practicable to the building berths, thereby reducing the area over which components had to be moved (figure 5). Among the best features of the design were that the steel structural material moved through the metal-working departments in a single direction, reducing handling to a minimum, and that both the wood and metal sections had easy access to the cranes at the head of the slip. In contrast, the shops at Clydebank were more spread out. The sheds of the shipyard smithy, forge, and angle-iron and beam-bending smithies stretched almost the entire length of the east side of the yard, making it necessary to carry the heavy sections over a greater distance.[11]

The effects that topography and preexisting plant had on yard organization are shown clearly by a comparison of the Barrow and Belfast installations. Both yards had two frontages, but the longer shoreline at Barrow made it possible to place all of the building berths on one side and the engine works and fitting-out dock on the other. As a result, the yard shops could be located at the head of the berths, convenient to all of them. At Harland and Wolff it was necessary to place the berths on both sides of the yard with the shops in between (figure 4). Again, most of the metal shops were at the head of the berths, but the smooth flow of materials that was possible at Barrow was precluded by space considera-

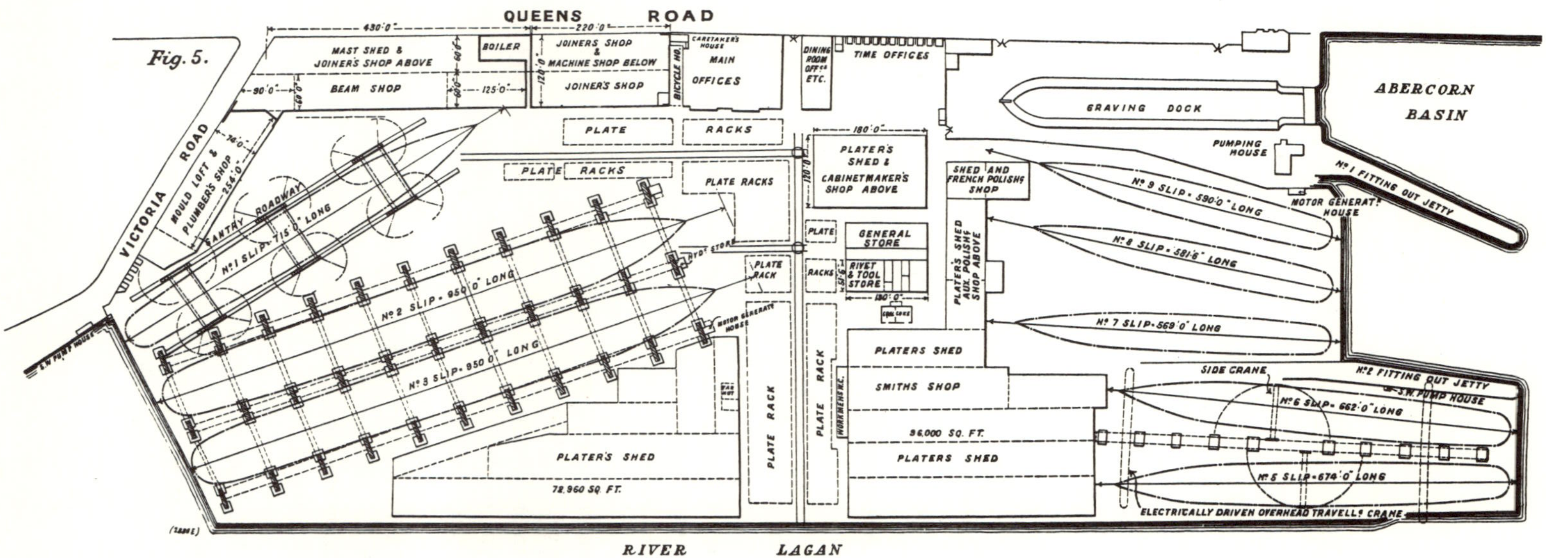

Figure 4. The shipbuilding department of Harland and Wolff, Belfast, 1912. (*Engineering*, July 5, 1912, p. 5.)

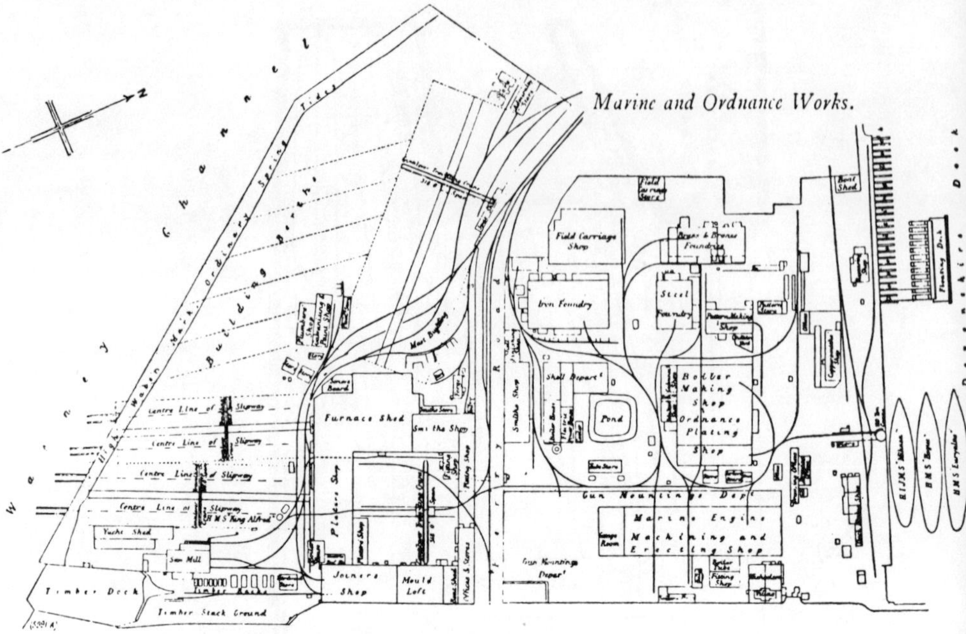

Figure 5. The Naval Construction Works, Barrow-in-Furness, 1901. (*Engineering*, August 2, 1901, p. 168.)

tions. The woodworking shops, in particular, were all on one side of the yard, a substantial distance from five of the eight berths. Most of the finishing and decorating shops were located across the Queen's Road, adjacent to the engine works. Since these were the least-used shops and their output the lightest,[12] this was the most logical location given the cramped area of the yard, but it again meant that materials had to be transported over a longer distance than was necessary in the more open layout at Barrow.

After the turn of the century Vickers benefited in a sense from the fact that Duncan's original plans called for buildings larger than were warranted by the conditions of the 1870s, thereby allowing for subsequent expansion without destroying the efficient unity of the design. Harland and Wolff had continually to contend with existing plant laid out for the needs of earlier decades, so that expansion often included the extra expense of removing or remodeling older structures. The platers' sheds at the head of the berths, for instance, had to be altered several times as the size of the ships was increased.

One reason so few new yards were built after 1870 was that most of the suitable land on the banks of the most important shipbuilding rivers was already occupied. When Yarrow's transferred its yard from London to Scotstoun near Glasgow in 1908, the problem was not overwhelming because the firm built small craft and required only 12 acres with a river frontage of 819 feet for both its shipbuilding and marine engineering operations. Until then, Yarrow's had rented facilities in Poplar but decided to move because of high rents, rates, and wages in the metropolis. Whereas the firm annually paid £100 per acre in ground rents in London, for instance, on the Clyde it was able to buy land for only £50 per acre.[13]

Sir W. G. Armstrong, Whitworth and Company, which built the largest classes of naval vessels and needed a much bigger plot, had a more difficult task in finding land for its new yard. In the early 1880s the firm had laid out a yard adjacent to its works at Elswick. The site had originally been chosen because of its suitability for armor and ordnance manufacturing, and the location of the shipyard made it necessary to launch vessels between the piers of a swing bridge spanning the Tyne. There was space to extend the building berths, but shortly before the outbreak of the war the firm was forced to seek a new site because the increasing beam of naval vessels made it impossible for them to pass safely between the piers of the bridge. Finally, it was decided to locate another yard at Walker-on-Tyne even though at that point the land rose steeply from the bank of the river. In order to obtain a level tract of 83 acres, the firm excavated over 4.1 million cubic yards of soil. Between 1910 and 1914 hills of up to 130 feet in height were leveled and a complete yard constructed.[14]

The tendency of yards to concentrate on certain classes of work was encouraged by the increasing size and complexity of vessels and by the widening range of facilities necessary to build ships of different types. Builders of large warships needed extensive shops for the preparation of armor and ordnance, while yards that specialized in liners had establishments for upholstering, paneling, and manufacturing furniture. Both classes required large building berths and fitting-out basins and heavy equipment for handling large plates and boilers. None of these installations was necessary for the construction of cargo steamers, however, and the yards of the builders who concentrated on their construction were generally somewhat different from those described above. Although they also increased in size and added shops for new trades in the decades before the First World War, they remained simpler, with fewer buildings, smaller tracts of land, and less elaborate equipment than was needed by builders of naval vessels and liners. The cost savings that this specialization permitted were particularly valuable in view of the cyclical nature of the industry and the consequent long periods when much capital stood idle.[15]

New Machinery

The increasing size of vessels and the severely cyclical nature of the industry were the most important factors in decisions of capital investment, but while the former (along with the increasing differentiation of the labor force) encouraged greater investment, the latter made yard owners wary of spending any more than was absolutely necessary. Thus, while new equipment was adopted steadily over the years, great care was taken to justify each expenditure. This did not reflect conservatism or a short time horizon, but was simply an accurate assessment of the character and needs of the industry.

By 1900 builders had begun to use machines to perform many of the jobs involved in constructing ships. At the Vickers-Maxim yard at Barrow, for example, more than 100 distinct types of equipment were in use in 1901, although not all were specialized to shipbuilding. Among others, there were 76 machines of 20 types in the mechanics' shops; 79 machines of 28 types in the machine shed; 93 machines of 9 types, including 76 fires and furnaces, in the smiths' shop; and 25 machines of 13 types in the plumbers' shop.[16]

Heavier components made necessary the provision of heavier and more powerful cranes. Power riveters were needed to cope with greater stresses since giant ships not only had more rivets than smaller ones but also had heavier platings and doublings, which were more difficult to close by hand. The substitution of mild steel for iron also made mechanization attractive, since the hardness of the material, in both forgings and castings, increased the labor involved in fitting, caulking, drilling, chipping, and cleaning. Finally, increases in vessel size and speed put a premium on accuracy of execution, which could only be achieved through the use of precision machinery. Tolerances that might be acceptable in a ship of 3,000 g.t. with a maximum speed of 12 knots could not be allowed in a steamer of 10,000 g.t. capable of 25 knots.[17]

Increased division of labor, or differentiation, in the shipbuilding process was also important. When wooden sailing ships were preponderant, a wide range of tools was neither necessary nor available. The shift to iron and steel as the principal materials of construction and the adoption of a panoply of shipboard appliances, however, led to the employment of 50 more or less skilled trades by 1914, all of which needed specialized tools and machines. While shipwrights and to a lesser extent iron workers specialized in shipbuilding, the remainder of the workforce—including many of the metal and woodworkers, painters, plumbers, electricians, millwrights, and engineers—practiced trades that were of significance in other construction and manufacturing fields. It was to serve these latter trades that much of the shipbuilding equipment was introduced. Lathes, saws, and drilling and grinding machines were all used extensively wherever metal and woodworking tasks were performed; only a few types of equipment for preparing and riveting plates or moving heavy equipment were designed to solve problems peculiar to ship construction.

Shipbuilders were of several minds on how machinery could be used to lower costs. In most instances, builders accepted that the high degree of skill of men in crafts already engaged in shipbuilding was necessary to obtain high-quality output, but machines were adopted to speed up work; this was most true of equipment used in trades that also operated outside the shipyards. At other times, machinery was added either because there were no men with the skills necessary to perform a new job as shipping technology changed, or because builders wanted to replace skilled laborers with lower-paid machine operators. Both trends affected relations between builders and their employees, for example, riveters. When iron shipbuilding began, it was necessary to train men to rivet the shells and frames. Shipwrights, who had assembled wooden vessels, refused to deal with metal. In any case, the skills involved in shipyard riveting were different from those in any existing trade. Shell riveters allegedly could not do the work of an inside riveter or vice versa, while boiler riveters could do neither class of work. To obtain riveters initially, firms imported workers from engineering shops as unskilled laborers and trained them. In time the men were organized into a powerful union, the Society of Boilermakers and Iron and Steel Shipbuilders, which insisted on working for piece wages. It was in part to replace the Boilermakers with a less expensive group of unskilled laborers that yard owners sought ways to mechanize riveting. The same tendency affected other trades when builders found that machines manned by semiskilled labor could turn out work as good as or better than that done by skilled men. Particularly after the turn of the century an increasing share of the labor input was executed by semiskilled operatives. Their range of abilities was narrower than that of the skilled men whom they replaced, but with the aid of machinery they could produce as high a class of work at lower unit cost.[18]

Sharp fluctuations in output, however, tempered the enthusiasm of yard owners for mechanization. Between 1903 and 1913, only about 70 percent of the capacity of the leading firms was utilized in an average year even by the most favorable definition, and in some years the figure fell below 50 percent (table 6.2). Because facilities might not be fully used for several years at a time, builders were particularly reluctant to put money into the most expensive types of machinery or other specialized plant unless it could be shown that they could not operate without them, or that the savings would be great enough to offset the additional overhead charges during periods of disuse. Hence, shipbuilders were often very conservative in their investment policies, and there were long lags between the introduction of an innovation and its general adoption.

In its lengthy accounts of various shipyards, *Engineering* usually devoted very little space to the machinery in the shops; and, aside from the

Table 6.2. Capacity and output of 20 of the leading shipbuilding firms, 1903-1913.

Year	Output for year (A) (000 g.t.)	Capacity as given to the *Directory of Ship Owners* (B) (000 g.t.)	(A)/(B) (%)	Capacity taken as the highest previous output (C) (000 g.t.)	(A)/(C) (%)
1903	759	1,220[a]	62.2	1,043	72.7
1904	696	1,260[a]	55.2	1,066	65.3
1905	907	1,295	70.0	1,119	81.1
1906	967	1,310	73.8	1,173	82.4
1907	935	1,360	68.7	1,188	79.0
1908	522	1,340	39.0	1,188	43.9
1909	493	1,410	35.0	1,188	41.5
1910	702	1,410	49.8	1,194	58.8
1911	1,050	1,410	74.5	1,218	86.2
1912	972	1,410	68.9	1,225	79.3
1913	1,068	1,555	68.7	1,284	83.2
Average, 1903-1913			60.5		70.3

a. Seven firms submitted no details of their capacity in 1903; figures for 1904 have been used instead. One firm submitted no details for 1903 or 1904; the figure for 1905 has been used instead.

question of the most efficient means of providing power, shipbuilders rarely discussed the adoption of this kind of equipment at meetings of their technical societies. It appears that the machines used were standard in most yards, that they were regarded as natural additions once the trades that operated them began to be employed in the yards, and that their value was almost always proven outside the industry before shipbuilders introduced them.

The *Transactions* of the Institution of Naval Architects, Institution of Engineers and Shipbuilders in Scotland, and North East Coast Institution of Engineers and Shipbuilders reveal that the types of machinery that yard owners were most concerned with were riveters and other kinds of machinery for working plates, large cranes, covered berths, graving docks, and means of providing power. Unlike shop machinery, all of this equipment was either very expensive, of little use except in shipbuilding, or both. In many cases, the only way of determining the actual usefulness of the equipment was to test it in the industry, because the conditions yard owners faced had no parallels in other fields.

Graving docks and covered berths are two examples of expensive facilities that, although useful, were never generally adopted because they could not be employed often enough to justify their high initial cost. Savings could be realized, for example, by nearly completing a vessel in a graving dock where it was possible to place most of the heavy equipment on board using movable cranes; conventional fitting-out basins usually had fixed cranes, which made it continually necessary to move the vessel at considerable effort and cost. In 1900, however, a graving dock cost approximately £100,000. Since yards could not count on a steady supply of orders for large, labor-intensive vessels, on which the savings were greatest, few replaced even one of their slips with the more economical dock. Moreover, the development of large movable cranes for fitting-out basins helped to solve the immediate problem and greatly reduced the apparent economic advantage of the graving dock.[19]

Covered berths were also acknowledged to be useful but, again, their high cost deterred shipbuilders from installing them. They had been used in the last days of wooden shipbuilding to protect the timber and gain a high class of insurance rating at Lloyd's, and later by builders of high-quality small craft such as yachts. Although a few iron shipbuilders, including Tod and M'Gregor of Partick, Robert Duncan of Port Glasgow, and Laird Brothers of Birkenhead, did erect small covered slips, the only extensive installations were those of Swan, Hunter and Wigham Richardson at Low Walker. They were said to save money by speeding work and allowing for work in bad weather despite complaints that they were drafty and excluded too much light. The berths at Low Walker must have proven profitable, for when work began on the Cunard Line's *Mauretania* in 1906 the firm was sufficiently impressed to add a cover 1000 feet in length to the two 500-foot covers it had already built in 1894. Cost considerations prevailed, however, and other builders failed to follow suit. In 1914 it was estimated that it would cost £150,000, exclusive of lifting equipment, to cover a slipway 1,000 feet long and 100 feet wide. Overhead costs would have approximated £12,500 per annum, not including maintenance. Most firms believed that the risk of obsolescence because of increasing vessel size more than outweighed any savings that might result.[20]

In the 1860s most shipyard equipment was powered by steam, and even in the early years of the twentieth century some British yards still had steam engines in their various departments, long after many foreign firms had installed more modern equipment. Steam had a number of disadvantages. The engines were often supplied with steam piped from a central boiler and were very inefficient, even in the yards of builders who went to great lengths to reduce fuel consumption. In other cases it was possible to power machines through the use of shafting, countershafting, and beveled gearing with the handicap that the engines then had to provide full power to the shafting even if most of the machines were not in operation. Where shafting was not used the situation was even worse, for then a separate engine was required for each machine. Finally, the loss of power in transmission was high and the engines, steam pipes, and joints were expensive to maintain.[21]

In some yards, natural gas or internal combustion engines were used to replace steam, but the adoption of hydraulic, pneumatic, and electric power was ultimately of greater importance. As early as 1846 Charles May patented a proposal for a hydraulic riveter, which was never used, and in the same year Sir William Armstrong designed his first hydraulically powered crane. Although elevated artificial reservoirs were used to provide water pressure in the early part of the century, the development of Armstrong's "accumulator"—essentially a large plunger that provided pressure by injecting water into the engine—raised the efficiency of hydraulic machinery to a commercially feasible level. In 1865, apparently inspired by a shortage of labor, three men working independently developed hydraulic riveters, which again were slow to be adopted. When Tweddell invented a portable riveter in 1871-1872 that was capable of driving 1,000 rivets in a 10-hour day, Clyde shipbuilders told him that its introduction was not warranted because riveters were plentiful at the time. By the 1880s, however, firms were more willing to invest in "accumulators" and pumps, and by the turn of the century, according to H. C. Gillmor, "Hydraulic power is to be found in practically every yard in which vessels of any size are built." It was used to power not only riveters but heavy flanging machines, manhole presses, and smaller cranes as well. After pneumatic and electric power became available, hydraulic units continued to be used, especially for heavy sustained work: the major advantage of hydraulic tools was that they drew no energy when not in use, but they were of little value in colder climates because they froze in winter.[22]

Pneumatic tools were first developed in America to avoid freezing. Although they were adopted more slowly in Britain than in the United States, by 1893 a pneumatic caulker and chipper had already been successfully tested on the Clyde. It performed more accurate work than a skilled laborer with hand tools. Although the vibrations of the machine were at first disturbing to the workmen, after a few days' practice it gave little further trouble. By 1905 pneumatic power was also employed for drilling, planing, and riveting. Pneumatic drills were preferred to electric ones because they could be used more quickly and were more portable. Pneumatic caulkers were capable of replacing at least four men with hand tools and pneumatic planers prepared a fairer surface than electric or hand-powered models. The primary objections to pneumatic power were that it was too wasteful and that production still depended largely on the skill of the operator; but the tools were particularly successful in raising the output of the slowest workers.[23]

The potentially most efficient form of power was electricity. It did not freeze, was used only when machines were in operation, and could be either purchased from outside suppliers or provided by a yard's own generators. Electric lighting had been introduced in several yards in the 1880s, and in 1893 Swan, Hunter installed electric power for driving machines. Furness-Withy followed in 1895, Alexander Stephen and Sons and the Thames Iron Works in 1896, and Harland and Wolff and Palmer's in 1898. By 1902 the factory inspectors could report that electric power was general in all up-to-date yards.[24]

Electricity proved to be half as expensive to provide as steam power and required only a slightly greater capital outlay. Furthermore, since each electric machine had its own motor, it was possible to place it wherever needed. As a consequence electricity was tried and adopted for a variety of uses in the 1890s. Electric lighting was found to be useful both on the slips where it could be used to illuminate the dark and less accessible sections of the hulls, and in the drawing offices. Although in the 1890s few machines were commercially available and builders were obliged to construct their own, electricity was also used to power punches, boring machines, pumps, winches, saws, fans, cranes, drills, cutting machines, and riveters. In most instances, electrically powered equipment was judged to be effective but hydraulic or pneumatic power was preferred for certain machines such as drills and riveters. While the Admiralty had used electricity successfully for all purposes, many private builders still favored hydraulic riveters and pneumatic drills with electromagnets to hold them in place.[25]

The advent of the all-electric shipyard had been predicted in the 1890s, but for many years hydraulic, pneumatic, and steam power remained in use. At the turn of the century, for example, all four types of power were used at Barrow.[26] The primary reason for the lack of uniformity was not that builders were loath to part with old equipment, but that they found different types of power more efficient and cheaper for different operations. The development of hydraulic, pneumatic, and electric machinery proceeded simultaneously, forcing builders who wanted to use the most modern methods to have all types of power available. Since most large firms had already installed facilities for all types by the early years of the twentieth century, further extensions were inexpensive. The costs of continuing to mix different power sources within the same yard were usually offset by savings derived from being able to power each kind of equipment most cheaply.

An entirely satisfactory machine for shell riveting had not been developed by 1914. Riveting remained one of the most difficult tasks in shipbuilding, in terms of both the skills involved and the physical energy required. The work was very specialized, hard to learn, and exhausting so that men often could not continue beyond middle age. Riveting was also a highly paid trade. There was incentive, therefore, on the part of both riveters and management to introduce machinery that would lengthen the working life of the one group and reduce costs for the other. A variety of technical problems, however, and the opposition of the men who worked for piece rates to the adoption of machinery that would have reduced rates to compensate for higher productivity helped to limit the use of power riveters.[27]

By the late 1870s power riveting was used to a limited extent for frames, floors, and beams. McMillan's of Dumbarton used a steam riveter for shells, but the machine was thought to be too hot when in operation and the workers did not approve of it—although their opposition was to some extent overcome by the higher wages they received. Use of the machine did not spread, however, and even hydraulic riveters, which generated less heat, were initially unpopular, partly because shipbuilders felt that they were uneconomical and partly because the quality of work they turned out was low. As early as the 1880s hydraulic power was used to rivet frames at Barrow and Fairfield's, but in the mid-1890s one speaker could still say that he "hoped before long to see mechanical riveting applied to the shell rivets."[28]

In 1881 A. Jamieson and F. J. Rowan both developed electrically powered riveting machines. Rowan's machine, which was held in place by electromagnets, met with opposition from workers who feared reductions in piece rates and threatened to boycott yards that used it. As a result it was never adopted. An electric riveter was used at Trieste in 1898, but British builders felt that it worked no more rapidly than hydraulic machines and had the disadvantage of exerting unequal pressure on plates of different thicknesses, bending the frames or beams. In addition, it was heavy: a model capable of driving 1¼-inch rivets weighed 1.25 tons; for 7/8-inch rivets, a machine of 3/4 ton was necessary.[29]

Pneumatic and hydraulic riveters were the two most popular types after the turn of the century. Hydraulic equipment had been improved until it could produce the highest quality of work, especially where great strains were placed on the rivets. The machines were heavy and clumsy, however, the largest weighing up to eight tons. Pneumatic machines were much lighter and more portable. A machine with a depth of gap of 9 inches weighed only 83 pounds and one with a gap of 70 inches, 220 pounds. They were modifications of pneumatic caulking hammers and were used extensively on framings but proved less satisfactory for shells, both because the plates had to be placed too close together, necessitating extra labor to bolt them tightly, and because it was difficult to place the rivet heads properly. It was also claimed that because of the wet climate pneumatic machines did not function well in Britain.[30]

Although it was objected that the rivets had to be placed too near the edges of the plates, hydraulic riveting was eventually adopted for shell work. Swan, Hunter and Wigham Richardson, for example, used hydraulic machinery for many of the more than four million rivets on the *Mauretania.* By 1914 many builders acknowledged that hydraulic riveters did work of the best quality although their weight remained an obstacle to their wider use. Pneumatic riveters, when the plates were properly held up, were capable of doing work as good as or better than hand riveting while greatly increasing output and enabling less-skilled laborers to do good-quality riveting. The adoption of pneumatic riveters was at that time limited only by the amount of power available, the willingness of the men to use them, and the extent to which holding-up tools were available and applicable.[31]

Electric welding, which was ultimately to replace riveting altogether, was adopted slowly in Britain. Before the First World War, welding was used to repair a number of ships, but the process was not extended to hull construction because the joints were too weak, causing a loss of strength and the buckling of plates and frame. In 1914 tests of the "Quazarc" process, using molten metal, showed that the elasticity in the region of the weld was reduced, increasing the risk of broken joints. Welding was both easier and cheaper to perform, however, than riveting.[32]

The adoption of cranes was fostered by the increasing size of ships' components and shipyard equipment. Ships larger than 6,000 g.t. were more expensive per ton to build, in large part because of the increased weight of beams and frames and the greater height to which such weights had to be lifted. The greater weight of plates and of hydraulic riveting machinery also provided incentives to introduce cranes on the slips.

In the early days, lifting was done through the use of poles, blocks, and tackle, but in the boom of the mid-1860s a number of firms including the Thames Iron Works and William Denny and Brothers installed sheerlegs. In 1868 Tod and M'Gregor had a crane capable of lifting 70 tons. These large cranes were employed to lift frame components and boilers. As the latter increased in size, so did the cranes. In the early 1870s cranes designed to lift more than 100 tons were installed in several yards, and by 1885 the largest crane at Clydebank could handle 130 tons. After the turn of the century Beardmore's erected a 200-ton crane in its new yard at Dalmuir; in 1914 two cranes capable of moving 250 tons apiece were in operation. Many yards possessed their own cranes, but in large shipbuilding centers such as the Clyde publicly owned cranes were provided for the use of nearby firms.[33]

The earliest large fitting-out cranes were sheerlegs, many of which were placed at the edge of the quay where they came into contact with the gunwales of vessels, pushing them away and limiting the effective reach of the crane. In addition, sheerlegs were slow, interfered with funnels and masts, and could only move loads vertically, necessitating frequent expensive and time-consuming movements of the ship.

In place of sheerlegs, several designs that could be moved in one or more directions were substituted, particularly for loads of less than 100 tons. Some models moved on rails and had jibs that revolved and could move in and out and up and down. Floating cranes to raise up to 150 tons were constructed for Swan, Hunter and Wigham Richardson, Armstrong-Whitworth, and Harland and Wolff. Because they could move around a vessel, they were more flexible than land-based cranes but had the disadvantages of being lower, thus increasing the distance over which loads had to be raised, of being easily disturbed by passing ships and rough weather, and of requiring expensive pontooning and machinery. A third model, the overhead traveling crane suspended from a covered fitting-out basin, was considered ideal. Because of the cost of building the cover, however, it was suitable only for yards such as Yarrow's, which built small craft.[34]

The designs adopted by most yards for heavier loads were all cantilever cranes. The earliest, German hammerhead cranes, were first used in 1889. They revolved and were able to move loads horizontally along the jib to any desired position across the beam of the ship. When placed between berths they could serve two ships except for the widest dreadnoughts. Traveling carriage cranes, which were attached to the sides or roof of a skeleton structure spanning an entire slipway, provided more hooks and were quick but cost much more than cantilever cranes to build and install.

The system most favored at the beginning of the war for lighter weights was the tower crane. It cost comparatively little, was quick, could work in high winds, and lent itself to extensions of the berths served, although it was expensive to maintain. Between 1912 and 1914 Sir William Arrol built tower cranes for Harland and Wolff, Workman, Clark, Armstrong-Whitworth, and Vickers. The installation at the Vickers yard at Barrow, illustrated in figure 6, had nine cranes in two rows able to serve four slipways of up to 900 feet in length. Each crane had a jib with a radius of 120 feet, and their ranges overlapped. Two of the cranes had a lifting capacity of 15 tons at the extreme radius and the other seven could handle 5 tons; the loads were doubled at half radius (figure 7). Congestion in the yard was reduced by the ability of the cranes at the lower end of the yard to pick up bulky materials from barges in Walney Channel.[35]

Table 6.3 gives an indication of the growth in value of shipyard machinery after the turn of the century. Much of the machinery that was

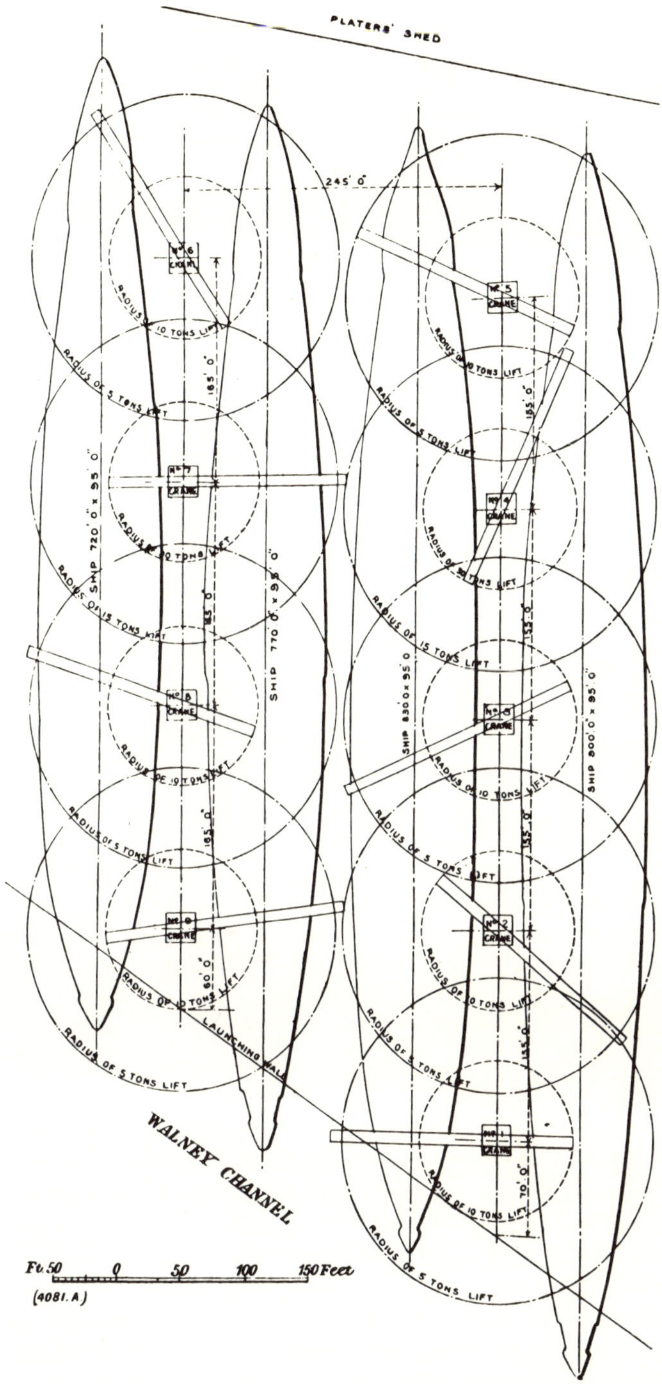

Figure 6. The positioning of the cranes at the Vickers Shipyard, Barrow-in-Furness, 1914. (*Engineering*, July 10, 1914, p. 70.)

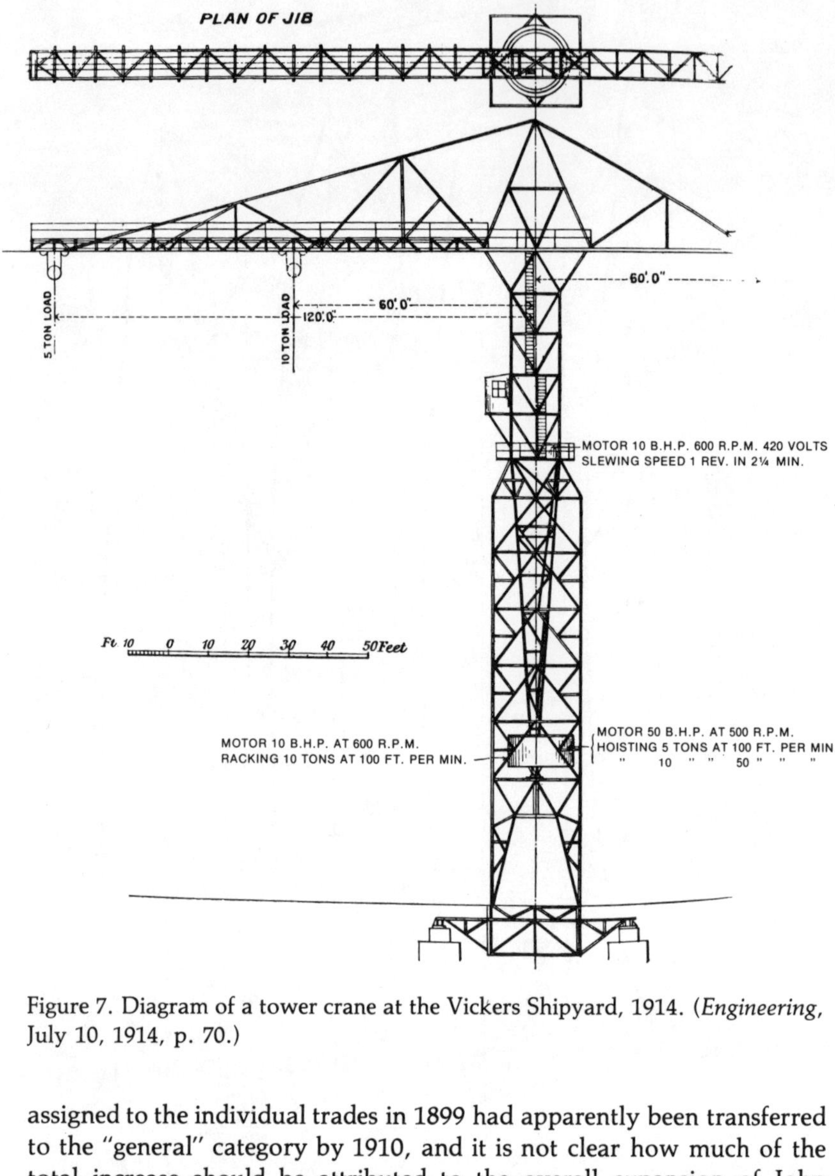

Figure 7. Diagram of a tower crane at the Vickers Shipyard, 1914. (*Engineering*, July 10, 1914, p. 70.)

assigned to the individual trades in 1899 had apparently been transferred to the "general" category by 1910, and it is not clear how much of the total increase should be attributed to the overall expansion of John Brown's operations and how much to increased capital per worker. Nevertheless, the trend toward greater use of machinery over the period is evident. Unfortunately, because there are no earlier breakdowns of the labor force by trade, it is impossible to calculate figures on capital per worker before 1910. At that time, however, as would be expected, there were wide variations, with the iron trades employing far more capital than did the older wooden trades (table 6.4). The absolute values assigned are, of course, highly tentative, since they would have fluctuated with the number of workers employed at each stage of the business cycle, but the ratios between the amounts of capital employed by each trade are probably fairly representative.

Table 6.3. Value of shipbuilding machinery by trade in the Clydebank yard of John Brown and Company, 1899 and 1910.

Trade	1899 (£)	1910 (£)
Iron workers	42,532	46,397
Ship smiths	14,151	8,114
Shipyard engineers	13,327	9,269
Joiners	3,996	3,793
Sawmillers	15,615	11,125
Patternmakers	275	243
Carpenters	1,827	2,182
Enginemen and cranemen	16,943	23,388
Electricians	165	610
Loftsmen	144	93
Sheet iron workers	2,319	1,734
Plumbers	519	522
Boat builders	203	118
Painters and redleaders	45	121
Riggers	15	9
Storemen, etc.	1,233	1,416
Electric lighting and power	12,411	6,251
Machinery, general	5,500	85,306
Total	131,220	200,691

Source: John Brown and Company, UCS1 25/6.

Table 6.4. Capital per shipyard worker, selected trades, John Brown and Company, 1910.

Trade	Capital/worker (£)
Ironworkers	38.89
Ship smiths	35.43
Engineers	32.87
Joiners	8.60
Sawmillers	144.48
Patternmakers	9.34
Carpenters	6.97
Electricians	5.45
Sheet iron workers	11.26
Plumbers	6.14
Painters	0.94
Riggers	0.11

Source: Calculated from data in John Brown and Company, UCS1 25/6 and 53/1.

Yard design and machinery in foreign countries often differed significantly from their British counterparts. Whereas few new yards were laid out in Britain after 1870, many foreign establishments, especially in the United States, were built in the 1880s and 1890s. Good river or ocean frontage was more readily available, and foreign firms were able from the beginning to plan their yards for the construction of large steel steamers. They avoided the burden of coping with cramped, irregularly shaped tracts and older buildings, which sooner or later plagued most British shipbuilders.[36]

The choice of equipment also differed abroad. Several extensive covered slips were built in the United States, though they became less popular in the years immediately before World War I. The Bath Iron Works in Maine, for example, had a covered slip 312 feet in length as well as a smaller shed for torpedo boats and other light craft. The Fore River Ship and Engine Company of Quincy, Massachusetts, had a shed 480 feet by 325 feet that covered four berths, and Moran Brothers of Seattle built a covered slip 850 feet in length for a battleship they were constructing. The largest and most expensive installation was at the New York Shipbuilding Company's yard at Camden, New Jersey, where practically the entire yard including the machine, boiler, smiths', and plating shops, molding loft, furnace, and several other departments as well as three large berths and a wet basin were connected under a single roof. In Germany, four of the older and smaller berths at the Krupp Yard at Kiel were covered.[37]

The Vulkan Yard at Hamburg and Harland and Wolff at Belfast used shiphouse frames similar in conception to the one first built at the San Francisco yard of the Union Iron Works. The original American frames were built of wood but later adaptations were of latticed steel to support both longer and higher structures and the larger cranes that traveled on them. At Hamburg the frame served three berths, one of which was 814 feet long. At Belfast two berths were served. The frame itself was 852.5 feet long and, including the outreach of the cranes at the ends, a total area 1,070 feet by 285 feet was covered. The considerations that encouraged builders to enclose their slips were heavily influenced by the climate and the nature of the work the firm undertook. Winters in northern Germany and parts of the United States were on average colder than in Britain, enhancing the value of covered berths. In a more moderate climate the heavy capital expenditures involved in building covered berths and shiphouse frames could be justified only if a firm could rely on a steady supply of orders for large ships. As a result only firms such as Swan, Hunter and Wigham Richardson or Harland and Wolff in Britain and the Vulkan Works in Germany, which built the largest liners, found it profitable to cover their berths or surround them with steel frames.[38]

The choice of other types of equipment was similarly influenced by climate and existing plant. Electricity for powering tools was adopted more quickly in the United States, France, Germany, Russia, and Austria than in Britain where steam power plants had been in operation for years. Despite the greater efficiency of hydraulic riveters for many purposes, pneumatic tools continued to be used in the yards on the Great Lakes and northeast coast of the United States because they would not freeze. This in turn affected the use of overhead cranes in America, since there was no need to transport heavy hydraulic machinery.[39]

Wage costs and the relative efficiency of labor were other factors affecting capital investment. In the United States, shipyard workers were often less efficient and more highly paid than their British counterparts. This gave American builders an incentive to adopt "all of the labor-saving devices possible," including cranes. The use of cranes was also fostered by the smaller number of shipbuilding firms in Germany and the United States. In Britain, where firms tended to be clustered together, it was sometimes feasible for several to share large wharf cranes, but in other countries the greater distances between yards made it necessary for each firm to have its own facilities.[40]

Clearly shipbuilding before 1914 is an instance of an industry in which, on balance, the nation with a head start had an advantage.[41] While foreign builders were able to choose better sites and design more efficient yards and shops, they were unable to overcome completely the greater efficiency of British labor, an efficiency that in part derived from Britain's longer tradition as a producer of iron and steel steamships. Hence, the tendency of British shipbuilders to invest less heavily in machinery than their foreign competitors was a sign of the foreign competitors' weakness rather than of cultural conservatism or technological interrelatedness preventing the British from sharing in the fruits of technological progress. British yard owners were willing to purchase any equipment that they believed would improve their profitability, and they did, in fact, try most of the innovations that originated abroad. The British found, however, that in many cases the benefits to be gained from increased capitalization would be erased by having to bear higher overhead charges during cyclical downswings. At the turn of the century, one American yard with five berths capable of housing large liners found them filled with a tugboat, a battleship, a ferryboat, a yacht, and a revenue cutter. It was the specter of precisely this kind of situation that made British shipbuilders wary of excessive capital investment and encouraged them to rely heavily on the skills of their workforce.[42]

7. Scientific Shipbuilding and the Evolution of Technical Education

British shipbuilders were able to invest less in equipment than their foreign competitors did, in part because of the superior skills of their workers. As long as the British persisted in a slower rate of innovation, therefore, their continued dominance in shipbuilding depended largely on their ability to maintain higher labor productivity than their competitors could achieve. Since the trend was clearly toward increasing technological complexity, this implied that traditional apprenticeship training, which primarily fostered manual skills, was no longer sufficient. Greater amounts of technical education were necessary as well. As with plant and equipment, however, the British also invested less in research and education than their foreign competitors. This tendency probably had few, if any, adverse effects in the years before 1914, but the legacy from this period of backwardness in research and training may have contributed to the relative decline of British shipbuilding in the interwar years.

The Science of Naval Architecture

On the continent, especially in France, mathematicians and physicists devoted considerable energy in the seventeenth and eighteenth centuries to the discovery of laws governing the performance of ships at sea.[1] Their findings (many of which were admittedly fallacious) had little effect on British shipbuilding practice, however, and even the more practical work of de Chapman (published in 1768 and 1775), himself a designer in the Swedish service, attracted no attention until the nineteenth century.

The first serious investigations in England were not made until the last decade of the eighteenth century when naval progress and superiority over the French became of major political importance. Between 1793 and 1798 Colonel Mark Beaufoy, the noted astronomer and physicist, conducted 1,700 experiments at a cost to himself of £20.000- £30,000. He followed up these tests, which were directed toward drawing up tables of the resistance of water to different shapes of solids, with comparative speed trials. As early as 1824 Beaufoy advocated model experiments to establish the speed and degree of stability of various hull forms, but his efforts were not followed up until John Scott Russell's investigations in the 1830s.

The shipbuilding trade treated the early experiments with contempt. The few rules laid down by continental mathematicians, which were used with effect abroad, were so little known in the British dockyards that every ship had practically an experimental design, except when the lines of captured French ships—which were admitted to be of superior design—were copied. In private yards the position was, if anything, worse. In Sunderland, around 1804,

> Science had comparatively little to do with the matter. A few general principles, no doubt, gave a basis, but the superstructure was greatly an affair of guess-work and eye-work. Frames were laid down from 5 to 10 feet apart, near the centre of the vessel. Others were then placed at a distance of 20 feet or so, from the bow and stern; and all the rest of the work was done at the arbitrary pleasure of the builder, by his individual caprice or conjecture, luckily or unluckily, as the case may be, and far more likely, of course, to be the latter than the former.

Fifty years later William Pile, the greatest of the Wear shipbuilders, still built according to his "shipyard training and wonderful judgement rather than to rule or plan," and in the following decade even "iron shipbuilding on the Wear was primitive—so primitive in fact that much of the construction work was done by rule of thumb." The drawing office of a yard building 4-5 ships had only one draftsman and two apprentices.[2]

By the time public opinion was at last aroused to the shortcomings of British naval architecture in the middle of the century, a number of pioneers had been at work for some time on the advancement of the science. Perhaps the most notable of them was Scott Russell, a brilliant Scottish scientist, engineer, and shipbuilder. In the 1830s and 1840s he devoted himself to observations on the performance of a hundred different vessels, from which he deduced a number of laws (most of which have since been disproved) on the proper form and length of hulls. Russell joined forces with I. K. Brunel in 1852 to build the monster *Great Eastern*. The methods of her design were as far in advance of the time as was her size.

Consciously and deliberately her builders set out to plan her from the bottom up in the greatest detail, according to the latest scientific knowledge, and to leave nothing to human skill, judgment, or error. Every component was designed beforehand; her speed, steadiness, and immersion in eastern waters were calculated; and her scientific instruments were specially designed by Airey, the Astronomer Royal, Professor Smith of the Edinburgh Observatory, and Captain Beeche of the Admiralty. Robert Fox, F.R.S., was consulted on the compasses, and William Froude on rolling at sea. The "machine"—Brunel preferred the term to "ship"—was intended to have even her route mapped out in detail in order to dispense with the human element of seamanship and reaction to weather conditions. A number of the latest innovations in naval architecture, such as the longitudinal system of framing and the cellular double bottom, were introduced into her design.[3]

The *Great Eastern* was not the typical ship of her day, but iron and steam shipbuilding of necessity brought with them a more scientific approach and shook the faith in the wisdom of tradition. Iron allowed longer and narrower ships. De Chapman's optimum in the eighteenth century had been three and a half beams for the length; in 1845 Fincham was allowed a maximum of five and a half beams at Portsmouth, and R. Napier thought six beams a minimum for ocean steamers in 1844. The advantage of length—which was thought to increase capacity without materially increasing resistance—continued to be exploited by iron shipbuilders. The *Warrior* had six and a half beams, and other ships in the 1860s went up to eight or nine.[4] In sailing ships the speed depended mainly on the wind, and the object was to spread the largest area of sail compatible with safety, but with steamers, once the power of the engine was given, speed depended to a great extent on the shape of the hull, and it was to this that investigation turned.

By 1860 science had entered the iron and steamship yards. In 1861 J. S. Russell wrote:

> In former times it has been reckoned enough for the constructive department of the Admiralty if they possessed science enough to calculate decently the proportions of a ship; and not many years ago it was too commonly found that even this great effort of skill exceeded the powers of the department of the Surveyor of the Navy. But the introduction of steam compelled also the introduction in to the Surveyor's Department of much higher skill and knowledge . . . [since the commencement of the iron fleet] it is manifest that the department of scientific construction, commonly called the Department of the Surveyor of the Navy, requires powers and functions vastly greater than have hitherto been conferred upon it.[5]

The apparent ease with which science was introduced into naval architecture in those years is subtly misleading. This minor scientific revolution was confined to the spearhead of the new iron and steam building, the handful of London and Glasgow engineers whose interests were not specialized and who, among their other activities, brought their scientific and mathematical experience to bear on the problems of shipbuilding. Occasional attempts were made to work out more detailed calculations, such as the designs of James Griffith of New York in the early 1840s, or the fully rigged five-foot models built by the Scotts in the clipper days for trial under ocean wind conditions on Loch Thom above Greenock, but most shipbuilders were not influenced by these new ideas. When the early engineers began to specialize and withdrew from shipbuilding, the industry settled back into its comfortable traditional groove, from which it had scarcely been disturbed. Apart from using the vaguely misleading and largely empirical "Admiralty coefficients," which were meant to indicate the power necessary for a given speed, most shipbuilders looked with unveiled hostility on the attempt to substitute "rule" for "instinct."[6]

Even after 1870 shipbuilders were by no means wholeheartedly committed to science or original research. Nevertheless, in some areas there was a spread of rational methods. The main direction of the change was toward more accurate calculations at an earlier stage in the process of production, until it became possible to predict speed, immersion, and stability, as well as the measurements and capacity of the vessels, even before their scale lines were drawn on the mold loft floor.

In this field, perhaps the greatest step forward was the idea of tests on small-scale models in experimental tanks. The credit for the discovery of the "law of comparisons," showing the similarity of reactions of hulls of the same shape, no matter what size, must go to William Froude, who derived his method of finding the resistance of ships from Rankine's streamline theory of resistance.

Froude's first experiments were in the late 1860s. He convinced E. J. Reed, at that time Chief Constructor of the Admiralty, that the behavior at sea of full-size vessels could be deduced from the behavior of exact-scale models in an experimental tank, a view not shared by most naval architects of the day. The Admiralty accepted his offer of 1868 to take over his station on the River Dart, and in 1871-1872 an Admiralty experimental tank, under Froude's supervision, was built at Torquay. After 14 years of successful operations it was removed to more satisfactory premises at Haslar, near Portsmouth, and continued to operate under R. E. Froude, the son of the founder.[7]

Over the next decades, foreign governments built tanks abroad. William Denny and Brothers was the first private firm to establish a tank, in

1883, but the second, John Brown, did not follow until 1903. Most private builders in Britain were still content to use rule-of-thumb methods or got their information at second hand from published results of the existing stations, since it was "but natural that the private firms possessing tanks do not consider it in every way expedient to draw public attention to the advantages they are deriving."[8] Cargo vessels in particular were built on established lines, and it was never known whether their hulls and engines were really suited to each other. Not until 1901 did Alfred Yarrow make the first serious proposal to the Institution of Naval Architects for a national experimental tank for the use of all private builders at a small fee. The proposal was supported unanimously, a committee was formed, and, in 1902, the council authorized the raising of the sum of £15,000 for the construction of an experimental station as part of the National Physical Laboratory at Bushy Park. The appeal for funds met with a disappointing response from private firms, however, and the project was delayed for several years. Only the persistent advocacy of Yarrow, who promised the capital sum of £20,000 for the project if the rest of the trade would guarantee £2,000 per annum for 10 years to cover the running cost, ensured its final realization. The experimental station was officially opened in 1911.[9]

In the 1870s and 1880s calculations on speed and stability had not yet become common among private builders. The decision of the commission of enquiry into the sinking of the *Daphne* in 1883 was significant: Stephen's of Linthouse had been determined at all costs to complete her in time for the Glasgow Fair. At her launch her engines were already installed, and shipwrights and others were working inside her hull in order to save time as she left her ways. The ship suddenly turned over, trapping all the men inside. The commission's "decision was that the builders must be wholly exonerated from blame on the ground that the estimation of stability on strictly scientific lines was at that date practised by nobody in the shipping line."[10] Neither the law nor enlightened self-interest did yet enforce the use of the data that were available for the safety of crews and passengers.

Around 1870 the simple relation between horsepower and speed, as given in the "Admiralty formula," began to be queried. Denny perfected the method of the progressive speed trial over the measured mile, at which the vessel's speed was measured at various levels of horsepower, and found in 1875 that the formula that power varied as the cube of the speed was inaccurate.[11] The formulae for determining resistance also improved and approached ever closer to the observed data, many of which were collected by the experimental stations. The shaping of hulls became increasingly a matter of calculation and experiment. Even before 1870, Harland and Wolff had challenged the accepted maximum length of steamers, itself changed since the eighteenth century, by lengthening the Bibby steamers in order to increase their capacity while keeping beam, speed, and power constant. An iron deck and a flat bottom and wide bilge (the "Belfast bottom") gave them stability and strength, the sharp bow gave them speed, and the straight stem and powerful steering gear allowed them to maneuver in narrow waters, docks, and channels. Their length of more than 10 or 11 beams became fashionable, and other companies, including the Cunard Line, had some of their ships lengthened. But the long ship had its opponents: a shorter ship had a smaller hull surface and was, therefore, cheaper to build, and had a smaller skin friction, an important factor at high speeds. It was also easier to maneuver, a trait that made it more suitable for warships, and broader beams, that is, more than one-eighth of the length, combined with finer ends, gave better results at high speeds. The optimum length lay somewhere between the extreme claims and could only be found by experiments in the model tank for each separate case.[12]

Other successful improvements in the 1870s and 1880s were bilge keels and the antirolling tank, which enjoyed a temporary popularity. Double-bracket framing, a system of combined transverse and longitudinal bottom framing, was introduced by Austin and Hunter of Sunderland in 1876, and applied to cellular bottoms, which had been fitted into colliers for water ballasting since 1860. Its application to ordinary mercantile steamers was an instant success, and between 1877 and 1880 steamers of a total of 200,000 tons were built or building according to that principle.[13]

Rationalized design was followed by rationalized workshop and yard practice. Denny's was again among the pioneers, with its complete set of rules—from the calculating tables for the offices to the details of the accident fund. The more complex the work undertaken on the ships, the more necessary did it become to have "perfect combination and supervision, that the completion of any work may, months or even years, be counted on in advance almost to a day."[14] This led to the greater emphasis on yard layout, which attempted to achieve a smoother flow of intermediate products, and to the more efficient use of machinery (see chapter 6).

The testing of metals before they entered the yard spread from the Continent and from the Dockyards to the private builders, and breakages due to faulty metals became rare. The Committees on Engineering Standardization, set up between 1901 and 1904, standardized the tests for the important metals. A Subcommittee for Shipbuilding Material Standardization was also formed, representing the Institution of Naval Architects, the Admiralty, the Board of Trade, the Iron and Steel Institute, the Scottish and North East Coast Institutions of Engineers and Shipbuilders,

Lloyd's Register, the Bureau Veritas, the British Corporation, the Steel Ingot Makers, and the Steel Merchants. Standard sizes were adopted in coordination with the Subcommittee for the Bridge and Railway Rolling Stock Sections of the Engineering Standards Committee.[15]

The presence of science-minded graduates in the yards was felt also in a new approach to experiments and inventions. The technical trade press was generally made available to the staff, together with the standard textbooks and general works on naval architecture. Incentive payments for inventions and improvements by the staff were found valuable in the more progressive yards. Although important inventions were introduced no more rapidly, it is significant that by 1913 they were to a much larger extent due to trained experts than had been the case 50 years previously.[16]

The emphasis on planning necessarily meant a larger staff, both for the drawing and the commercial offices, so that the number of black-coated workers in the industry grew rapidly in the twentieth century. It was a long move from the three draftsmen whom Hunter found in a large yard on the Wear in the 1860s, and from the office of Parsons who as late as 1894 had to write all his letters himself, to the shipyards of the twentieth century. An important shipbuilding firm in the United States, in advance of Britain in this matter, in 1901 employed a commercial and technical staff of 183 out of a total personnel of 5,206. Of these 183, the engineers and draftsmen in the shipbuilding department numbered 55, the marine engineering department 40, and other, including the scientific (research) department, 8, a total of 103; the other 80 were in commercial, managerial, and clerical positions.[17]

These developments changed the shipbuilding industry fundamentally between 1870 and 1914. In the former year, a vessel was the product of a shipbuilder whose design was empirical and unscientific if not entirely intuitive; who relied mainly on the traditional skill of his men for the accuracy and quality of the work; who possessed little capital, controlled few horsepower or machines, with a yard that was a haphazard collection of workshops that he could supervise easily. In 1914, at least in the larger and more progressive firms, a ship was designed collectively by a group of specialists, carefully coordinated in their functions, trained in a common body of advancing scientific knowledge; the shape, dimensions, and performance of the ship were calculated and tested in detail before the first keel plate was laid; her progress was planned, her material tested, her components turned out by reliable machines in a shipyard where massive power and equipment were minutely geared to produce the required results.

Despite these advances, as late as 1914 science and research were still the poor stepchildren of the industry. Many of the improvements described here, such as the testing of materials, had been developed in other industries. Others, including the use of experimental tanks or the measured mile, were empirical responses to empirical problems. Pure research had as yet little place in the industry; few builders maintained even rudimentary laboratory facilities or employed any scientists. It would be misleading to say that naval architecture had ceased to be a craft and had become a true science.

Technical Education

The increasing scientific sophistication of naval architecture and marine engineering, combined with the decay of the apprenticeship system (see chapter 8), forced yard owners and engine works managers after 1850 to find ways of providing the proper amounts of training and/or education for all groups of workers, be they unskilled, semiskilled, skilled, or professional. In the era of wooden sailing ships, when almost all the men engaged in shipbuilding had been either shipwrights or laborers, it had been relatively simple to provide training for the workforce. Shipwrights went through rigorous apprenticeships, which had evolved over the decades, and laborers were able to learn their jobs with a few days of practice. As ships became more complicated and more machines were introduced into the yards, the spectrum of required skills broadened considerably and builders were faced with ensuring that all of their employees received the mix of on-the-job training and formal technical education that would allow them to perform their jobs most efficiently.

At one end of the spectrum were the professional naval architects and mechanical engineers. Not only did their jobs require more formal education than the old shipwrights had needed, but they involved less manual effort and agility. The new materials and machines likewise increased the variety of artisan trades, in many cases craftsmen who needed the same lengthy training in the yards as the shipwrights had received, since for them a high degree of manual dexterity was still essential. Other tradesmen were in reality semiskilled machine operatives who required a shorter training period. As scientific knowledge increased, at least a modicum of technical education became desirable to help both of the latter groups of workers understand the general principles on which the new ships and machinery were based.

Aldcroft's account of the conservative reactions of British industrialists toward research and education has already been reported (see Introduction). Whatever the accuracy of the stereotype for British industry as a whole, it does fit a large proportion of shipbuilders and engine works managers, although there were some who were well aware of the value of technological improvements, knew that workmen abroad often were better educated, and deplored the situation. Yet there remained many, perhaps a majority, who like Aldcroft's typical industrialists had little faith in applied science and scientific instruction. Although this latter group declined in importance as the First World War approached, they were influential in blocking plans to bring British shipbuilding and marine engineering workers up to the same level of technical competence as workers in foreign nations.

In their attempts to establish programs that would make it possible to supply all types of workers with the proper training and education, British shipbuilders and engine works managers were hampered not only by the decline of the apprenticeship system but also by (1) the inadequacies of the national system of education, including a scarcity of facilities for technical or scientific instruction, and (2) the unwillingness of many builders to believe that the benefits to be gained from technical education justified significant adjustments in the established patterns of work in their yards or factories, or investments of time and money on their part. The efforts of those who supported increased education to overcome these obstacles absorbed much of the attention of the builders in the half century before the First World War. The transactions of their professional societies were full of schemes for improvements, and the issues involved were discussed from all points of view. Their statements, however, give the modern reader a strong sense of *déjà vu.* They frequently lamented the existing arrangements and offered suggestions for improvements, yet their comments have a curiously static quality. For more than 50 years each new generation found the same conditions, proposed the same broad remedies, and cited much the same reasons for failure; and each time their remarks were greeted as original and salutary. Because of the failure of the entrepreneurs to agree on a plan of reform, however, very little was accomplished.

The point on which in their public statements shipbuilders and engineers were almost unanimous was the necessity for increased technical education. To be sure, those who spoke at meetings of the societies were unrepresentatively progressive, but they were often the leaders of the largest firms and their opinions could not be easily ignored. For the most part, they felt that reform was urgent yet were unable to decide on who should be educated, what form the education should take, or how much should be imparted. A few even doubted whether there should be any technical education at all.

The primary theme running through their discussions was the need to preserve the "practical man" while at the same time providing enough technical training to meet the competition from abroad. This adoration of the "practical man" influenced the decisions of builders and engineers and led ultimately to the unsatisfactory compromises that characterized all British technical education.

Much thought went into the question of who was to be educated. It was acknowledged that naval architects and marine engine designers needed more theoretical training than mechanics or other artisans, but the constant fear was that on any level too much education might spoil the boy by making him too proud for the dirty work of the yards and shops. From the beginning two sets of workers were distinguished: "The man who will be a hewer of wood and a drawer of water all his days," and the future "captains of industry." Under the system that eventually evolved, both groups began their apprenticeships in the same program—as in the state-supported system of general education in Britain—but a system of rewards and punishments was used to separate "from the ruck the brilliant boys and . . . raise the standard of the rest who must, perforce, remain what nature intended them to be, namely, industrial units." Long hours of work (6 a.m. to 5 p.m.), which made daytime instruction impossible, and overtime, which interfered with evening classes, were even glorified by some:

> I think that many of our best men who can doubtless look back with a great deal of pride on the education and training which they have obtained at evening classes would consider the matter in precisely the same way. I think that if their path had been made altogether smooth by having facilities only for day instruction they would not have made perhaps such good men as they appear to be at the present time.[18]

There was general agreement on the training of artisans. Their jobs required manual adeptness and a knowledge of conditions that could be obtained only through experience. Any further education, technical or otherwise, that they received was welcomed, but it was most important that they serve their time in the yards and shops. On the one hand "workmen must also advance with the others," but on the other, "when all is said and done, for the largest proportion of the ordinary apprentices . . . a very high degree of technical education is not an essential in their after life."[19]

In the decade before the First World War, the North East Coast Institution of Engineers and Shipbuilders issued several sets of recommendations on education, which a number of firms in the area put into operation. Under their plans apprentices were given bonuses for good timekeeping and discipline and conduct as well as for examinations passed in the local technical colleges. Significantly, though, an apprentice could obtain enough marks to earn a bonus without passing any of the examinations or even attending a class.[20]

Throughout the period, education for artisans was regarded as having as much a social as a vocational mission. The Science and Art Department examinations administered under the auspices of the national government were originally designed for "those who can scarcely spell two words correctly," "to benefit the working classes, by inducing working men to devote their spare time to the study of the principles as well as the practical part of their professions . . . to make better men of them, i.e. more intelligent." In short, evening classes were to provide a channel of career mobility by compensating for deficiencies in the state-supported system of elementary education. As education improved, the evening classes became more difficult, but their social "uplift" side remained. Sir William White, a long-time proponent of technical education, looked on evening schools almost as correctional institutions:

> There have been hundreds of young men going through the evening classes who have been helped to fight evil habits, whose minds and bodies have been wholesomely occupied, who have been saved from becoming Hooligans and made better men by these evening classes—I do say, although the classes are by no means substitutes for higher and complete technical training, they are one of the glories of this country, and I hope the time will never come when those who can only find time to attend evening classes will not have evening classes to go to.

Another felt that "the moral effect on the boy who put in this 6 to 5 day's work, spent three nights at classes, and did a little homework was so beneficial that he sincerely trusted evening classes would not be given up." Others had different but equally nonprofessional reasons for supporting the education of artisans. When it was proposed to broaden the curriculum to include the social studies and humanities, Sir Benjamin Browne, the chairman of R. and W. Hawthorn, Leslie and Company, one of the few firms to insist on strict apprenticeship programs at all levels, remarked that "viewed as hard money-making subjects he found [economics and history] more valuable than anything else."[21]

In all, technical education for skilled workmen was deemed beneficial because it discouraged vandalism, promoted moral strength, and broadened a man's outlook as well as giving him a better grasp of his job. It was important not so much because it imparted a better knowledge of the principles of shipbuilding and engineering, but because it helped to inculcate industrial discipline and habits of good conduct.

There was no question that technical education was of professional value for the higher grades of apprentices, the potential naval architects and engineers. Only at this level was university or other higher education required. Furthermore, the training of naval architects and engineers was in practice closely connected, in part because of the similarity of many of the subjects they studied, but also because the number of students in both areas was so small that it was often necessary for financial reasons to run their courses in tandem. This was true both in local evening classes and in full-time courses such as those at the Royal Naval College, Greenwich, and Armstrong College, Newcastle. In the case of training these higher-level apprentices, the problem was to devise a curriculum satisfactory to a sufficient number of employers so that suitable educational institutions could be established. Since this agreement was never forthcoming, the system remained a patchwork of compromise programs in and out of which students drifted according to their own inclinations or those of their employers.

The most difficult barriers involved questions of when such higher-grade apprentices should attend classes and how much schooling they needed. Should they go directly from secondary school at age 15 or 16 into a technical college and then begin their apprenticeships at perhaps 19, or should they spend one to three years as apprentices, attending evening classes, before entering a college or university in a B.Sc. curriculum? Was daytime education superior to evening? Should full-time students be allowed to lose touch with their jobs or should they follow a "sandwich" course of six months at school alternating with six months in the yards or shops? These questions were debated throughout the period and never resolved.

To some extent the answers were provided by society. Most boys left school before they were 15 with practically no education in science or in mathematics beyond simple arithmetic. Often one or two years were then wasted before boys were accepted in apprenticeships, allowing them to forget what little they had learned. This was a national problem.[22] It meant that immediate full-time study in a technical college was out of the question because the boys were neither old enough nor well enough prepared to undertake further schooling. In order to obtain a basic education preliminary to higher technical instruction, boys required further secondary schooling, a need that, unless they were bright enough to win scholarships at approximately 10 or 12 years of age or their families were comparatively well-to-do, forced them to attend evening classes while working. When, especially after 1900, owners of shipyards and engine works committed themselves to drawing more talent from among the poor,[23] they also committed themselves to finding a way to compensate for the deficiencies in the national school system.

It cannot be said that they attacked the problem with vigor or imagination. By forcing apprentices to work from 6 a.m. to 5 p.m., they made the road for naval architects and engineers more a test of endurance than of aptitude. Over the years numerous complaints were registered on the impracticality of expecting boys who rose at 5 a.m. to study until 9.30 or 10 p.m., and some firms did shorten the hours for students. "One or two" firms on the Clyde, for instance, allowed apprentices to come in later on the mornings following classes, but the fact that the complaints persisted showed that little was done.[24]

One alternative was to allow students to attend classes during the day. When in 1915, Armstrong, Whitworth and Co. began teaching boys during the afternoons at their Walker shipyard, the plan was received by some as original although it had been in operation in the Royal Naval Dockyards since 1843 and had been included in the 1806 recommendations of the Commissioners for Revising the Civil Affairs of the Navy. Moreover, despite 70 years of success in the Dockyards complaints were still raised. It was objected, for example, that afternoon instruction would disrupt the yard by laying off others from their work while benefiting only the more intelligent apprentices; that in large shipyards with 1,500 apprentices, if boys were to attend classes two afternoons a week at least 600 would be absent from work on any day; that the cost to a firm would be as high as £15,000 to £30,000 per year.[25]

Advanced daytime instruction for the higher-grade apprentices in technical colleges and universities was also controversial. Some contended that practical training was more important and that students who had received full-time instruction suffered from "college conceit," that they were arrogant and unwilling to dirty their hands. In the decade before 1914, this attitude became less prevalent and the emphasis shifted to "sandwich" courses. When Alfred Yarrow began propagandizing for them in 1903, it was said that his plan would "give us in our industries an advance such as we have not had for a good many years" even though similar courses had been in effect in the engineering department of Glasgow University since 1840 and in the Admiralty for 40 years. By 1914 a number of firms had adopted "sandwich" programs, but the drawback remained that full-time education was expensive, especially in England. While several scholarships existed, they were generally too small to cover all of the costs, so the majority of students were forced to remain in evening classes.[26]

In summary, while the principle of technical education for all levels of apprentices was accepted, many shipbuilders and engineers were unwilling individually or collectively to take the steps necessary to create a workable system. Considering their indecision and lack of a basic commitment, it is not surprising that the results were disappointing.

The attitudes of private shipbuilders and engineers should be contrasted to those of the Admiralty. Official concern was first turned toward naval architecture during the Napoleonic wars. Captured French vessels were found to be superior to British, and as a result a school of naval architecture was established at Portsmouth in 1811. Under Dr. Inman, a Cambridge mathematician, the school trained 41 Admiralty students between 1811 and 1832. The teaching had to be carried on from foreign books, as no English ones were in existence. At first its graduates were cold-shouldered in the service, but by the momentous changes of the mid-century several of Inman's former pupils attained key positions. These included Isaac Watts, the Chief Constructor, Lloyd, the Engineer-in-Chief, Creuze, the Chief Surveyor of Lloyd's Register, and Moorsom, the author of the new Tonnage Law. After the school was closed, training for Dockyard apprentices was resumed in 1843, at first at a lower level. Schools were opened at Chatham, Portsmouth, and Pembroke in that year, at Sheerness in 1844, and later in Devonport, Plymouth, and Woolwich. The teachers for these compulsory classes were drawn from the Dockyard officials, and the apprentices still received all their practical training in the yards. The courses lasted for four years, with an examination at the end of each year. In the fourth year the most promising students were given instruction in draftsmanship.

When the first batch of pupils had passed through the Dockyard schools, their talents were not allowed to run to waste. A central college was founded to train the best pupils from all the Dockyards up to the highest level of their science. The School of Mathematics and Naval Construction at Portsmouth was placed under the direction of Dr. Woolley, another Cambridge graduate. It was in existence only from 1848 to 1853 and its annual intake was a mere eight pupils, but the small group of talented naval architects trained there carried through the modernization of the merchant marine and the Royal Navy;[27] created the armored navy; kept alive scientific advance in naval architecture; and was largely responsible for the foundation of the Institution of Naval Architects and the third School of Naval Architecture in the 1860s. Dr. Woolley's school, like Inman's two decades earlier, was closed by the First Lord of the Admiralty, Sir James Graham, "whose conduct had the same merit of consistency in putting a stop to the growth of knowledge."[28]

The Royal School of Naval Architecture and Marine Engineering at South Kensington was founded in 1864. The members of the Institution of Naval Architects, concerned over the lack of facilities for advanced education in Britain, joined with the Admiralty in sponsoring the school, and from the outset provision was made for private students. In 1873 the school was absorbed into the Royal Naval College at Greenwich, but it continued to be accessible to private students. Scholarship funds were available to those who did well on the entrance examination.[29]

In 1883 Mrs. John Elder endowed a chair in naval architecture at Glasgow University in memory of her husband, a well-known Clyde shipbuilder. A combined chair of naval architecture and marine engineering, separated in 1906, was established at Armstrong College, Newcastle, in 1890, and, after years of urging by Sir William White, a chair was finally endowed at Liverpool in 1908. The ancient English universities were less accommodating. Around the turn of the century, a would-be donor offered £100,000 to establish a school of naval architecture at Cambridge but withdrew his pledge in disgust when he discovered that students would have to qualify in Greek.[30]

Despite their roles in establishing these courses, the private shipbuilders and engineers did not support them by sending students. From the beginning, very few private students enrolled at the Royal School of Naval Architecture and Marine Engineering. Of those who did attend, most were foreigners or had no previous connection with shipbuilding. For several years after the school was transferred to Greenwich, no British private students enrolled, and private support barely increased in subsequent years.[31] Of the 166 students who completed the course from its founding down to 1904, only 60 were private students, and of those only 6 were known to have been employed by private firms in Britain. In all only 23 graduates of the school went into private establishments, many with Lloyd's Registry, as compared to the 91 who remained in the Admiralty service and 17 who went abroad. Of the 28 private students who are listed as having left the profession or unknown, a large proportion were foreigners who were also probably engaged in shipbuilding abroad.[32]

Eight naval architects and eight marine engineers were sent annually to the school from the Dockyards. The list of the first eight students was topped by W. H. White, later Director of Naval Construction and designer of Armstrong's shipyard at Elswick; it also included H. E. Deadman, later his assistant; Francis Elgar, later Professor of Naval Architecture at Glasgow and head of the Fairfield yard and Cammell-Laird; William John, later of Lloyd's Register; and W. J. Bone. In the following nine years the school also qualified P. Watts, Director of Naval Construction after White; J. R. Perret, later of Armstrong's; S. J. P. Thearle, Chief Surveyor of Lloyd's; and J. J. Welch, first Professor of Naval Architecture at Newcastle.

Various reasons were given for the failure of private students and firms to patronize the school. In the early years, the school was oriented too much toward naval construction without much consideration given to merchant ships, and too much time, which many felt could have been better spent in practical work, was devoted to theoretical training. The required mathematical skills were too great for many private students. Admiralty students, who had passed through the afternoon classes in the Dockyard schools, were better prepared. A shortened course for private students, with less emphasis on mathematics and theory, was approved in principle but never put into effect.[33]

Poor preparatory training also proved to be a barrier to students in the newer university programs. After the turn of the century, about 40 students enrolled each year in the naval architecture classes at Glasgow University, but on average only about 2 completed their degree requirements. Most of the candidates had simply not been able to prepare adequately during the five-year apprenticeships they had served before entering the university. Armstrong College at Newcastle was not empowered to administer entrance examinations and found as a consequence that it had to admit pupils who were incapable of following a curriculum almost identical to that of the Royal Naval College.[34]

Besides full-time study, there were other means available for obtaining advanced instruction. In the earlier years, mathematics was taught in the yards and shops themselves. There were also advanced technical classes in many of the important towns. Dumbarton had 25 naval architecture students as early as the 1870s. Newcastle established classes in the early 1880s and Sunderland, West Hartlepool, and other industrial centers had facilities for the course offered under the supervision of the central government, first by the Department of Science and Art and later by the Education Department. The conditions were not as good and the instruction took longer, but the classes provided a reasonable facsimile of the university programs. Approximately 85 to 90 percent of the topics covered in the syllabus of Armstrong College, for instance, were available in Newcastle's evening courses.[35]

Nevertheless, facilities in Britain lagged behind those of Germany and the United States even though the British were the more important shipbuilders. In 1907 Britain turned out 1.6 million gross tons of ships, the United States, 460,000 gross tons, and Germany, 280,000 gross tons. But the Germans had 280 full-time students of naval architecture at Charlottenburg and Danzig, and the Americans more than 100 at Cornell University, the Massachusetts Institute of Technology, the University of Michigan, and Webb's Academy—more than in Britain. Furthermore, foreign institutions were better staffed, and if the Germans were producing more naval architects than demand warranted and were subsequently forced to reduce their enrollments, the poor support that British institutions received throughout the period is all the more striking.[36]

Part of the success of the Americans and Germans was due to their more extensive systems of secondary education. The Germans carefully prepared their students for higher technical training, whereas the British relied largely on voluntary evening classes. In the United States, no practical experience was required before admission to a university and only a year was required before entering the equivalent German *technische Hochschule*.[37] Although there were advantages to requiring some experience, the lengthy apprenticeships in Britain, combined with the tepid backing for daytime instruction (in spite of the success of the Dockyard schools), made it more difficult for British students to obtain the preparation necessary to benefit from rigorous technical training.

Complete statistics are not available, but the fragments we have make it obvious that even the secondary education facilities that existed were not fully utilized. A number of firms on the North East Coast found that despite inducements only about 20 percent of their apprentices passed any evening examinations. In Sunderland in 1913-1914 only 38 percent of the engineering apprentices and 12 percent of the shipbuilding apprentices attended evening classes. In Hebbern the figure for shipbuilding apprentices was only 9.3 percent, and although the population of Wallsend had tripled over the years the number of students had increased very little. Armstrong, Whitworth had difficulty in persuading even 60 apprentices to join the afternoon classes at its large Walker shipyard.[38]

Not all of the blame can be attributed to the shipbuilders and engineers. Some of the trade unions, especially the Boilermakers, were cool toward technical education. The experience of Armstrong, Whitworth shows that apprentices themselves were not always eager to take advantage of even the most attractive opportunities for training. In some areas, the education authorities bear much of the onus. In Newcastle, for example, more public funds were spent on schools of correction than on the technical colleges. But although modern technical secondary schools for 1,200 boys could have been provided at an estimated cost to the city corporation of less than a penny rate, owners of local engine works had neither cooperated with a project to establish the schools nor demanded that the funds be appropriated.[39]

British shipbuilders and engineers were aware of developments and had a healthy fear of foreign competition. John Scott Russell's plea—which in the 1860s had led to what was probably the most important joint action ever taken by the profession, the establishment of the Royal School of Naval Architecture and Marine Engineering—was included in an article describing the École d'Application du Génie Maritime in Paris.[40] On several occasions members of the Institution of Naval Architects and the North East Coast Institution heard papers on education in Germany and the United States. They were impressed by the facilities available and the number of students trained. Although they did not entirely approve of the curricula, they felt that foreign education was on balance superior to British. They noted that tuition costs were lower in other countries, allowing a broader range of students to attend, that the state provided greater support, that secondary education better prepared students for university-level training, and above all that foreign employers were willing to hire graduates with very little workshop experience and train them despite their comparatively advanced age. British shipbuilders and engineers knew that in other fields technical training had already helped the

Germans and Americans to overtake the British.[41] And yet they did very little to encourage the spread of technical education at home. They continued to regard technical schools as deterrents to hooliganism, to discourage evening instruction by prescribing long hours and overtime for their apprentices, to squabble over the proper course of training rather than join to form an effective system, to cling to the cult of the practical man. Why?

There is evidence that suggests that in the short run, at least, British shipbuilders and engineers were correct in investing less in technical education than their American or German competitors. British shipbuilding appeared to be holding its own in world markets: as late as 1900-1914, about 60 percent of the world's new tonnage was produced in British yards. Secondly, because shipbuilding was a construction industry that relied heavily on handicraft skills that could be gained only through practice, scientific or technical training was of marginal importance for the lower grades of apprentices. It is possible that the Germans, in particular, with their extensive system of schooling for all grades of workmen, may have overinvested in education at this level. If data on real wages are a reliable indicator, the productivity of British shipbuilders rose steadily between 1860 and 1902.[42] Evidently the craft apprenticeship system, with all its imperfections, was sufficient, at least until the early years of the twentieth century, to allow workmen to make good use of the new machinery that became available during the period. This finding is buttressed by the knowledge that, despite their relatively poor education and outmoded equipment, British shipyard workers consistently produced more than their German or American counterparts.[43]

It is conceivable that the lack of facilities for the technical education of naval architects and marine engineers, whose work required a grasp of scientific principles, had more serious consequences. Even here, however, it appears that down to 1914 the British were as competent and inventive as their better-educated competitors abroad. An interesting example is the German-built *Imperator,* the largest passenger liner completed before the outbreak of the First World War. The vessel was originally scheduled to be built by Harland and Wolff, and only after being pressured by the German government did the Hamburg-Amerika Linie transfer the order to a German yard. The original British design was retained and the *Imperator* was basically an enlarged version of the *Olympic* and the *Titanic* without many of the innovations incorporated in the Cunarder *Aquitania* of 1914, which was designed and built in Britain.[44]

Nevertheless, it seems probable that the "practical" attitudes of British shipbuilders toward research and education did impede technical progress and, ultimately, the competitive edge of the industry itself. Machines were concrete objects whose efficiency could be accurately tested in a relatively short time; research was more suspect because the end product was often a body of ideas that took years to be translated into anything marketable; education, a long and costly process that trained people for research, was perceived as the riskiest of all.

There were, to be sure, tremendous improvements in vessels between 1870 and 1914, some of which—such as the manipulation of deck designs, the discovery of new lines, and the development of new ways of strengthening the structure while keeping down the weight of the steel used—were developed within the industry. Many improvements, if not most, however, were developed outside the industry, in the steelmaking, electrical products, or engineering industries. Many of the innovations in marine engineering did come out of firms that were associated with shipbuilders, but the most important, the steam turbine, was developed by an independent researcher, Charles Parsons. In a great many cases, it was only necessary for the shipbuilders to adopt the innovations after the basic research had been done elsewhere. Few laboratories were established in the yards, and as the reluctance to use experimental tanks demonstrates, builders were not even very interested in investing funds to solve problems peculiar to their industry.

In short, British shipbuilders did not believe that the benefits to be gained from increased research and education justified their expense. Most progress consisted of many small changes produced by men of little or moderate education. Great improvements, based on a solid knowledge of scientific theory, were few, and the attitude of the builders was: why should we equip laboratories and hire scientists to produce some basic breakthrough that may never come? Furthermore, they had little faith in the ability of scientists to contribute anything of much value, because scientists did not know how to build ships, and it was felt that only people who had passed through a proper apprenticeship in the yards and shops were likely to know enough of the problems involved in shipbuilding to be able to arrive at satisfactory solutions. The odd scientist or two working in a corner might be useful, but otherwise progress was best left to men who understood ships and shipbuilding. The position of most builders, though it was beginning to change by 1914, is perhaps best expressed by Pirrie's remark that "I can get all the brains I want for £3 a week."[45]

The high degree of competition in the industry, in the context of a laissez-faire society, also strongly influenced the attitudes of the builders. By 1914 most of them no longer disdained applied science or technical education as such; the difficulty was that they were unwilling to put their own money into them. They favored a state-supported system as existed in Germany, but the British government was unwilling to assume the responsibility. As the general mood of British shipbuilders and engineers was against any investment that might work to the advantage of their competitors—domestic as well as foreign—rather than themselves, education remained anemically financed. As one engineer noted, "It is not reasonable to expect private firms or even limited companies to equip technical colleges and give exceptional advantages to their apprentices unless they reap the reward. The instruction so given will probably accrue to firms competing against them."[46] This discourse on "externalities" suggests that shipbuilders and engineers were probably more fearful of competition at home than from foreign builders. The situation should be contrasted with that in the United States, where public aid to higher education was also scanty, "but fortunately for Americans they produce a race of millionaires over there. We do not seem to do that at all—at any rate not millionaires who make a hobby of technical education."[47]

This is a valuable clue to the general attitude of private British shipbuilders, which may be described as empirical and guarded. Despite the ties that grew from being associated in the same industry, yard owners always regarded each other as competitors, to whom it would be foolhardy to give away information unless something tangible could be gained in return. It is suggestive that the most important professional society, the Institution of Naval Architects, was initially sponsored primarily by employees of the Admiralty or other official bodies, or by men who were closely linked to the Admiralty as contractors.[48] These were men who were not affected by competitive markets in the same way as were private builders, and they played a most useful role by bridging the components of what was otherwise a highly fragmented industry. The institution was in existence for two decades before most private builders began to participate actively, and right down to the present officials of the Admiralty and of such quasi-public bodies as Lloyd's have continued to be among its leaders. The keeping of one's distance is normal and sensible behavior in highly competitive industries, even when, as in shipbuilding, many of the sectors are in practice oligopolistic, but it is at least conceivable that at times the attitudes of the builders surpassed the limits of reasonable caution and that the good sense of preserving individual initiative may in the long run have impaired the prosperity of the industry by restricting the opportunities for improvement.

The cumulative effects of the British neglect of research and technical training may have finally become disastrous during the interwar period, when the British share of world shipbuilding output declined dramatically. Although British builders and engineers had begun to make greater provision in these areas in the years just before 1914, a high proportion of the labor force in the 1920s and 1930s was the product of the earlier era when formal technical training had been disdained by the men and their employers alike. Yet the problems of the interwar period must be seen in the perspective of the immediate postwar years, which left a glut of tonnage that lasted well into the next decade; the decline in the volume of world trade and the further decline of Britain's share; and the restrictive practices of foreign buyers, which, more than ever before, discouraged their nationals from purchasing British-built vessels.[49] Against this background, it is difficult to determine even for this later period how large a share of the British decline may be attributed to reductions in labor force productivity brought on by poor education, but it seems certain that it was of some importance.

8.

Labor and Labor Relations

THE TRAINING of the workforce was just one facet of labor organization and control. Even when the workers have the requisite skills, the proper coordination of labor with equipment may be difficult to achieve in a period of changing technology. The introduction of new talents or machines that can perform multiple functions, particularly if the functions have hitherto been assigned to more than one man or group of men, disturbs existing work routines, upsets prevailing cost relationships, and forces management to reorganize entire sections of the productive process. If there are rigidities within a firm or industry that limit the capacity of management to reorganize effectively, the potential productivity of new talents and machinery may not be fully realized and the rate of technological or economic progress may be reduced. As a consequence, the competitive ability of a long-established firm or industry, as against a newcomer without the burden of tradition, may be jeopardized.

One of the central problems confronting the management of British shipyards between 1870 and 1914 was precisely how to coordinate workers and equipment. Management had first to determine that enough (but not too many) men with appropriate skills were available, and then to deploy them in relation to each other and to the available machinery so that ships could be built as efficiently and profitably as possible. To be sure, this is virtually a definition of the managerial function in any era, but it is a particularly vital function in periods of rapid technological change, as with the increasing complexity of ships after 1870.

Trades and Apprenticeship

In the days of wooden shipbuilding, the shipwrights, assisted by unskilled laborers, formed the main workforce in the yards, and other trades were ancillary. As ships became more complex and machines were introduced, the spectrum of required skills broadened considerably. In time almost all of the metal trades came to be represented as well as many of the building and engineering trades and several trades peculiar to shipbuilding. The division of labor thus proceeded largely through the addition of crafts new to the industry rather than through the subdivision of traditional shipbuilding trades. Moreover, the cyclical nature of the industry, together with the fact that it produced for the most part a large, custom-made commodity that was not susceptible to many of the techniques of mass production, ensured that a premium continued to be placed upon skilled labor. This stratification of the workforce along trade lines was reinforced by the presence of a strong network of traditional craft unions, many of which operated primarily outside the industry. In a period in which mechanization was increasingly destroying the functional barriers that had divided one trade from another, these factors inhibited the ability of management to make the adjustments in labor force organization that would have been necessary to take full advantage of technological improvements.

The development from the few clearly defined trades of the early years to the great number of skilled and semiskilled crafts represented in a modern shipyard might form a study in itself. The iron workers, frequently promoted from unskilled laborers, developed into such separate trades as platers, angle-iron smiths, riveters, drillers and their necessary holders-up and assistants; the engineering trades developed independently, yet on a parallel line, into fitters, turners, and drillers; other metal trades, including foundrymen, brass and copper workers, and finishers, found employment in the yards; in the larger centers the shipwrights split early into riggers, mast and block makers, and caulkers; and finally, improved construction and amenities brought painters, cabinetmakers, electricians, and plumbers into the shipyards. Not only did the trades come into conflict with each other on demarcation questions, but the fact that some unions were confined to the yards or had their main strength and interests in the yards while others looked on shipbuilding as a minor source of employment at times made collective bargaining nightmarish.[1] The inclusion of engineering shops, foundries, and other diverse facilities in some yards added further confusion and made comparisons of employment in the industry over long periods almost impossible.

The trades employed in shipbuilding were generally highly skilled (see table 8.1), but almost all of them needed a number of semiskilled and unskilled assistants, so there was always an appreciable amount of unskilled labor employed in the yards. The proportion of skilled to unskilled labor remained fairly constant from 1870 to 1900. At the end of the century, however, the influx of automatic machines weakened the

Table 8.1. The percentage of skilled labor in shipyards.

Trade	Eight English yards 1892	Six Scottish yards 1892	Scotland 1911	North East Coast 1913
Skilled woodworkers, "amphibians"[a]	58	66	21	47.04
Skilled metal and engine workers	—	—	46	12.58
Unskilled	29	22	18	25.16
Miscellaneous[b]	13	12	15	15.22

Source: Royal Commission on Labour, 1892-94, Group A, Replies to Trade Union Questionnaires (274-287); Board of Trade, *Industrial Survey of the North East Coast*, 1934, p. 273.

a. Amphibians are tradesmen who work both in the yards and on land, including plumbers, painters, etc.

b. Apprentices, etc.

position of the skilled men. In some cases the men won the right to retain the machines for skilled laborers only, as happened at Earle's in Hull following a nine-month strike in 1897; in other cases the masters were able to introduce unskilled labor for the machines. After 1900 the industry generally accepted as semiskilled the men who tended the machines in the shops and on the slips.[2] As the field of machine work widened, the men at the machines—hydraulic and pneumatic riveters, drillers, and so on—were again gradually recognized as fully skilled.

The shipwrights suffered most from the technical changes in the industry. Their refusal to handle iron in the early years of iron construction deprived them of their main building employment, since little timber, if any, was used in the hulls of iron and steel ships. By the time it became obvious that most of the larger ships were to be built of iron, the metal trades were firmly established and the shipwrights were unable to regain their position. They were restricted to drawing the lines of hulls, shaping the frames, and launching the vessels as well as performing heavy timber work on board. They also supplied molds for the components, plates, and bars, which iron workers then shaped and assembled.[3]

Other skills were brought in as shipping technology grew in complexity and many firms began to build their own marine engines. Platers, angle-iron smiths, riveters, drillers, and caulkers were hired from engineering shops, and unskilled laborers from within shipbuilding were trained to perform metal work. Although some workers, such as riveters, became specialized to work in shipbuilding, most of the 90 crafts represented in 1914 were identical to those needed in machine tool making, light and heavy metal fabrication, engine shops, or general construction. Only the trades that evolved from the shipwrights were peculiar to shipbuilding, and they were responsible for perhaps 10 percent of the work on iron or steel vessels.[4]

The privileged position of the shipyard craftsmen rested largely on their long and careful apprenticeships. During this period, which usually lasted for seven years, the boys were taught the manual skills of their trades, but the aim of the apprenticeship was also to instill traits of reliability and responsibility on the job and an overall sense of pride in one's craft.

On the main rivers the shipwrights' indentures were never observed very strictly. The migration of the industry to the North, where the shipwrights' position was less secure, weakened the strongly entrenched unions on the Thames and the Mersey, and the end of wooden shipbuilding imperiled their position still further. The boilermakers' apprenticeship faced different dangers. Like the engineers', it lasted five years only, but it always was in a state of flux stemming from changes in techniques and materials. In time it became subdivided in various ways, with a ladder established on which a worker climbed from the lesser-paid stages of iron shipbuilding, from being holders-up, caulkers, riveters, to the better-paid skills of platers and angle-iron smiths.[5]

Some employers were notorious for the large number of apprentices they engaged. In 1897 Waterman Brothers were said to employ 65 apprentices and 8 journeymen, and Reads of Portsmouth, 44 apprentices and 8 journeymen. In 1894, a slack year, the Greenock firms differed significantly in their ratios of apprentice shipwrights to journeymen: while Caird and Company employed 40 of each and Russell and Company had 40 apprentices and 30 journeymen, Scott and Company employed 84 apprentices but only 6 journeymen.[6] As late as 1877, the shipwrights in Scotland had so far found it unnecessary to restrict the number of apprentices, but in other areas relations were not so peaceful. On the Wear the marine engineering members of the Amalgamated Society of Engineers were out for two and a half years, between 1883 and 1885, on the apprenticeship question before they declared themselves beaten.[7] By the end of the century, however, the average number of apprentices among shipwrights and engineers amounted to about one-sixth of the labor force, a proportion that the unions had been struggling to ensure in earlier years.

The Boilermakers, the union embracing most of the iron and steel shipbuilding trades, had insisted on a five-year apprenticeship since their foundation in 1834. It was accepted by nearly all districts, although in the outports at the end of the century the apprenticeship period was frequently shortened and only one special trade acquired. At the same time, the limitation of the number of apprentices became an important item on the union's agenda. In 1891 a national agreement was attempted to ensure a five-year apprenticeship, and in the following year a limitation of one apprentice to five journeymen was laid down. In the National Shipyard Agreement of 1893 the ratio was raised to two apprentices for every seven journeymen, and most important yards accepted these figures from 1894 on.[8]

The shipwrights' seven-year apprenticeship remained substantially unchanged in the main ports until the end of the century. It is difficult to state exactly to what extent the reluctance to touch iron work and the undisguised contempt with which the shipwrights viewed the work of the boilermakers were caused by their long period of training in timber shipbuilding; it is certain, however, that they refused to recognize any apprenticeship in iron work and forced the iron men to form their own competing unions. Toward the end of the century even this old-fashioned and strongly entrenched trade could not prevent the decline in the duration of its apprenticeship, and the Associated Shipwrights' Society itself became diluted with drillers and others.[9]

In fact, by that time apprenticeship had almost disappeared as a valid means of imparting skills. Testifying before the Royal Commission on Labour, Robert Knight of the United Society of Boilermakers and Iron and Steel Ship Builders contended that by 1892 the old system of indenture had practically died out and that few boys were then bound. Apprentices would stay in a yard for a few years at half a journeyman's wages and then move on when they received better offers. The new system, Knight believed, produced inferior workmen and, more importantly, overstocked the labor market and led to reduced wages. Other trades, including the blacksmiths and engineers, were similarly concerned.[10]

The cheapness of apprentice labor, combined with the wide range of trades that came to work in the shipyards, inevitably disrupted the carefully controlled apprenticeship programs inherited from the preindustrial era. This change was not confined to shipbuilding and marine engineering, of course, and it occurred long before any publicly organized system of technical education arose as a replacement. Around 1880, a typical five-year apprenticeship in engineering had been described as a year of odd jobs in the pattern shop—lighting the fire, sweeping the floor, and so on—followed by two years of desultory training in woodworking. A year or two in the fitting shop, working with metals, might then, if the apprentice were lucky, lead to entrance into the drawing office, where he would have to begin again. Along the way, he would receive little help from his foreman, and at the end of his apprenticeship he would qualify only as a journeyman pattern maker or fitter and not as an engineer. Although he would be skilled in "practical" work he would have received little or no theoretical instruction and would have forgotten what he had learned before beginning his apprenticeship. "It had always been assumed, when engaging an apprentice, that the employer undertook to teach him the trade which was to be the business of his life; in reality, however, he received no teaching but was allowed to learn by his own observation and practice as best he could."[11]

There is no doubt that both employers and craft unions were of two minds on the question of training apprentices. Employers recognized that apprentices could be a cheap source of labor, but when they were indentured they could also be very expensive since they could not be laid off during depressions. They were often the only hands kept on, under the supervision of foremen, and there are indications that some of the speculative building in slumps was carried on mostly to keep the apprentices busy. The unions regarded formal and controlled apprenticeship as the basis of their strength, but as long as the apprentices required indentures they threw the entire burden of unemployment upon the journeymen and made it possible for apprentices to be used as strikebreakers. To gain tighter control over apprentices some of the unions including the Boilermakers and the Shipwrights took them in as half members and looked after their interests until they qualified as journeymen.[12]

Trade Unions

The origins of the shipbuilding industry in a skilled craft and the way in which it evolved through an accretion of skills borrowed from other industries ensured that stratification by trade would continue despite mechanization. Evolving trade unionism in the shipyards came to be based almost entirely on the division of crafts, and not upon a distinctive "industrial" or general consciousness. The shipwrights' societies were traditional craft unions, jealous of their prerogatives and determined to exclude outsiders from their ranks. As a result, when men with different skills entered the yards they could not associate themselves with unions that were already active in shipbuilding. In order to obtain representation, the new trades had either to form societies of their own or to join ones that had been set up earlier to represent their trades in other fields of manufacturing or construction.

Even before the repeal of the Combination Acts in 1824, the skilled men engaged in building and repairing wooden vessels had formed illegal societies. In subsequent years, the shipwrights founded unions of varying importance in a number of ports. They were generally strong enough to force a closed shop and to demand union membership and apprenticeship for foremen. One of their main preoccupations was to restrict competition for jobs by eliminating geographic mobility. The various local unions began to consolidate in the 1850s with the formation of the United Kingdom Amalgated Society of Shipwrights. The Amalgamated Society had no central funds, and the local unions continued to act independently until the mid-1880s, when its loose organization was finally shown to be inadequate. It was then superseded by the much stronger Associated Shipwrights' Society, which guaranteed the right of its members to work in all ports and provided strict centralized control. By 1897 all but three of the local unions had been absorbed, and after the turn of the century, when drillers and some iron shipbuilders assimilated, the name was changed to the Shipconstructive and Shipwrights' Association.[13]

A second great union in the shipyards was the Amalgamated Society of Engineers. Like the shipwrights, mechanics and engineers had formed local societies throughout the kingdom in the early years of the nineteenth century, in which the various branches of the trade were generally represented in different unions, which joined together to conduct strikes. The Amalgamated Society of Engineers was formed in 1851 from a merger of several of the London and provincial societies. Most of the engineering unions joined, but a few important ones, including the Boilermakers and the Steam-Engine Makers, remained aloof.[14]

The largest union of metalworkers, the United Society of Boilermakers and Iron and Steel Ship Builders, was founded in Manchester in 1834. Over the years, local societies were absorbed and the union began to admit less-skilled men in order to prevent blacklegging. Eventually caulkers, holders-up, light plate workers, and a few shipwrights who had been performing iron work were taken in. Drillers were not accepted, however, and in 1900 they merged their own unions with the Shipwrights, perpetuating old antagonisms that would have been dampened if they had been admitted to the Boilermakers.[15]

Every shipyard employed a large number of skilled men, general laborers, storemen, and others. There were also a number of assistants to the craftsmen—helpers, holders-up, rivet heaters—who were not entirely unskilled and who succeeded in the course of time in gaining entry into the craft unions. The helpers were the only large group to be forced into their own societies. Their main grievance stemmed from their employment by the platers rather than directly by the yard owners. The platers organized their work to maximize their own wages, which were based on piece rates, to the detriment of the helpers. After long struggles, the helpers in some ports managed by the late 1890s to gain better benefits including the right to be paid directly by the builders.[16]

Between the 1860s and 1914, the scope of collective bargaining had widened from the yard to the river, and from the river to the national level. At the same time, some trades had coalesced, and others had become crystallized as separate skills, so that national uniformity was almost achieved. After 1890, a last step remained to be taken: a common organization of all the trades in the yards, so as to settle the demarcation battles in amicable national agreements, regulate their wage movements and their hours together, simplify collective bargaining with the employers, and prevent the employers from playing off one trade against another.

The first Federation of the Shipbuilding and Engineering Trades was formed in 1889 at the initiative of the Boilermakers. Its primary goal seems to have been the desire to end the demarcation difficulties and raids among the unions. Most of the shipbuilding unions belonged, but the A.S.E., which attended the preliminary meetings, did not join until 1905. The federation was instrumental in reducing the demarcation chaos of the early 1890s but was not designed for any common action toward the employers. The A.S.E. fought their great dispute of 1897-1898 over the eight-hour day alone.

The first real efforts at unity in trade matters were not made until the bitter strike and lockout of the Shipwrights in 1907. This led to the national organization of employers and workers in the industry on the terms of the Edinburgh Agreement of 1908. The employers had deplored the refusal of the men to accept the agreements reached by their leaders in the dispute of 1907 and pressed for the establishment of mechanisms for preventing and settling strikes with a minimum of disruption. The subsequent agreement had two main sets of provisions dealing with wage changes and means of settling disputes. While preserving local differences in wages and hours, "general fluctuations" in the activity of the industry as a whole were to be met by alterations in the wages of all trades and all districts at the same time. This represented a victory for the employers, who desired the national market for ships to be matched by national wage movements, and was accepted by some union leaders only under protest, as they realized the difficulties of creating strong pressure in all ports and among all the trades at the same time when a rise in wages was justified.[17]

As for the settling of disputes, a system of conferences and mediation was set up to allow work to continue while final agreement was being negotiated. The terms were praised in the national press as a contribution toward industrial peace, but they also limited the ability of the men to enforce any of their demands, as they arose in the shipyards, without national backing. In fact, from 1909 on, nearly all the negotiations in the industry took place in camera, beyond the light of day. Although the agreement was not universally successful in eliminating unofficial stoppages, it was generally effective in achieving its ends and was renewed in 1912 with only slight modifications.[18]

The national development of unionism in the industry was matched by a parallel growth of employers' associations. In 1889 the Federation of Shipbuilders and Engineers of England, Scotland, and Ireland was formed, but the engineering employers founded their own national federation in 1896 and the shipbuilders followed in 1899 with local branches in the more important shipbuilding regions. With both engineers and shipbuilders the formation of the national body led to immediate aggressive trade policies and national lockouts, which resulted, on the initiative of the greatly weakened unions, in the standing national agreements.[19]

Despite this movement toward national consolidation, there continued to be a multitude of unions operating in the shipyards. Between 1892 and 1896, 130 "general" shipbuilding unions were in existence, not including the societies of engineers, carpenters, joiners, plumbers, and other trades that were active in other industries as well. In addition, there were 7 unions of riggers and 17 of sailmakers, most of which were confined to one port. In the period between 1906 and 1910, the number of general shipbuilding unions was reduced to 18 while the riggers and sailmakers consolidated into 5 societies. Fragmentation remained common, however, and at times a trade was represented by more than one union in the same yard.[20]

Working hours and wages were two of the questions that led most frequently to strikes in the decades before 1914. There were considerable local differences in working hours before 1870, ranging from as many as 64 hours per week at Cardiff to as few as 54 hours at Liverpool. On the North East Coast, full hours were worked only in summer, with dawn to dusk the rule in winter. Liverpool, too, worked as little as 48 hours or even 44½. Scotland, on the other hand, worked the same hours the year round except that Saturday was a half day in winter while alternate Saturdays were taken off in summer.[21] Actual hours worked per day varied widely, depending upon the length of mealtime breaks.

The first serious dispute over shorter hours occurred on the Clyde in 1866, when the men won a modified demand of 57 hours with a proportionate reduction in wages following a lockout by the newly formed Employers' Association. In the early 1870s the 54-hour week spread in all shipbuilding areas, and a 51-hour week, the shortest reached anywhere before the 1890s, was conceded on the Clyde in 1872. By 1890 the Tyne and Wear engineers had won their 12 o'clock Saturday after a short struggle. The agitation for an 8-hour day made considerable headway in many unions, and the Trades Union Congress was induced to press for parliamentary action to establish it. The Royal Dockyards adopted a

48-hour week in 1893, and they were followed by some employers who realized that they could get more work out of their men in 8½ hours than in 9½ by abolishing the working hours before breakfast. The Engineers' struggle over the 8-hour day in 1897-1898 was defeated by a national lockout. By 1906 the average number of hours worked in a full week was slightly under 54.[22]

The working week seldom corresponded to the actual hours worked, however. In depression years it served as a standard of reference for the half or three-quarter time worked; in boom periods, it fixed the hours to be counted as overtime; but conditions were rarely normal in the industry. Work was not to be had every day, even in active times, since most craftsmen were needed only at certain stages in the progress of a vessel. Piece workers who earned higher wages would stay away for some of the time, to the detriment of their helpers. Bad weather, especially in winter, also kept the workers idle for days on end. Even some of the employers agreed that wages in the industry were not too high, if the fact were borne in mind that the men were unemployed for long periods. P.J. Robson, a trade unionist, thought in 1867 that a daily wage of 6/6d. meant an average of only 27/- a week; in 1892, Seaton, an employer, thought that a 53-hour week meant in practice 50-51 hours, while Robert Knight held it to be less than 48 for the boilermakers. In 1903 the Glasgow Commission on Housing of the Poor found that boilermakers lost 10-15 percent of their possible time. Tyne helpers were stated to earn 26/- out of a possible 30/-.[23] These proportions refer to periods of employment and omit the drain on the men's resources during the long stretches of unemployment. Thus the workers bore many of the risks of the industry.

Overtime working was common in the years of good employment. Competitive bidding for rapid delivery, penalty clauses, and the need to take advantage of good weather all justified to the employers the payment of higher rates. Furthermore, the full use of costly machinery required overtime and sometimes night shifts. The unions permitted overtime in emergencies, such as the docking of a vessel or the erection of staging, but they were firmly opposed to the "pernicious" system of regular overtime—which extended the working day from 6 a.m. to 8 p.m.—because they felt it created unemployment.[24] The agreement to end regular overtime work on the North East Coast, signed in November 1891, had the support of many of the employers who had found that their wages bills were going up because of the higher rates payable, and also that it "taxed the men beyond their strength."[25] In most areas the men succeeded in abolishing systematic overtime by 1893.

After the signing of the Edinburgh Agreement in 1908 an attempt was made to settle the question of hours of work and overtime on a national level. The employers desired a maximum of 32 hours overtime per man every four weeks. They also proposed to the standing committee in 1909 to abolish the weekly wage and pay hourly rates with a quarter-hour minimum instead, as a means of forcing all districts to return to the 54-hour week, and to pay for only 48 hours in winter in the districts where shorter seasonal weeks were established. The men, many of whom had been receiving their full week's pay in the winter months, countered with a proposal for a uniform 51-hour working week throughout the year. Discussions continued through 1910 and 1911 and at the end of that year the unions were prepared to allow four hours overtime per man per week and also agreed to overtime up to the full amount of the reduction if the hours were reduced to 51. This question remained unsettled in 1914.[26]

The shipbuilding industry boasted of successful conciliation schemes and of close contacts between well-organized unions and employers' associations at an early date, yet some of its disputes were fought as bitterly as any on record. The closely organized skilled trades could threaten to inflict crippling damage on the masters during periods of full employment, so that their wages generally rose rapidly when the yards were full. In slack times the masters could easily dispense with their tradesmen, and wages slumped as quickly. Only with the unskilled men, however, was it ever found worth while to settle disputes by importing blacklegs. The masters found, with A. F. Hills, that "the men you get when a strike is on are not up to the mark as a rule," and that when they arrived they were, like Harland's blacklegs from Glasgow in 1857, "intimidated, or enticed, or feasted, and sent home again."[27]

Although it is doubtful whether the unions had much effect on the long-term trend of wages, a high proportion of their efforts was naturally devoted to that issue. On the Clyde in 1877, for example, the shipwrights asked for an increase of 15 percent or one penny per hour to make up for a reduction three years earlier. A strike of six weeks was followed by a lockout of 35,000 workers in 22 yards. The strike and lockout were almost ruinously expensive for both the men and the unions: it is estimated that they cost £156,000 in strike benefits. Nevertheless the unions held together and the question of their recognition and survival was settled at that time. It was also with the strike on the Clyde and other disputes in the 1870s that the focus of negotiations shifted from the local to the district or river level.[28]

From the 1890s, the employers' federations began to use national lockouts as a tool to impose their will upon the unions. In 1895 engineers in Belfast struck to obtain higher wages. The local Engineering Employers' Association linked up with the Clyde Association and insisted on treating both districts together. Engineers on both rivers were locked out. An agreement was reached that imposed an unpopular settlement upon the membership in Belfast, and the general secretary was replaced by a militant. Similarly, the lockout of engineers in 1897 resulted from a dispute in one district, which spread throughout the kingdom via the medium of the employers' reprisals. It proved to be the most costly dispute in the industry for the whole century, and its effects can be seen clearly in the annual output figures. For 30 weeks up to 47,500 highly skilled men were idle, of whom 28,000 were members of the A.S.E. (out of a total membership of 92,000). The conditions of settlement were very hard, coloring relations in the industry until 1914, but there were no longer any attempts to eliminate the unions by the "document."[29] As the employers had enforced district agreements in the 1870s, so, in the 1890s, did they inaugurate national collective bargaining.

The irresponsible use of the immense power wielded by the national associations of employers in the engineering and shipbuilding trades had begun to jeopardize the prosperity of the industry and the livelihood of its workers. Within a short period, the associations had turned local disputes into national lockouts four times (in 1897, 1907, and twice in 1908) and had turned a sectional dispute into a general lockout once (in 1906). The unions were forced to combine not only on a national but also on an industrial level. As the disputes in the 1870s had led to local agreements, and those in the 1890s to national agreements, the disputes in 1907-1908 led to a national industrial meeting ground, the Shipyard Standing Agreement, which included all of the important trades. At the same time, the marked effect of disputes in one shipyard trade on employment in the others made a medium for union consultation imperative.

The hopes of the union leaders for a more moderate attitude on the part of the federated employers proved unfounded. In September 1910, after all the unions had agreed to the employers' requests to postpone their wage rise to October, the employers suddenly decided to use the pretext of two small unofficial strikes at Elswick and Meadowside to lock out the whole membership of the Boilermakers' Society in the federated yards at a day's notice. Only after two months did the employers agree to meet the union's representatives. Although the executive and a specially elected delegate conference were willing to accept the employers' "promises" of good behavior, the men decisively rejected them with growing majorities. The lockout, lasting 15 weeks, affected 15,300 men directly and 20,000 indirectly and was calculated to have cost £800,000 in wages. It crippled the Boilermakers to such an extent that they had to suspend the payment of unemployment benefits for three years.[30]

In an industry blighted by a high and fluctuating level of unemployment, the restriction on the speed-up was, next to the restriction on overtime, one of the major union aims. In 1867 J. Allen, representing the London riveters, was already complaining about the speeding-up process. "We find by a strict examination of the accounts (and I have had the accounts from the offices of the various shipbuilders in London), that in place of an advance of wages of 10% being given, we have fallen back about 50% when we consider the amount of work done." Thirty years later the process had gone much further: riveting "is a difficult trade to learn, a hard and exhausting one to follow, wearing a man out in his youth, for no one ever saw an old riveter . . . wages are inevitably high, and most of the work is done by the piece. When the work is to be had, therefore, the riveter makes a great deal of money, but at the expense of his vital energies."[31]

Throughout the period, union policy was made easier by the high degree of geographic mobility of labor in the industry that followed the establishment of national trade unions. The unskilled labor was commonly Irish, not only on the Clyde, but also on the Tyne, and migrated fairly easily when a dispute of skilled men put them out of work. Skilled labor migrated equally freely, driven by long-term prospects or by temporary local unemployment and disputes. The decision of the Barrow and Belfast branches to follow the wages and conditions of the Clyde is easily understood when it is realized that their members had nearly all had their training there, but even on the Tyne and the Wear the men had "very largely come as young shipwrights from Dundee, Montrose, Aberdeen."[32]

This mobility enabled the unions to finance long disputes on slender means. During the 9-hour strike on the Tyne in 1871, more than 4,000 workers found employment elsewhere, so that the number "on the funds" was small and actually decreased during the dispute. On the Clyde in 1877, of the 1,200 shipwrights out, only 400 remained on the books and the rest found employment in other districts. They had left their families behind and planned to return at the earliest opportunity. Other disputes were marked by similar temporary migration. Ease of migration was among the forces that led unions and employers to coordinate their trade policies over wider areas.[33]

Demarcation Disputes

Of all the types of industrial strife, the one most closely associated with shipbuilding is the demarcation or jurisdictional dispute. Not only was it especially common in the industry, but it illustrates well both the technological changes in shipbuilding and the course of trade union evolution.

Labor historians have often accepted the argument that demarcation disputes were especially common because shipbuilding "from the wooden sailing ship to the steel steamship had undergone in a short span of years a thorough-going technical revolution without a corresponding regrouping of unions."[34] This is to misunderstand both the nature and the development of the industry. The multiplicity of crafts and unions active in the shipyards was, at least to a large extent, the *result* of the technical revolution. It survived because, although the proportion of unskilled and semiskilled machine operatives did increase, shipbuilding was still in 1914 a construction industry that required men with varied skills of a high level and was not susceptible to many of the techniques of mass production. The division of unions along craft lines did correspond to the actual functional division of the labor force; the introduction of unions enrolling members from all sectors of the workforce would not have been as appropriate as it was in other branches of heavy industry in which the workforce was predominantly unskilled and workers were largely interchangeable. While subsequent changes in the technology of shipbuilding tended to blur the boundaries between trades and resulted in interunion conflict, the stratification of the industry along craft lines remained on balance the most appropriate way of organizing the highly skilled workforce and its trade unions.

Although practices varied from port to port, the wooden shipbuilding industry was organized in much the same way as house building or other sectors of construction. In London, for example, yard owners, themselves master shipwrights, acted as general contractors who negotiated with customers and supplied facilities and materials in return for a percentage of the cost of the work performed. The shipwrights, who provided their own tools, then acted as subcontractors. In many yards, they grouped themselves into "gangs" or "teams," which were hired by the job for a fixed price based on an estimate of how much they would have received for the same amount of work if paid individually by the day. The price of each job was determined separately, and as the work progressed yard owners advanced wages depending on the proportion of the job completed or the time elapsed. If any money remained at the end, it was paid in a lump sum, which the gang members divided among themselves. It was essentially a contract system of piece wages, which encouraged the shipwrights to complete the work more quickly than they would have done if hired and paid by the day or week. Particularly after 1824, when the Shipwrights Provident Union of the Port of London was founded, master contractors were unable to dictate the membership of the gangs, and in some cases apprentices were bound to gang members rather than to the yard owners.[35]

In some areas, yard owners had even less power to regulate their workers. At Liverpool, building was performed for day wages but in the important repair sector of the industry shipwrights were paid directly by shipowners and could refuse to follow the orders of the yard masters.[36]

As the number of trades and workers grew and the yards became more heavily capitalized with the advent of iron steamships, the contracting builders found subcontracting to be increasingly intolerable. The complexities of negotiating with each trade made it difficult to arrive at prices that were satisfactory to buyers, and placed the fate of the builders' large fixed investments beyond their own control. In some cases, helpers were paid by and responsible to skilled tradesmen rather than to yard owner-builders. The latter could not be sure of how much their employees were receiving and were sometimes plagued by friction among workers created by unequal wages for the same tasks. Subcontracting persisted down into the 1860s on the Thames, where in at least one yard even riveters were covered under contracts granted to shipwrights. It never became firmly established in the newer iron shipbuilding centers on the North East Coast and the Clyde, however, where most unions objected to piece wages and to paying their assistants themselves. As the newer builders gained the right to pay the entire labor force through the central office, the older contract system eventually disappeared.[37]

Only one major group, the members of the United Society of Boilermakers and Iron and Steel Ship Builders, continued to work on a variation of the gang and contract system throughout the shipbuilding industry. The difficulties that resulted have already been alluded to. Platers and angle-iron smiths worked in teams with one skilled worker and six or eight helpers. As the Boilermakers were paid by the piece but the helpers received day wages, the skilled men were able to sweat the unskilled by forcing them to work at a maximum pace for no extra pay beyond an occasional bonus. Conflicts arose between platers and angle-iron smiths and the employers over control of the helpers, but by the 1890s Boilermakers no longer wanted to pay the helpers themselves and generally acceded to the employers' demands to pay them directly through the central office. From the standpoint of the members of the Boilermakers Society, the gang and contract system worked well: by 1906 shipyard platers were probably the most highly paid trade in all British industry.[38]

Greater mechanization and capitalization increased central authority, but the demise of the contract system did not bring an end to the organization of yards along craft lines. Since most vessels were custom-built, there was little scope for techniques of mass production, and tradesmen could retain their identities as skilled workers even though they were employed as wage laborers in large concerns. Especially before the turn of the century, large-scale mechanization of work processes was not feasible, although the workforce was differentiated into a large number of skilled trades. Between 1870 and 1900, the greater complexity of vessels led to increases in the contribution of skilled labor to the shipbuilding process. Whereas in 1873 hand labor accounted for 24 percent of the cost of a hull, by 1887 the proportion had risen to 34.5 percent for large vessels and to between 36 and 38 percent for smaller ones. Similarly, skill differentials in pay increased until approximately 1900, suggesting that constructional techniques in the late nineteenth century emphasized skill; and although increased use of capital equipment helped to reduce the importance of skilled labor somewhat after 1900, much of the machinery introduced was intended to enhance the productivity of existing trades rather than to displace them.[39]

Two factors usually suffice to distinguish one trade from another: the nature of the materials worked and the types of tools used. In the middle of the nineteenth century, for example, engineers worked on cold metal with hammers, files, and lathes while boilermakers handled hot iron to rivet hull plates, bend angle irons, and perform similar tasks.[40] As long as technical improvements fell within the province of only one trade, there were few problems. The craft system of organization became increasingly anomalous in the decade before 1914, however, with the development of equipment that could be used by several trades, and of unspecialized machines designed to perform several activities without regard for craft boundaries.

The problem of reorganizing labor inputs to suit the needs of changing technology was accentuated by the mood of the unions. Not only had they inherited a guild outlook on the need to restrict entry to the trades, but they were immediately alarmed by the periodic unemployment and by what they early perceived as a growth in the stock of labor relative to the demand. Under the circumstances, the job security became the paramount concern, and each union was bound to resist changes that might result in the loss of opportunities for its members.[41] Every technical change that disturbed the existing structure of trades and seemed to give one craft an advantage over another was greeted by a jurisdictional or demarcation dispute in which the unions fought each other for work. Because unions were strongly organized, it was awkward for management to interfere. As a consequence both increased mechanization and efficient labor allocation became more difficult than might have been the case if the industry had not been organized so closely along craft lines.

Demarcation disputes were common even before iron was adopted, and they grew in importance as the number of trades multiplied. Some disputes, such as those between platers and their helpers, were products of the gang and contract system. Others, including attempts by the A.S.E. to destroy the Steam-Engine Makers and the United Pattern Makers as independent societies, were examples of what might be termed trade union imperialism. There were even cases in which members of the same union, the Boilermakers, fought each other. Most disputes, however, were the result of changes in shipping or shipbuilding technology that created tasks that either could be performed by two or more crafts or could involve entirely new machinery. Traditionally, for instance, joiners were confined to wood of one and one-half inches or less in thickness and used hammers and planes. Shipwrights performed heavier woodwork with adzes and mallets. Both types of tools, along with machines that were unknown when the trades had developed, were necessary to complete the decks of modern passenger ships, hence joiners and shipwrights both claimed the "right" to do the work. Logically, neither side was right or wrong. Since both trades were essential in other stages of construction, however, the employers were powerless to divide the work without arranging a compromise acceptable to both unions. Any arbitrary decision could lead to a strike, which might cripple the entire yard.[42]

Disputes could arise wherever the work of one trade bordered on that of another. The engineers quarreled with the boilermakers, shipwrights, joiners, brassworkers, and tinplate workers; the boilermakers with the shipwrights, smiths, chippers, and drillers; the shipwrights with the caulkers, boat and barge builders, mast-and blockmakers, and joiners; and the joiners with the millsawyers, patternmakers, cabinetmakers, upholsterers, and French polishers. On the Tyne, there was an average of one major strike per month over questions of demarcation between 1890 and 1893.[43]

Strikes caused by disagreements over jurisdiction might be costly. Since losses were suffered by the unions involved, by employers, and by the men in other trades who were thrown out of work when yards were closed down, the incentives to find peaceful means of settlement were great. In the early years, there were attempts to avoid disputes through interunion agreements, but they generally worked for short periods if at all. In 1865, for example, the joiners and shipwrights on the Clyde drew up a list of work to be performed by each trade in accordance with the custom on the river at that time. The agreement was followed for a few years but dissolved when the practice of the firms began to change.[44]

After 1880, unions and employers turned to more formal conciliation and arbitration to settle disputes. In 1883 a court of conciliation was established by builders and unions on the Wear. Although not originally intended to deal with matters of demarcation, the court functioned well for several years and was instrumental in settling a dispute between the shipwrights and joiners. Panels established to settle two conflicts on the Tyne were less successful, however, as their decisions were initially rejected by one or another of the opposing parties.[45] In 1892 the rules of the Federation of Engineering and Shipbuilding Trades provided that

> If any dispute takes place between any of the societies forming this Federation, unless amicably settled, such dispute shall be referred to a court of arbitration selected by the parties affected

> by the dispute. When a court of arbitration is required each party to the dispute shall appoint one or two arbitrators to appoint an umpire. Failing the arbitrators to agree his decision shall be final and binding. The court when formed to decide as to place of meeting, method of procedure, &c.; each party to pay half of the expenses, unless otherwise ordered by the court.[46]

As with other methods of arbitration that were tried, the courts of the federation failed because in practice there was no way to force either side to accept an agreement that it felt to be unjust.

When the trades proved unable to settle conflicts between themselves, the employers, who had long claimed the right to decide, obliged the unions to accept the concept of local boards of conciliation. By 1900 seven local boards were in existence in the Tyne and Tees districts, at Middlesbrough, at Birkenhead, and on the Clyde. Courts of arbitration similar to those of the Trades Federation were formed with an independent chairman who could issue a morally binding award in case of deadlock. Because of the problems of different customs of work on the various rivers, local boards continued to be the primary tribunals down to 1914, and the otherwise comprehensive national agreement of 1909 between the Shipbuilding Employers' Federation and the Shipyard Trade Unions explicitly delegated any questions of demarcation to the local bodies. Nevertheless, they were unsuccessful in providing a lasting framework for compulsory and binding arbitration. By 1911 the shipping journal *Fairplay* was asking why some means could not be found to eliminate demarcation disputes altogether "by a joint tribunal of all trade unions concerned, with the employers in co-operation," and the following year an official of the Boilermakers Union pronounced the Tyne Demarcation Agreement to be "atrophied." The reasons for the failure of conciliation boards and courts of arbitration to cope with all types of industrial conflict are perhaps best summed up by Ernest Aves' remark of 1897: "The experience of the [engineering] dispute has afforded yet another illustration of the impossibility of having recourse to any regulative machinery for the solution of industrial disputes if either party is averse to them or thinks them unnecessary."[47]

Despite their notoriety, it is difficult to find a crime, much less a villain, in the story of shipyard demarcation disputes. Irresponsible and even irrational as the disputes may appear, from the standpoint of the workers they were necessary. In an era of endemic unemployment brought on by fluctuations in trade, job security was certainly a legitimate goal. Demarcation quarrels may be viewed as essentially an investment in the future in which unions forsook current employment by striking to protect themselves against incursions by other trades and, if possible, to gain the greater part of any new work that changing technology might bring. Nor is it wholly correct to say that the interests of the employers were "remorselessly sacrificed."[48] Workers as well as employers recognized that some alterations in organization were necessary to accommodate changes in shipping and shipbuilding technology if Britain were to retain the preeminent position in the industry of which both groups were so proud. Ways were found of adjusting to each shift in routine. Demarcation disputes were definitely annoying and, in the short run at least, costly to yard owners, but there is little evidence that this, or any other form of union activity, significantly retarded technical change or the rate of growth of the industry.

Dockyard Labor

The conditions of labor in the Royal Naval Dockyards differed widely from those of the rest of the country. Perhaps the most important distinction was their division into "established" and "hired" men. While the number of established men remained almost constant, the number of hired men grew rapidly over the period (table 8.2).

The trades were not divided along the same lines as those in the private yards. In particular, the shipwrights did all the iron, steel, and timber shipbuilding, so that the boilermakers were confined to the building and repair of boilers, and the so-called skilled laborers, drawn from the ranks of the yard laborers, took the place of a number of semiskilled trades in the private yards.

It is not easy to assess the relative levels of earnings in the royal and private yards. Wage rates were certainly considerably lower in the Dock-

Table 8.2. Employment in the Royal Naval Dockyards (in thousands; annual averages).

Period	Established men	Hired men[a]	Total
1870-1873	6.6	5.7	12.3
1873-1880	6.5	8.9	15.4
1880-1888	6.7	11.0	17.7
1888-1892	5.7	12.9	18.6
1892-1899	5.9	14.8	20.7
1899-1903	6.0	21.2	27.2

Source: Bowley and Wood, *Journal of the Royal Statistical Society*, 1905, p. 706, table 2.

a. Including men in the steam factories.

yards, yet the hours were fewer too: 51 until 1894 and 48 from then on. The established men also enjoyed pension rights unavailable to men in the private yards, and as a result of their steadier employment their actual earnings seem to have compared well with those in the private yards (table 8.3).[49]

The upheavals in the Dockyards in 1859-1861 caused by the introduction of the iron screw and armored warship led to changes in labor relations as well. Apprenticeship no longer meant an "establishment," but a hired status first. Three means of paying were current in the Dockyards: day pay under occasional (check) measurement, ordinary day pay, and "task and job" with unlimited earnings. The latter became by far the most common after 1858 and threatened to throw the clerical work into utter confusion, since there were said to be 94,762 different prices payable for various items of work in 1861, and 51,400 different prices for the shipwrights alone. Moreover, the prices appeared to be unjust, as the same gangs varied widely in their earnings from week to week.[50]

The system was somewhat improved and rationalized after the enquiries of 1860 and 1861, but the accent on piece payments remained. Where the men worked in gangs, rates for the job were paid to the gangs, who divided them according to their daily ratings. If they managed to earn more than 30 to 50 percent above their daily rates, however, the piece rates were lowered, since the total annual wages bill was limited by the naval estimates.[51]

The nationwide union activity in the early 1870s found an echo in some of the Dockyards, particularly in Devonport, but soon petered out

Table 8.3. Earnings in private shipbuilding firms and in the Royal Naval Dockyards (average annual incomes in £).

	Private shipyards			Royal Naval Dockyards		
Year	Iron and steel ship-building	Wooden ship-building	Engi-neering	Established men[a]	Hired men	Average
1886	76	56	59	75.7	63.8	68.1
1906	70.5		69	85.0[b]	66.5[b]	70.4[b]

Source: Bowley and Wood, *Journal of the Royal Statistical Society*, 1905, p. 706; Board of Trade, *Report on Wages*, 1893, p. 477; *Enquiry into Earnings*, 1911, p. IV.

a. There were very few unskilled men on the establishment.

b. Figures for the year 1902-03.

again. There were no further signs of a general labor agitation until the payment reforms of 1888 and 1891, and the growing interest of the outside unions in their Dockyard members in those years led to the formation of Dockyard unions and labor councils.

In 1888 a new "tonnage" payment system, a group piece rate, was introduced in the Dockyards. The men were unanimously opposed to it (though it permitted higher earnings) mainly because its workings remained obscure and because they had no voice in its arrangement. After a trial period even the official responsible for its introduction had to admit its failure, which he explained by the absence of individual incentive in a group payment. Some gangs, indeed, decided to divide their tonnage payments equally among all the members.[52] The system was abolished, and there remained a straight piece rate, which sometimes led to earnings below the daily rate, and "check" payments, that is, piece payments with a minimum daily rate.

In 1891 the Admiralty decided to overhaul the whole structure of Dockyard wages. The principal changes were the establishment of proper wage differentials between the different trades and a graded system of payment, whereby men of each trade were eligible for a range of different pay rates, according to a rating by their officers. The scheme was accompanied by a general rise in pay, bringing the earnings of the hired men close to the standard of the established men and of the men outside.[53]

Much to the surprise of the Board, this new scheme met with strong resistance from the men. They were willing to accept the higher rates, estimated to cost £90,000 per annum, but in the graded payment they saw not an incentive to promote merit over the claims of seniority but a means of nepotism at the hands of the officers. "There must be no relaxation of our efforts," the Devonport branch of the Shipwrights' Constructive Association resolved in 1892, "until this unjust scheme is removed from us." Resolutions were passed against it in all the Dockyards, and an effective attack was made on it before the Royal Commission on Labour, then sitting, both by the unions limited to the Dockyards and by others.[54]

The Admiralty was sufficiently impressed to revert to uniform payments for most skilled trades, and the men also secured other improvements in wages and conditions. In 1894 their hours were reduced to 48. Soon after, the need to attract more seagoing shipwrights into the Navy induced the board to improve their pay and conditions. In 1906 the Admiralty agreed to a series of pay rises, estimated to cost £60,000 per annum, while trying to establish a premium bonus system in the Dockyards. In 1911 the shipwrights were again successful in raising their wages. The efforts to achieve a uniform minimum pay were defeated, but the men succeeded in securing the withdrawal of the premium bonus system in 1911.[55]

In the Dockyards, as in the rest of the country, the tendency was evident of an ever closer approach to national uniformity of wages, hours, and conditions. Sometimes the causes were fortuitous, as in the Fair Wages Resolution of the House of Commons of February 13, 1891, when Sydney Buxton and Sir J. Hibbert pledged the Government to insist on full district union wages to be paid by their contractors. This diverted all contracts to the lowest-paid districts and reduced wages to the same level elsewhere, where it did not strangle the industry altogether.[56] More generally, economic forces of widening markets, cheapened transport, and improved means of communication equalized working conditions over the whole country by unifying the scope of bargaining, associations, unions, and disputes. At the same time, the beginnings of iron shipbuilding and its primitive machinery crystallized the various skills into separate trades, until the introduction of electric, pneumatic, and hydraulic power transmission and automatic machines reduced the skills again to the level of semiskilled machine operators.

How far the workers in the industry, both in the private yards and the Dockyards, benefited by all the changes described in this chapter can be judged, if at all, only by a study of labor conditions in the whole country, not in a single industry in isolation. It may be questioned, however, whether rising real wages and shorter hours compensated for the greater intensity of work, the lack of freedom, independence, and mobility, the long stretches of unemployment and short time, and the dehumanizing character of labor relations in the shipbuilding industry.

9.

Wages, Employment, and Productivity

The efforts of British yard owners to educate and train their workers and to achieve a proper coordination between labor and machinery were only means toward the end of efficiently building ships. Ultimately they affected the level of productivity in the yards, which in turn helped to determine the levels of wages and employment and the rates of profits that prevailed. Although they were perhaps unnecessarily reluctant to invest in education or research, on balance British builders were highly successful. Increasing productivity, much of which was passed on to consumers in the form of lower prices for finished vessels, allowed the British to maintain their dominance of world shipbuilding down to 1914.

The establishment of appropriate wage and salary scales and the control of variations in wage rates were among the most critical tasks facing shipbuilding management. As output fluctuated, wage rates also tended to change, making it difficult for builders to estimate costs when bidding for contracts and thereby jeopardizing expected profits. Furthermore, when requisite new trades were introduced into the yards ways had to be found of setting their wage rates in relation to both those of older shipbuilding trades and those of workers in the same trades employed in other industries. Both problems were complicated by the actions of the trade unions and by changes in the relative skill levels and productivity of trades brought about by increased mechanization.

Fluctuations in wages, salaries, and employment, and their effects on the business strategy and tactics of British shipbuilders are examined here on the basis of two broadly consistent sets of statistics. The first comes from a series of articles published by A. L. Bowley and George Wood in 1905-1906 and purports to cover all the major shipbuilding ports of Great Britain.[1] The data are based on lists of trade union standard-time wages and indicate the general trend of wages throughout the island. Although they are useful for certain interport comparisons, the Bowley and Wood data give an invalid picture of the amounts of wages paid out by shipbuilders, that is, the actual earnings of the workers. Bowley and Wood do not take into account changes in the number of working hours over the course of the business cycle, including overtime, or variations in wages owing to the adoption of piece rates. Finally, the national wage index hides other differences. "Old work"—repair—was paid at higher rates because it was dirtier and made heavier demands on the men's tools, and because there was no fear of competition from other ports. The "amphibian" trades were a special source of difficulty. The painters on the Clyde, for example, took advantage of the seasonal demand for housebuilding to "almost regularly" ask for a rise in the spring, but wages were reduced again later in the year.[2]

A second set of data, drawn from the unpublished records of the important Dumbarton shipbuilding firm of William Denny and Brothers, is in several ways more useful. Although the Denny statistics refer to only one firm, they give the exact month of changes in the wage rates of more than 20 trades. This makes it possible to correlate fluctuations in output with alterations in wage rates more accurately than the annual Bowley and Wood data will allow. Moreover, the Denny records include two series of actual earnings: one for all nonsalaried employees at the firm's Leven Shipyard, and the second for shipjoiners and joiners machinemen.[3] The series were calculated by dividing the total annual wages bill by the number of fortnightly pays and then by further dividing the result by the average number of workers per pay to arrive at a figure for "average earnings per man per pay."[4] In table 9.1, the figures, which have been halved to give average weekly earnings and converted to indices, are presented together with Bowley and Wood's index of average time rates for all workers in shipbuilding on the Clyde.

The series follow varied paths. All show frequent fluctuations depending on the stage of the business cycle. The Bowley and Wood index of time wages, however, in most instances fluctuated less violently than did the index of earnings of all workers at the Leven Shipyard. This is as would be expected since variations in earnings would have reflected changes in the number of hours worked as well as differing wage rates. The long-run tendency of earnings to increase faster than wages is more difficult to account for. The trend over the period was toward shorter working hours, which should have been reflected in reductions in earnings relative to wage rates. Several explanations are conceivable. As transportation facilities improved, Dumbarton became less isolated from other shipbuilding towns on the Clyde. This could have forced Denny's to raise wage rates more rapidly as employment opportunities increased for workers living near Dumbarton. Builders in larger towns, where the

Table 9.1. Indices of average wages and earnings of workers in shipbuilding and engineering on the Clyde and at the Leven Shipyard, 1851-1914 (1860 = 100).

Year	(1) Time wages of all workers on the Clyde	(2) Earnings of all workers, Leven Shipyard	(3) Earnings of joiners and joiners machinemen, Leven Shipyard
1851	89	—	—
1852	91	87	103
1853	101	93	99
1854	102	99	102
1855	105	104	106
1856	106	100	133
1857	104	100	109
1858	100	95	108
1859	98	95	87
1860	100	100	100
1861	103[a]	101	106
1862	106[a]	100	113
1863	110	103	119
1864	116[a]	109	128
1865	116[a]	112	128
1866	115	114	134
1867	108[a]	111	123
1868	108[a]	108	113
1869	108[a]	113	122
1870	111	155	123
1871	118[a]	153	136
1872	126	158	134
1873	130	147	135
1874	131	149	148
1875	128	146	118
1876	126	147	126
1877	125	130	200
1878	124	157	163
1879	121	150	160
1880	125[a]	150	171
1881	131[a]	157	186
1882	135[a]	172	177
1883	141	176	—
1884	135[a]	166	197
1885	130[a]	149	154
1886	124	158	157
1887	124[a]	146	142

(Cont.)

Table 9.1. (cont.)

Year	(1) Time wages of all workers on the Clyde	(2) Earnings of all workers, Leven Shipyard	(3) Earnings of joiners and joiners machinemen, Leven Shipyard
1888	136[a]	163	167
1889	147	167	166
1890	151	183	194
1891	150	180	191
1892	145	176	175
1893	144	180	191
1894	144	180	205
1895	144	179	189
1896	154	180	187
1897	156	178	188
1898	164	187	199
1899	164	208	214
1900	164	204	216
1901	164	209	214
1902	164	216	235
1903	160	192	—
1904	160	191	202
1905	—	187	201
1906	—	196	—
1907	—	209	211
1908	—	199	—
1909	—	199	—
1910	—	200	—
1911	—	230	—
1912	—	236	—
1913	—	255	—
1914	—	262	—

Source: Column 1, Bowley and Wood, *Journal of the Royal Statistical Society*, March, 1906; columns 2 and 3, appendix B, table B.1.

a. Figures based on incomplete data.

competition for labor was greater from the beginning, may have been forced to pay higher wages initially, but improved transport would not have affected them as severely. Alternatively, the composition of Denny's workforce may have tended more than that of other yards to trades that had greater leverage in the labor market and were able to raise their wages faster than the average. Finally, the Bowley and Wood index may not be representative and may understate the rate of wage increases along the river.

Wages and Employment

The struggle to obtain reductions in the length of the working week has been described in chapter 8. Holidays and the state of trade, however, were of greater importance than the official hours of work in determining fluctuations in the level of earnings.

Holidays imposed regular annual reductions in wages in a period in which paid vacations were unknown. At Charles Connell's yard on the Clyde, for example, between 1902 and 1909 the wages bills for January and July—the months of the traditional Scottish New Year's holiday and of Glasgow Fair—were significantly below the average. This was in part compensated for in that pays were the largest in the preceding months of December and June. While the timing of local celebrations would have varied, similar fluctuations would have appeared in other centers. Perhaps surprisingly, however, poor weather and the shorter days of winter seem to have affected the wages bills very little. Not only was December the busiest period at Connell's, but February and March were more active than the autumn months of September, October, and November.[5]

Another important influence on earnings that is not reflected in lists of wage rates is that different trades were employed in the various stages of a ship's construction. Tables 9.2 and 9.3 present an analysis of Fairfield's progressive wages books for the years 1904-1914. The tables, which are based on data for 10 ships of a variety of types, show the proportion of work performed by each trade during each quarter of the vessels' construction, and the proportion of each trade's work performed during each quarter. During the first half of construction, more than 60 percent of the work was performed by the iron workers; in the last quarter, only about 20 percent. Two-thirds of the iron workers' wages was earned in the first half, but barely 10 percent in the final quarter. On the other hand, as might be expected, the greatest portion of the wages of the joiners, cabinetmakers, sheet iron workers, and other trades who worked inside the hulls was earned in the final quarter of construction and comparatively little in the early stages. Even in years in which substantial tonnage was launched and average employment was high, large numbers might be laid off from work during periods in which their trades were not needed.

Finally, employment varied tremendously with the business cycle. Denny's employment figures (table 9.7) reveal decreases of more than 60

Table 9.2. Percentage of work performed, by value, by each trade in each quarter of a vessel's construction.

Trade	First quarter	Second quarter	Third quarter	Fourth quarter	Total[a]
Angle-iron smiths, time	0.4	0.3	0.2	0.1	0.2
Angle-iron smiths, piece	2.5	2.2	1.8	0.5	1.8
Fitters and helpers, time	5.5	3.8	2.8	2.4	3.6
Fitters and helpers, piece	24.0	11.8	7.3	2.6	11.5
Riveters, holders-on, and boys, time	5.3	7.2	4.2	2.2	4.7
Riveters, holders-on, and boys, piece	14.1	18.7	11.0	2.0	11.5
Caulkers, time	2.3	4.7	3.3	2.4	3.2
Caulkers, piece	2.9	5.5	4.2	3.0	3.9
Borers, time	0.5	0.9	0.7	1.0	0.8
Borers, piece	3.7	5.9	5.0	4.2	4.7
Total iron workers[b]	61.1	61.1	40.6	20.5	46.1
Smiths, strikers, and finishers, time	1.1	1.5	2.4	2.6	1.9
Smiths, strikers, and finishers, piece	1.3	1.4	1.8	1.3	1.5
Mechanics and helpers	4.4	6.9	9.3	10.0	7.6
Carpenters and helpers	8.0	6.7	7.5	6.3	7.1
Patternmakers	1.2	0.6	0.4	0.2	0.6
Boatbuilders	0.5	0.4	0.1	0.2	0.3

percent in the number of men in only three years, followed, sometimes very slowly, by increases as trade recovered. Employment at the Leven Shipyard dropped from 1,577 men in 1873 to only 588 men in 1876, and it was not until 1883 that the previous peak was surpassed. Other firms showed similar patterns,[6] and, given that all builders tended to be affected simultaneously, it is clear that there were stretches of months, or even years, over which many men either remained out of work or found jobs in other sectors. In areas dominated by shipbuilding, however, the latter was difficult, forcing families to migrate. Although the population of Barrow rose from 47,300 in 1881 to 63,800 in 1911, for instance, the loss of population by migration from the Ulverston registration district was 16.3 percent in 1881-1891, 13.4 percent in 1891-1901, and 16.4 percent in 1901-1911, a total permanent loss of 11,800 persons from Barrow in those 30 years.[7]

Table 9.2. (cont.)

Trade	First quarter	Second quarter	Third quarter	Fourth quarter	Total[a]
Joiners and machinemen	4.4	4.9	10.2	16.2	8.9
Cabinetmakers and machinemen	0.8	1.1	1.5	2.7	1.5
Polishers	0.8	0.7	0.8	1.5	0.9
Coppersmiths (yard)	—	—	—	0.1	—[c]
Plumbers	0.1	1.8	3.9	4.0	2.5
Tinsmiths	0.1	0.1	0.1	0.2	0.1
Sheet iron workers and helpers	0.4	1.8	4.8	10.8	4.4
Painters	0.4	0.4	4.3	5.5	2.6
Redleaders	0.1	1.3	2.6	4.0	2.0
Cementers	—	0.3	0.4	0.4	0.3
Riggers	1.3	0.9	1.2	2.4	1.4
General laborers	5.6	4.7	4.5	6.4	5.3
Draftsmen	5.9	2.1	1.8	2.5	3.0
Miscellaneous	2.4	1.4	1.8	2.2	1.9
Total[a]	100.0	100.0	100.0	100.0	100.0

Source: Fairfield Shipbuilding and Engineering Company, Progressive Wages Books. The figures are based on data for vessels 439, 445, 450, 458, 467, 468, 469, 485, 490, 494.

a. Because of rounding, the individual items may not sum to the totals given.

b. Includes iron measurers.

c. Less than 0.05 percent.

It is clear that variations in both wages and earnings and in unemployment were closely related to fluctuations in output. In figures 8 and 9, tonnage launched on the Clyde and the earnings of Denny's employees are compared with the Bowley and Wood index of time wages and with figures on the rate of unemployment for all workers in engineering, shipbuilding, and metal working in the United Kingdom. With the exception of the unemployment data, all are expressed as percentages of trend values to highlight the basic cyclical movements.[8] As can be seen from the charts and from table 9.4, the series are all similar, with their major peaks and troughs coming in approximately the same years.

The difficulties shipbuilders had in coping with changes in wage rates that occurred after contracts for vessels had been signed were accentuated by the tendency for the changes to be bunched in periods of rapid change in output, rather than evenly distributed. In all, the time wage

Table 9.3. The percentage of each trade's total output, by value of wages, performed in each quarter of a vessel's construction.

Trade	First quarter	Second quarter	Third quarter	Fourth quarter	Total[a]
Angle-iron smiths, time	36.3	28.2	23.6	11.9	100.0
Angle-iron smiths, piece	35.3	31.0	26.4	7.3	100.0
Fitters and helpers, time	37.6	26.5	19.6	16.3	100.0
Fitters and helpers, piece	52.2	26.1	16.3	5.4	100.0
Riveters, holders-on, and boys, time	27.6	38.5	22.7	11.2	100.0
Riveters, holders-on, and boys, piece	30.4	41.1	24.4	4.1	100.0
Caulkers, time	17.6	37.3	26.7	18.4	100.0
Caulkers, piece	18.2	35.7	27.5	18.6	100.0
Borers, time	15.8	29.5	22.6	32.1	100.0
Borers, piece	19.6	31.7	27.1	21.6	100.0
Total iron workers[b]	33.1	33.6	22.6	10.8	100.0
Smiths, strikers, and finishers, time	14.7	20.0	32.2	33.2	100.0
Smiths, strikers, and finishers, piece	22.3	23.5	31.8	22.4	100.0
Mechanics and helpers	14.4	22.9	31.1	31.6	100.0
Carpenters and helpers	27.9	23.9	26.9	21.4	100.0
Patternmakers	49.5	24.1	17.2	9.2	100.0
Boatbuilders	42.9	30.1	12.2	14.8	100.0
Joiners and machinemen	12.4	13.9	29.4	44.2	100.0
Cabinetmakers and machinemen	13.5	18.1	25.4	43.0	100.0
Polishers	20.5	19.7	21.5	38.3	100.0
Coppersmiths (yard)	0.5	12.1	21.5	65.9	100.0
Plumbers	1.5	19.0	40.3	39.2	100.0
Tinsmiths	15.3	14.5	27.3	42.9	100.0
Sheet iron workers and helpers	2.1	10.5	28.0	59.4	100.0
Painters	4.1	4.0	41.3	50.5	100.0
Redleaders	1.4	16.6	33.7	48.4	100.0
Cementers	1.2	25.4	37.0	36.4	100.0
General laborers	26.5	22.4	21.8	29.2	100.0
Draftsmen	48.2	17.4	14.9	19.5	100.0
Miscellaneous	30.6	17.8	24.2	27.4	100.0
Total[a]	24.9	25.3	25.6	24.2	100.0

a. Because of rounding, the individual items may not sum to the totals given.
b. Includes iron measurers.

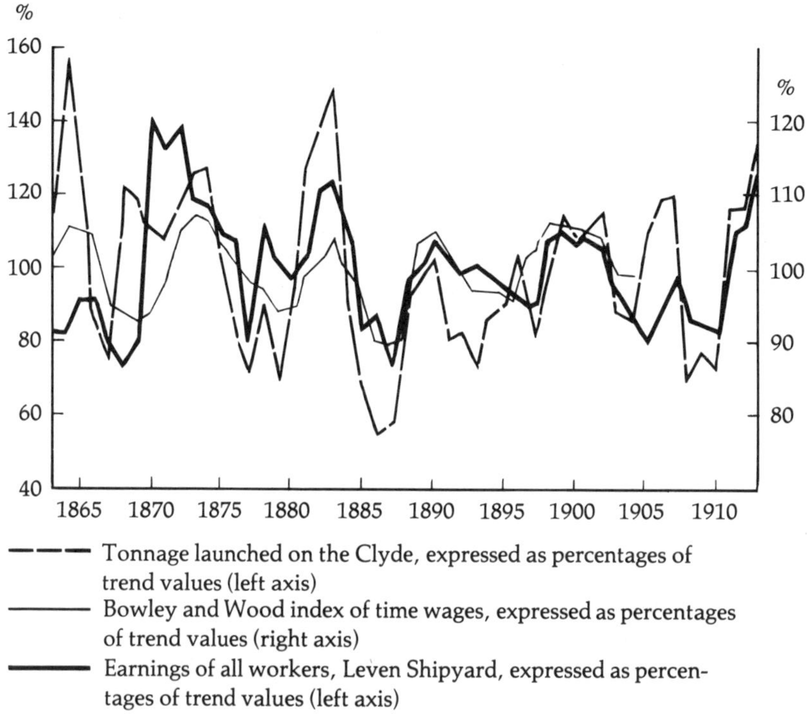

Figure 8. Money wages and earnings, and tonnage launched on the Clyde, 1863-1913. (Table 9.1 and appendix table B.9.)

rates of William Denny and Brothers employees were altered on 73 occasions between 1880 and 1912 (although it was rare for the wages of any single trade to change more than 20 times over the interval). However, of the 19 changes in rates for engineers in the 32 years after 1880, for example, 3 were in 1882-1883, 2 in 1885, 2 in 1889, and 3 between December 1908 and May 1910; 13 of the alterations occurred in roughly 7½ years. By contrast, there were only 2 changes in the 10 years between 1898 and 1908. As most trades sought increases in rates in the same short intervals, builders who had not foreseen an approaching boom in orders could easily underestimate wage costs when tendering bids (table 9.8).

Because skilled laborers in shipbuilding were well organized and delays caused by strikes could be costly, yard owners often had no choice but to accede to trade union demands for higher wages during prosperous periods. It is difficult, though, to sort out the effects, if any, that union activities had on either short- or long-term changes in wage rates and earnings between 1860 and the outbreak of the First World War.

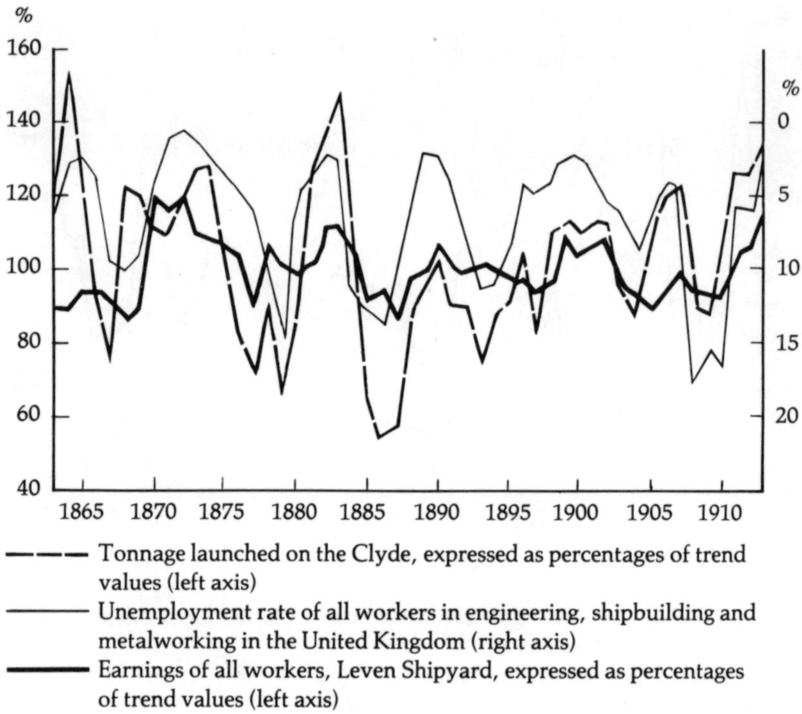

Figure 9. Money earnings and tonnage launched on the Clyde, and the unemployment rate of all workers in engineering, shipbuilding, and metalworking in the United Kingdon, 1863-1913. (Table 9.1 and appendix tables B.3 and B.9.)

Much of the ability of trade unions to increase wages during booms can be traced to the normal operations of a competitive labor market. Unemployment was endemic in shipbuilding before 1914, as we have seen. In good periods, labor became scarcer and unions could demand raises; in recessions, when the demand for labor declined, employers were able to reduce wage rates, although generally not by as much as they had been increased during preceding booms. Although trade unions may perhaps have been strong enough to gain larger increases in prosperous periods and to keep rates from falling as far or as fast in depressions as they would have if the workforce were unorganized, they did not have the power to counteract the forces of supply and demand.

Similarly, it is doubtful that unions were able to influence strongly long-run changes in wages and earnings. In fact, the unions were sharply divided on piece work, one of the major organizational changes that helped to raise wages. Among those who opposed it were the shipwrights, the shipjoiners, the patternmakers, and all the engineering trades. The trades that worked successfully on the piece rate system were the boiler-

Table 9.4. Cycles in time wages, unemployment, tonnage launched on the Clyde, and earnings of all workers at the Leven Shipyard, 1863-1913.

Peaks			
(1) Wages	(2) Unemployment[a]	(3) Tonnage launched	(4) Earnings
1864-65	1865	1864	1865-66
		1868	
			1870
1873-74	1872	1874	
		1878	1878
1883	1882	1883	1883
1890	1889-90	1890	1890
	1896	1896	
1898	1899		1899
		1901 & 1903	1902
	1906	1907	1907
	1913	1913	1913

Troughs			
(1) Wages	(2) Unemployment[b]	(3) Tonnage launched	(4) Earnings
1869	1868	1867	1868
		1871	
		1877	1877
1879-80	1879	1879	1880
1886-87	1886	1886	1887
	1893	1893	
1895			
	1897	1897	1897
1903-04	1904	1904	1905
	1909	1908 & 1910	1909-10

Source: Column 2, appendix B, table B.3; column 3, table B.9 for 1870-1913; "Glasgow Herald Shipbuilding and Engineering Supplement" for 1863-1869.

a. Trough.

b. Peak.

makers and the sailmakers. The shipwrights had, of course, had "task" contracts in the wooden days, but as their power and influence waned, they became bitterly hostile to payment by the piece. It led to sweating and irregular employment, especially in the badly organized areas, such as the minor ports on the Bristol Channel; work was subcontracted by competitive bidding, making it impossible to keep wage rates up in times of unemployment and destroying collective bargaining. Above all, the increased output per man created unemployment. In 1888 the shipwrights finally succeeded in ending all piece work on the Tyne, and in 1891 it was abolished on the Clyde after some struggle.[9]

Other trades, such as the joiners and the A.S.E., also secured abolition of piece work. Platers and other iron workers on the Thames had been paid by time since the 1860s.[10]

The Boilermakers' Society was strong enough to prevent abuses to its members, by the "common rules" under which all piece work was carried out, and subcontracting where it did exist meant the employment and sweating of unskilled helpers who were not their charge. The close connection between piece work and the installation of machinery in the shipyards was obvious to a number of observers: while the constant improvements in the equipment raised the earnings of the men, the more intensive rate of work was profitable to the employers, and it enabled them to calculate their costs beforehand. The union was strong enough to ensure piece rates of 25 or 30 percent more during overtime—a frequent occurrence as the iron had to be worked hot and it was not always convenient to break off work at the normal time. It was the employers who complained about the abuses of piece work among the boilermakers, the irregular working hours, and the lack of discipline. The boilermakers worked on piece rates only in maritime areas, not in the inland boilershops. In the ports they took on piece work whenever possible; estimates of the proportion of piece work done by them in the shipyards vary from 90 to 98 percent.[11]

In shipbuilding as a whole, 25 percent of the wages paid in 1886 were piece wages, and in wooden shipbuilding, 5.5 percent (see table 9.5). In 1906 the figure for shipbuilding had risen to 33.5 percent, approximately the same proportion as was performed at Fairfield's between 1904 and 1914. It is evident that this fact partly accounts for the increase in earnings in the industry between those dates and is merely the counterpart of the greater intensity of work.

In both 1885 and 1906, apprentices worked for time wages, though the practice was spreading of allowing apprentices to earn piece rates, less deductions, in the last years of their indenture. The deductions usually amounted to 30 percent in the second year, 25 percent in the third, 20 percent in the fourth, and 15 percent in the last year. In 1909 the Factory and Workshop Act of 1901, which made prominent advertisement of piece rates obligatory, was applied to some of the boilermaker trades, and in 1912 it was made applicable to all trades in the yards.[12]

The piece rates were fairly high, and allegations of platers' earnings of £150 per annum and more were common among the builders, despite the irregularity of employment. These statements, and statistics showing weekly earnings to be more than double the time rates, were misleading. Information supplied to the Royal Commission on Labour in 1892 by the

Table 9.5. Weekly time and piece rates in the shipbuilding trades, 1886-1906.

Trade	Boilershops and engineering (workers on full time)				Shipbuilding and boatbuilding and repair, 1906		All industry, 1906	
	1886		1906					
	Time workers	All workers	Time workers	All workers	Workers on time	Full-time piece	Time	Piece
Smiths	28/9	30/1	34/-	37/2	35/6	53/-	33/11	51/5
Smiths' strikers	18/10	19/11	21/11	24/1	21/11	31/3	21/8	29/9
Platers	31/10	34/1	38/11	42/3	42/5	71/3	35/4	64/7
Platers' helpers	—	—	—	—	27/-	30/2	20/10	24/4
Riveters and caulkers	29/-	29/9	36/7	39/4	—	—	—	—
Riveters	—	—	—	—	34/9	55/7	34/6	47/10
Caulkers	—	—	—	—	32/11	54/8	33/8	50/7
Holders-up	22/5	22/8	27/-	28/10	27/10	42/-	26/7	35/-
General laborers	17/10	17/11	20/5	21/-	21/1	29/5	20/11	28/-
Shipwrights	—	—	—	—	36/4	46/2	35/2	45/11
Ship painters	—	—	—	—	35/9	38/11	32/11	31/8

Source: Board of Trade, *Enquiry into Earnings*, 1911, pp. XXXII, 107-108.

employers showed that in West Hartlepool platers' helpers and blacksmiths' strikers were paid by the craftsmen, and in Scotland, the North of England, and Belfast piece workers generally paid their own helpers. Actual earnings, although still substantial, were shown by the Board of Trade inquiries to have differed by much smaller fractions from those of time workers.[13]

Moreover, it should be remembered that the high wages earned in the rush of night shifts, overtime, and weekend work had to last over the days, weeks, and sometimes years of unemployment and short time. The industry or skill of very few men could have secured them a steady income or enabled them to plan their future when shipyard earnings varied as widely as they did.

Productivity and Real Earnings

Three sources of productivity data have become available in recent years. Although an appropriate annual series on capital employed is not extant, we do have figures on labor costs and costs of hulls for the Clyde firms of William Denny and Brothers, Charles Connell and Company, and the Fairfield Shipbuilding and Engineering Company. The first two date from 1881 and 1883, respectively, but the Fairfield data do not begin until 1898. From them, it is possible to compute series on manpower productivity (output by value of hull per man-week expended) for merchant vessels for all three firms, and for warships constructed at Fairfield's (table 9.6). Our analysis, however, will concentrate on a composite series of productivity based on more than 700 vessels built by Denny's and Connell's between 1883 and 1913.

Any calculation of real productivity has to take account of the fact that the total wage bill for each vessel cannot simply be deflated by the wage index of the launching year in order to provide an index of real labor input. It has to be deflated by an index that includes wage rates of the years before and after the launch in which some of the work on any vessel may have been performed.

The ratio we have chosen is a double weighting for the year of the launch and a single weighting for the year preceding and the year following.[14] Thus if the average weekly money earnings were 28s. in year $(x - 1)$, 29s. in year x, and 34s. in year $(x + 1)$, the index would be $[28s. + (2 \cdot 29s.) + 34s.]/4 = 30s. = £1.50/week$. This wage index ($w$) would then be applied to all the labor costs of the ships launched in year x. Assuming that the labor cost involved in constructing the vessels launched during the year was £150,000 (L), a total of 100,000 man-weeks would have been expended (L/w).

For launchings, we simply apply the price index for year x. Thus if the total value of hulls launched by Denny's and Connell's in year x was £600,000[H] and the price index number for that year was 120[p], the real value of hulls launched was £500,000, or $[H \cdot 100]/p$.

By the simple formula $(H \cdot 100)/p \div L/w$, that is, by dividing the real (deflated) value of the total launchings by the composite wage index, we derive the real value of hulls launched per man-week expended, which in our example is £500,000/100,000 = £5.00.

The series is obviously highly sensitive to the price index used to calculate the real value of the hulls constructed. This index must correct not only for variations in the selling price of the vessels but simultaneously for changes in their quality and complexity. This is especially important in an era of rapid technological change such as that between 1880 and 1914.

Two alternative series of productivity are employed here in an attempt to get around these conceptual problems. The first is calculated with the price index for ships given in C. H. Feinstein's *National Income, Expenditure and Output of the United Kingdom, 1858-1965.* This is based on Maywald's cost data for new vessels and does not take into account most changes in quality or complexity.[15] The second series uses an index that we have calculated. The figures for 1898-1913 are based on the *Fairplay* price of a "representative" ship in those years. The figures for 1883-1897 are derived from a GNP deflator with the variations accentuated because of the extreme fluctuations in shipbuilding output, which, it has been assumed, would result in greater fluctuations in the price of ships than in the economy as a whole.[16] The *Fairplay* observations are for a basic cargo ship without frills or extra equipment and can therefore be assumed to eliminate most variations in price due to changes in quality and complexity, although their applicability to other types of vessels such as mail liners and warships is conjectural.

In general, the two price indices agree quite well for the period from 1883 to 1897. From 1898, however, they diverge widely, with the *Fairplay* index registering a much steeper drop in prices after the turn of the century. Between 1900 and 1908, the Pollard-Robertson (*Fairplay*) index shows a decline of 33 percent in the price of ships, whereas the Feinstein index never falls by more than 10.3 percent from its peak of 1900. The productivity of Denny's and Connell's increased by 29.6 percent between 1901-1903 and 1908-1910 when calculated using the Pollard-Robertson index, but by only 1.3 percent when the Feinstein index is used. This is largely because, for the years before 1910, Maywald's figures are based on the costs of inputs and therefore, as he points out, tend to fluctuate less severely than an index based on the actual selling prices of the finished vessels.[17]

The pattern of Denny's and Connell's productivity, using the Feinstein

Table 9.6. Fluctuations in the price of ships and in productivity in shipbuilding, 1883-1913.

Year	(1) Denny and productivity (Feinstein price index) (£/man-week)	(2) Feinstein price index (1913=100)	(3) Denny and Connell productivity (Pollard-Robertson price index (£/man-week)	(4) Pollard-Robertson price index (1900=100)	(5) Fairfield merchant ship productivity (Pollard-Robertson price index) (£/man-week)	(6) Fairfield warship productivity (Pollard-Robertson price index) (£/man-week)
1883	4.32	113.8	4.35	113	—	—
1884	4.30	110.2	4.64	102	—	—
1885	4.42	105.1	5.05	92	—	—
1886	4.71	92.9	4.97	88	—	—
1887	5.30	85.4	5.26	86	—	—
1888	5.26	86.6	5.30	86	—	—
1889	5.22	87.7	5.09	90	—	—
1890	5.45	91.0	5.51	90	—	—
1891	5.78	88.2	5.79	88	—	—
1892	5.71	85.3	5.41	90	—	—
1893	5.63	79.1	4.97	88	—	—
1894	5.61	79.6	5.59	80	—	—
1895	5.63	79.1	5.71	78	—	—
1896	5.35	80.3	5.51	78	—	—
1897	5.35	81.9	5.35	82	—	—
1898	5.42	84.2	5.13	89	3.83	3.29

1899	5.43	90.3	5.16	95	4.22	3.29
1900	5.37	96.2	5.16	100	4.40	3.25
1901	6.32	91.4	6.21	93	4.81	3.44
1902	6.20	89.1	6.50	85	4.78	3.79
1903	5.62	87.7	6.08	81	5.09	3.84
1904	5.84	86.3	7.10	71	5.31	5.25
1905	5.86	86.3	6.83	74	5.54	4.30
1906	5.98	88.2	6.94	76	5.48	—
1907	5.87	90.7	7.30	73	5.65	8.08
1908	6.43	89.0	8.71	67	6.40	—
1909	6.10	89.4	8.02	68	5.72	5.93
1910	5.97	91.1	7.88	69	6.15	4.85
1911	5.62	94.1	6.61	80	5.49	—
1912	6.37	99.5	6.60	96	5.50	4.31
1913	6.38	100.0	6.65	95	6.12	3.97

Sources: Columns 1, 3, 5, and 6 are derived from data on labor costs and total costs of hulls in the records of William Denny and Brothers, Charles Connell and Company, and the Fairfield Shipbuilding and Engineering Company. Column 2, Feinstein, *National Income, Expenditure and Output of the United Kingdom 1855-1965,* table 63, pp. T136-137.

Note: These figures refer to hulls only.

index, is similar to that of British industry as a whole, with increase down to the turn of the century and stagnation thereafter.[18] Long-term variations in the series, however, reflect changes in the amount and quality of capital equipment employed by the workforce as well as changes in the quality of labor as such. The Denny and Connell productivity series using the Pollard-Robertson index has the advantage of according with the increasing use of capital equipment that took place after the turn of the century (see chapter 6). The series yields a compounded rate of increase of productivity of 2.5 percent per annum between the peaks of 1894-1896 and 1908-1910, but this seems excessive in light of the minimal increase for industry as a whole. As the decline in prices after the turn of the century in the Pollard-Robertson index may well be exaggerated, it seems most reasonable to conclude that productivity did increase after 1902, but at a lower rate than the use of that index would indicate.

The pattern of fluctuation of neither series agrees very well with changes in the money or real earnings of shipyard employees over the period (table 9.7 and figures 10 and 11). To be sure, the recorded decline in real earnings after 1902 is in part an illusion. As the introduction of machinery in this period led in many cases to the replacement of skilled workers by semiskilled operatives, it is not surprising that average earnings did not increase; improvements in the real earnings of all groups of employees could have been masked by these changes in labor force distribution. The greater part of any productivity increases, though, probably contributed to higher profits for the builders or were passed on to the consumers in the form of lower prices, or both. As prices declined in both indices—and had they remained constant, our calculations of productivity would have shown an actual fall—buyers rather than the builders or their employees seem to have been the primary beneficiaries of any improvements in productivity in the dozen years before 1914.

It seems likely, in fact, that the greatest influence on long-term increases in real wages lay outside shipbuilding itself. The data given in table 9.7 are consistent with the general index of wages-earnings in composite units of consumptables for Great Britain calculated by Phelps Brown and Browne. Both indices show an almost linear increase from 1860 to around 1900 and then fluctuations around a level trend until 1913.[19] Since so many of the trades involved in shipbuilding were also active in other industries, variations in wages throughout the economy would have set a pattern that shipbuilders would have been forced to follow to attract workers employable in other fields. Interunion rivalry within the shipyards would have then tended to spread wage increases to trades limited to shipbuilding as well.

Finally, in their attempts to raise wages, trade unions in shipbuilding

Table 9.7. Employment and real earnings of all workers, Leven Shipyard, 1852-1913.

Year	(1) Number employed	(2) Index of number employed (1860 = 100)	(3)[a] Index of real earnings (1860 = 100)
1852	657	118	111
1853	632	114	100
1854	649	117	95
1855	641	115	100
1856	717	129	97
1857	743	134	95
1858	1,122	202	102
1859	408	74	100
1860	555	100	100
1861	959	173	106
1862	661	119	100
1863	1,334	240	102
1864	1,651	297	111
1865	1,670	301	116
1866	1,451	261	116
1867	1,044	118	113
1868	1,665	300	113
1869	1,438	259	127
1870	1,122	202	170
1871	1,266	228	160
1872	1,332	240	149
1873	1,577	284	140
1874	1,541	278	149
1875	1,321	238	151
1876	558	101	154
1877	851	153	143
1878	1,035	186	187
1879	1,085	195	184
1880	1,274	230	178
1881	1,463	264	190
1882	1,526	275	206
1883	1,627	293	211
1884	1,790	323	211
1885	1,371	247	203
1886	1,152	208	222
1887	1,116	201	217

Table 9.7. (cont.)

Year	(1) Number employed	(2) Index of number employed (1860 = 100)	(3)[a] Index of real earnings (1860 = 100)
1888	1,808	326	235
1889	1,668	301	240
1890	1,863	336	254
1891	1,361	245	252
1892	1,262	227	259
1893	1,787	322	265
1894	1,585	286	294
1895	1,794	323	300
1896	1,741	314	298
1897	1,709	308	289
1898	1,808	326	282
1899	1,965	354	298
1900	2,143	386	270
1901	2,102	379	294
1902	2,230	402	303
1903	1,950	351	270
1904	2,073	374	278
1905	2,134	385	262
1906	2,016	363	254
1907	1,991	359	260
1908	1,714	309	276
1909	2,088	376	263
1910	1,993	359	249
1911	2,571	463	273
1912	2,526	455	273
1913	2,622	480	290

Source: Column 1, unpublished book of "wages notes," William Denny and Brothers; column 3, appendix B, table B. 2.

a. For the Rousseaux overall price index used in computing the raw data on which column 3 is based, the average of prices in 1865 and 1885 was taken to equal 100.

appear to have been victims of "money illusion." When real earnings are calculated as percentages of trend values, only three major movements appear: a decline from 1853 to 1868; a slow, uneven rise from 1868 to 1894-1895; and a more rapid decline from 1895 to 1910. The relationship between output and real earnings was much weaker than that between output and money earnings.

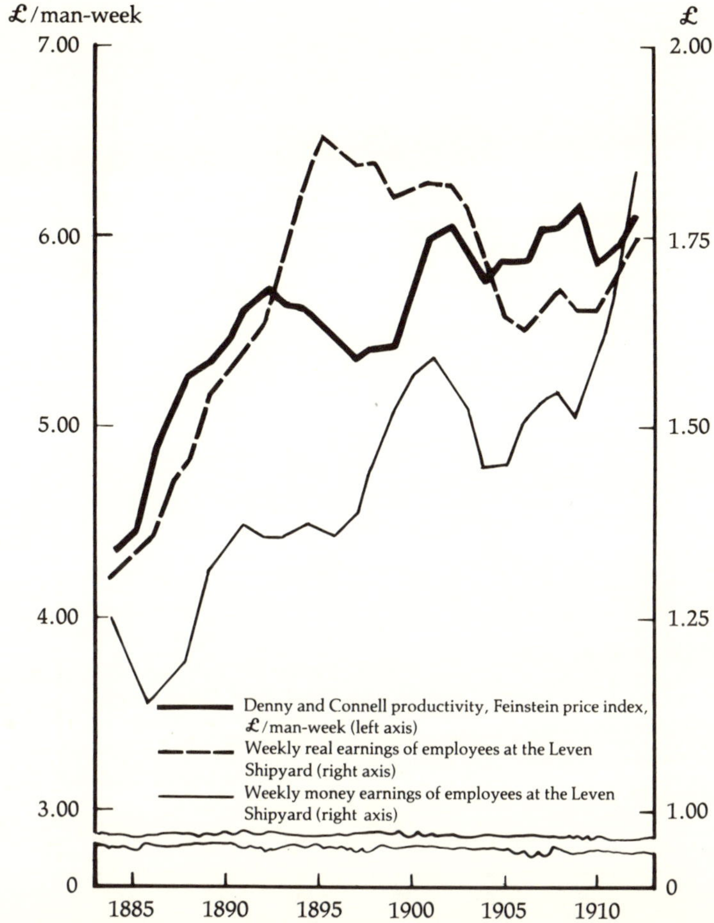

Figure 10. Denny and Connell labor productivity (Feinstein price index) and money and real earnings of employees at the Leven Shipyard, 1884-1912. (Table 9.1 and appendix tables B.1 and B.2.)

Wage Differentials, Salaries, and Pensions

Increased mechanization in shipbuilding, the changing nature of vessels, and the growing strength of national trade unions all had significant effects on wage differentials. The increased importance of the skilled crafts before approximately 1900 and the subsequent erosion of their position are shown in table 9.8, in which the time rates of five trades are given for the period 1880-1910. Four of the trades are skilled and the fifth, blacksmiths hammermen, semiskilled. All figures are expressed as percentages

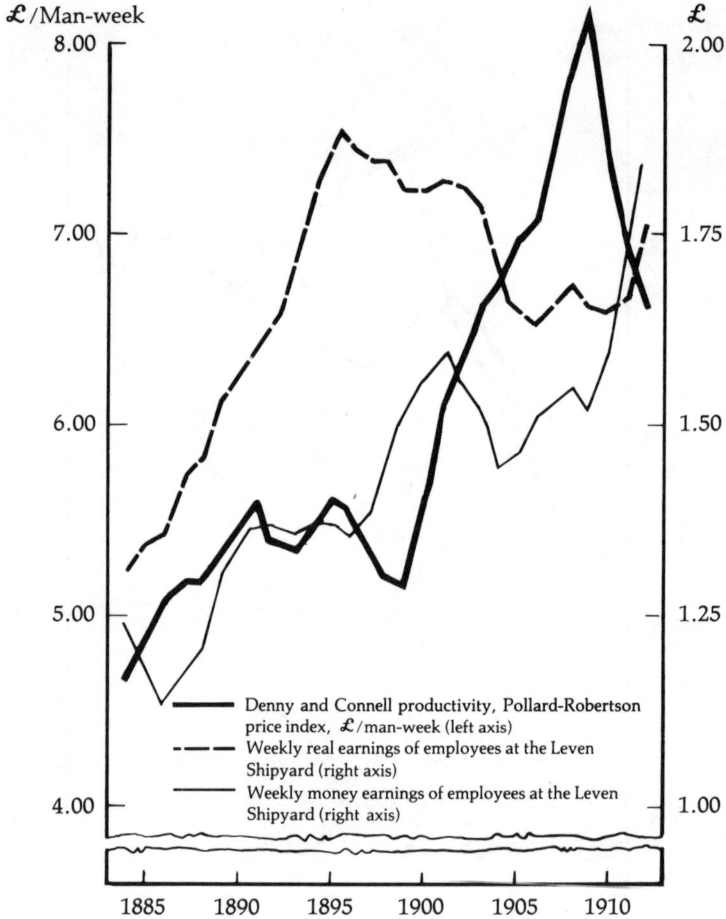

Figure 11. Denny and Connell labor productivity (Pollard-Robertson price index) and money and real earnings of employees at the Leven Shipyard, 1884-1912. (Table 9.1 and appendix tables B.1 and B.2.)

of the wages of unskilled laborers. The percentages of all the skilled groups increased during the booms of 1881-1888 and after 1900. During the uneven period of the 1890s, they remained constant at a high level.

The enhanced wage position of the craftsmen before 1900 illustrates the important role the skilled trades played in the construction of modern metal steamships. Each innovation in this period led to new uses for skilled labor. After 1900, however, the situation was altered as increased mechanization made it possible to substitute unskilled for skilled workers, thereby lowering the bargaining power of the latter and increasing that of the former. In addition, as was suggested above, the reduction or

Table 9.8. Wage rates of selected trades, Leven Shipyard, expressed as percentages of laborers' wage rates, 1880-1910.

Date	Blacksmiths	Carpenters	Joiners machine-men	Engineers	Blacksmiths hammermen
Dec. 1880	178.9	185.7	210.6	185.7	110.6
Jan. 1881	186.6	186.6	216.6	186.6	—
Jan. 1882	186.2	186.2	216.2	186.2	—
Mar. 1882	—	193.6	—	—	—
Apr. 1882	192.6	—	—	—	—
May 1882	—	—	—	199.5	—
Mar. 1883	—	206.1	—	—	—
May 1883	—	—	222.6	—	—
June 1883	—	212.8	—	—	—
July 1883	—	—	—	206.1	—
Feb. 1884	186.6	199.5	216.2	—	106.4
July 1884	178.9	192.9	210.2	—	106.7
Jan. 1885	178.9	192.9	210.2	213.6	106.7
Mar. 1885	178.8	192.9	214.3	206.7	100.0
Oct. 1885	—	203.7	—	—	—
Oct. 1888	178.9	—	213.8	206.7	106.7
Apr. 1889	179.6	199.5	216.2	—	109.8
Aug. 1889	—	—	—	205.0	—
Sept. 1889	—	212.8	—	—	—
Jan. 1890	—	—	—	206.4	—
Feb. 1890	191.4	—	—	—	—
Nov. 1892	193.1	212.7	216.0	206.9	109.8
June 1893	—	—	219.5	—	—
Feb. 1896	—	—	—	206.4	—
Apr. 1896	193.3	—	220.2	—	110.1
June 1896	—	—	—	213.5	—
Aug. 1896	—	213.3	226.6	—	—
Sept. 1896	199.1	—	—	—	113.3
Dec. 1896	—	—	—	212.8	—
Apr. 1897	199.5	212.9	226.0	—	112.8
May 1897	—	—	—	—	116.1
Sept. 1898	200.0	219.3	238.9	212.8	125.7
Mar. 1903	193.8	206.3	203.2	—	114.6
June 1903	—	—	—	200.0	—
Jan. 1908	187.5	—	—	—	114.6
Feb. 1908	—	200.0	—	—	—
Dec. 1908	—	—	—	207.7	—
Jan. 1909	—	—	—	196.8	—
May 1910	168.5	180.0	175.4	183.3	109.0

Source: Appendix B, table B.4.

stagnation of real wages that took place throughout the British economy after 1900 may have helped to erode the position of craftsmen in shipbuilding by lowering the wage rates they could expect to receive in other industries.

Except for the rapid relative decline in the wages of joiners machinemen after 1903, the wage rates of the various skilled crafts changed only slightly in comparison to each other between 1880 and 1910. Moreover, in this period the metal and engineering trades, which were newer to shipbuilding and whose jobs were more technologically advanced, seem to have had no advantage over the older woodworking trades. The wages of engineers and ships carpenters (shipwrights), for example, remained almost equal to each other over the 30 years. Although the shipwrights' proportion of the total workforce declined as metal workers came in, they remained a highly skilled trade, which performed essential tasks in drawing the lines for vessels and undertaking interior woodwork. As table 9.9 shows, the status of the trades that became outmoded with the decline of the wooden sailing ship began to deteriorate long before 1880. As early as 1851-1855, the relative wage rates of the riggers declined significantly while the rates of other skilled trades were improving.

One area of wage regulation in which trade unions achieved considerable success was in the elimination of regional differentials. After 1850 time wages on the Clyde, Tyne, and Wear increased more rapidly than in the more highly paid Thames and Mersey districts until virtual parity was achieved around 1900.[20] By 1904 time wages on the Clyde were 60 percent higher than in 1860 as compared to increases of 40, 34, and 17 percent respectively on the Tyne, Wear, and Thames. Since Bowley and Wood provide only indices of wage levels, it is not possible to determine precisely how money wages changed on the four rivers. It is clear, how-

Table 9.9. Wage rates of selected trades, Leven Shipyard, expressed as percentages of laborers' wage rates, 1846-1880.

Period	Carpenters	Smiths	Fillers	Riggers
1846-1850	197.7	190.9	223.5	227.3
1851-1855	204.8	211.8	231.6	184.7
1856-1860	203.2	207.7	226.3	175.3
1861-1865	217.0	210.9	223.1	180.2
1866-1870	201.5	215.8	225.0	192.9
1871-1875	206.1	200.0	210.0	182.8
1876-1880	185.3	191.9	182.4	178.5

Source: Unpublished book of "wages notes" of William Denny and Brothers.

ever, that over the period the differentials decreased. By 1913 the conciliation board established as a result of the 1909 agreement between the Shipbuilding Employers' Federation and the shipyard trade unions had nearly equalized wage rates for each trade throughout Great Britain—regardless of productivity differences—by raising the rates in the lower-paid districts.[21]

In 1861 joiners on the Clyde received between 59 and 64 percent as much as those on the Thames and between 88 and 92 percent as much as those on the Tyne; Clyde shipwrights received between 64 and 67 percent as much as Thames shipwrights and between 88 and 92 percent as much as Tyne shipwrights. By 1890 joiners on the Clyde were getting 87 percent as much as joiners on the Thames, 92 percent as much as joiners on the Tyne, and 94 percent as much as joiners on the Wear. Shipwrights on the Clyde were paid 88 percent as much as those on the Thames, 94 percent as much as those on the Tyne, and 99 percent as much as those on the Wear. Wage differentials remained at approximately those levels down to 1904.[22]

There are two main explanations for the narrowing of regional wage differentials. With the breakdown of the restrictive craft system in the second half of the nineteenth century, shipyard workers achieved a high degree of mobility and often moved from port to port when labor unrest or lack of orders led to unemployment. The shipwrights of the new and expanding centers such as Glasgow and Newcastle favored complete mobility, and by the 1860s migration was common. Most of the unskilled laborers on the Clyde and Tyne were Irish and moved freely whenever work became scarce. Nearly all of the men at Barrow and Belfast were said to have been trained on the Clyde, and many of those on the Tyne and Wear had migrated as young shipwrights. From its inception the Boilermakers' Society provided traveling or "tramping" funds for the unemployed, and in later years when builders found that there were not enough workers available locally they could apply to the Society's national office, which would arrange for men to come from other districts.[23]

This newer mobility, when combined with the development of national trade unions, worked to standardize wage rates and even out levels of productivity throughout the country. Wage rates on the North East Coast and the Clyde tended to move together, especially after 1880. Bowley and Wood found that it was also common for employers and workers in the smaller ports to tie wage rates to rates on the main rivers. Wages of Barrow engineering workers followed those on the Clyde after one month while the wages of the shipbuilders in Belfast and on the East Coast of Scotland also followed those on the Clyde, and platers and riveters at Southampton received 5 percent less than those on the North East Coast and followed their changes after an interval of a month.[24]

Data on salaries are not as plentiful, but from the lists that have survived and from Denny's salary books it is possible to determine how salaries were assigned.[25] There were three categories of salaried employees in Denny's yard: (1) the officials in the counting house—the manager, secretary, clerks, and timekeepers; (2) the women in the tracing department who copied blueprints; and (3) the employees in the drafting office—designers and architects as well as draftsmen. With the exception of the tracers, their salaries were on an annual basis and they were not subject to layoffs during slack periods.

Unlike the craftsmen and laborers in the yard, each salaried employee had his wage rate individually fixed on the basis of his position, age, experience, and years of service with the firm. Each of the timekeepers or draftsmen might receive a different salary, and the range of salaries of men in the same classification was sometimes as great as 100 percent. In 1897, for example, 21 draftsmen, not including the apprentices or head draftsmen, received an average salary of £115 per annum. The range of salaries, however, was from £90 to £190.

Between 1882 and 1897, the average salaries of employees in both the counting house and the drafting office dropped slightly while the average earnings of employees in the shipyard increased by about 3 percent. This may be accounted for by the growth in the office staff, which increased the proportion of young and inexperienced, and therefore low-paid, workers. No employee who was with the firm over the entire period had a reduction in salary, and several received raises of 25 percent or more.

By 1897 foremen and group leaders in the yard received between 16/6 and 80/- per week, the lowest wages being paid to women in charge of the polishers and decorators. Most of the men were paid between 40s. and 50s. per week.

In addition to their salaries or wages, many of the draftsmen and foremen received tonnage bonuses, a form of profit sharing. The bonuses ranged from 1/80d. to 1 5/16d. per ton. In 1897 this amounted to between £1.6.2 and £137.7.6. Countinghouse officials, messenger boys, and apprentices, although ineligible for bonuses, received year-end gratuities of 10s. to £10. Instead of salary increases, officials sometimes received raises of 1/16d. or 1/10d. per ton in their bonuses.

Denny's had no regular pension fund in the nineteenth century, so workers who became disabled owing to age or injury were forced to rely on the generosity of the management for their support. The financial status of petitioners was always carefully investigated. From the reports that have survived it appears that employees were expected to have saved a portion of their earnings and that their children, not the firm, were responsible for maintaining them whenever possible. Generally, although there were exceptions, the firm would only grant a pension when no other source of support was available.

Most of the pensions were for 5s. or 6s. per week. In a few cases in which an employee had died leaving small children, the widow was given a small pension, perhaps 2/6 per child per week, until the children were able to work. A typical letter from a retired employee and the report of E. S. Twigg, the head clerk, are given below. On the bottom of the report, one of the partners has written, "Recommended for a pension, say 5/- a week from 11/2/95 onwards."

> Dumbarton 8/2/95
>
> Messers Wm Denny & Bros.
> Gentlemen:
>
> Excuse me troubling you, but as my means are exhausted I take the liberty of asking if you would kindly Assist me. I have been in the employment of Mr. Arch^{d} Denny, Mr. Alexr Denny and your own firm for upwards of 43 years, and wrought up to about 6 years ago when I was afflicted with Paralysis and since then have been unable to work. I was a Sawyer in the old Sawmill for years and latterly wrought as a labourer, since the Steam Saw was started. If you would look into my case you would confer a benefit on an Old Servant.
>
> William Walker

> Dumbarton, 12 Feby. 1895
>
> William Walker 79 years of age at one time a Sawyer came to the employ in April 1865 and wrought at his trade until the Sawmill started. Since then he had been employed as a labourer and so continued till about 6 years ago when he took a shock of Paralysis since then he had done no work.
>
> He had previously been in Mr. Alexr. Denny and Mr. Archd. Dennys yards. he says he had been over 40 years in the Denny employment. I can find that he had been over 30 certain.
>
> He had three of a family
>
> 1 son a fitter
> 1 " " A. I. Smith
> 1 son a brass finisher out of employment and a useless man. He (the father) is assisted by his family but I learn his own savings are now exhausted. He was a well doing man.
>
> E. S. T.

A highly skilled workforce and the efficient deployment of labor combined with increased mechanization to permit great increases in labor productivity in British shipbuilding after 1870. Before the turn of the century these increases were translated into higher real earnings in the shipyards; subsequently, shipowners seem to have been the principal beneficiaries as the price of new vessels declined significantly until 1911. Skill differentials increased until 1900 and then began to narrow again, perhaps because greater mechanization decreased the relative bargaining power of the skilled trades after the turn of the century. Regional differentials decreased steadily over the period as trade union pressures helped to enhance the position of the worst-paid ports. Unlike the wages of skilled and unskilled labor, salaries and pensions were individually determined. Salary levels depended on such factors as age, rank, experience, and years of service with the firm, and they were sometimes supplemented by tonnage bonuses. There was no institutionalized pension system, and aid was granted only when it could be shown that an individual's other sources of income were no longer sufficient.

10.

The Influence of the State

Although the half century opening with the repeal of the Corn Law and the Navigation Act has good grounds to the claim of an era of free trade and laissez-faire, it was also one of increasing state intervention, control, and regulation of society and economy.[1] As one authority has noted, "The twin themes of . . . laissez-faire and state intervention . . . were equally characteristic developments of the middle quarters of the nineteenth century, and it is not necessary to assume that they were in contradiction to one another."[2] While the solid achievements of the Gladstonian budgets and of the commercial treaties were backed by confident economic and political doctrines that have continued to exercise their spell to the present day, the state never abandoned its right to mediate and arbitrate in situations in which it was felt that the interests of the public could not be left to the mercy of individual initiative.

The interaction between laissez-faire and state intervention is perhaps nowhere better illustrated than in the history of the shipbuilding industry. Through the navy, the state had for centuries been the largest single consumer of ships and shipping services. National defense and the maintenance of the empire required large fleets of both naval and merchant vessels throughout the nineteenth century. Moreover, the Royal Naval Dockyards had long possessed major shipbuilding and repair facilities. On the other hand, the increasing strength of the private shipbuilding sector after 1860 was based on as rigorous an adherence to the principle of economic freedom as was demonstrated by other pressure groups in similar circumstances. Private shipbuilders, who could point with pride to an international superiority in the quality as well as the quantity of their output, at times complained bitterly of state interference, as in questions of legislation on unseaworthy vessels. The encouragement, assistance, and guidance of the state remained constant, however, and while private builders in some cases accepted this support only grudgingly, in others, when it suited their interests, they imperiously demanded governmental help.

The Evolution of Naval Design

A detailed account of changes in the design of naval vessels is beyond the scope of this work. Here, it is necessary only, first, to point out the ways changes in naval vessels paralleled those in merchant ships; and, second, to discuss the role of the British state in advances in naval architecture and marine engineering.

Between 1860 and 1914, naval vessels underwent considerable changes in armor, firepower, means of propulsion, and materials of construction. In combination, these led to the development of ever-larger, faster, and more lethal ships, which cost more and, like merchant vessels, required for their construction shipyards of greater size and capitalization, the purchase of components from other industries, and workmen with a wider range of skills.

Progress in naval design was spurred largely by a perceived threat from foreign sea-power, especially from the French and, latterly, from the Germans. From 1862 to 1865 the Admiralty worked frantically to build new vessels and convert timber three-deckers to overtake the French lead in armorclads. Tests at Shoeburyness between 1861 and 1864 showed the superiority of wrought iron over copper, steel, and other metals as armor plating—at a thickness of 4.5 inches, it could withstand the shells made at that time. By 1865 the British lead over the French, whose program remained largely on paper, was unmistakable, but it was not long before heavier guns forced the use of even thicker armor and "the see-saw contest between the guns and the plates was on in earnest."[3]

Two important developments were shaping British naval construction between 1871 and 1884. On the one hand, the introduction of heavy shells, the torpedo, and the torpedo vessel made large ships too vulnerable; the tendency was to build shorter, faster, and more highly protected capital ships. On the other hand, the demands of the navy included not only an equality with the skillfully built and powerful French navy in home waters, but also fast cruisers to protect the merchant navy on the high seas, and heavily armed vessels for the stations along the distant outposts of the empire, where other nations had outgrown the tutelage of mere gunboats and frigates. Until these problems were finally solved by the evolution of triple expansion and turbine engines, by Krupp and Harvey steel, and by protected cruisers and torpedo boat destroyers, naval construction lacked a general plan.

At first, under the impact of improved ordnance, the original *Warrior* type of vessel, with its broadside armament and overall center and bulkhead armor, was merely enlarged. The armor was extended in the *Hector* and *Valiant* and spread over the whole side of the vessel in the *Minotaur* types. Its thickness and that of the teak backing were increased, but as shells and guns advanced in weight and power, any further extension would have made the armor too heavy for the buoyancy of the hull.[4] The simple rolled wrought iron armor of the 1860s, made by the Parkhead Forge and others, reached its limit at the decisive trials at Spezia in 1878, when the 100-ton Armstrong gun completely perforated a 22-inch armor plate and its timber backing. There followed the period of compound armor—hard steel outside to keep out the shells and soft iron inside to bind the plates together—produced by either of the two methods evolved simultaneously by A. Wilson of Cammell and Company and J. D. Ellis of John Brown and Company. Compound armor was approved by most admiralties and was found satisfactory in the trials at Okhta in 1882, Copenhagen in 1884, and Pola in 1885. But then the chromium-steel armor-piercing shell brought back the all-steel plates, produced by Schneider (Creuzot) since 1877 and by Whitworth since 1887. After its successful trials at Spezia in 1882-1884, Portsmouth in 1889, and Annapolis and Okhta in 1890, the all-steel plate, improved by additions of nickel and chromium, held the field. The Harvey and Krupp processes of armor plate manufacture, invented in the 1890s, were the last great changes in armor before the official demands for secrecy disturbed the march of technical advance within the close international confraternity of armament combines.[5]

Three types of naval cruisers were developed: fast, unprotected cruisers built to take the place of frigates; "protected cruisers" with a low vaulted armor deck built from 1878 on; and "armored cruisers" with a complete armor deck, designed by Armstrong for foreign countries and accepted in Britain only after 1890, when better armor and engines allowed the speed and range necessary for protecting the empire. The Whitehead torpedo, with its depth-controlling mechanism introduced in 1868, led to the building of torpedo boats, growing in size and speed from 7.5 d.t. and 15 knots in 1873 to 80 d.t and 21 knots in 1885, followed by "catchers" and "destroyers," larger and faster vessels carrying torpedos themselves.[6]

Battleships, too, increased in size and armament, particularly after the Battle of the Yalu River (September 17, 1894) in the Sino-Japanese War had shown the deadly power of shells. A specially long and narrow type of battleship, the *Canopus* class, was designed to pass through the Suez Canal and meet the possible challenge of Japan. The naval building program of 1893 had also introduced faster torpedo boats and destroyers to protect the merchant fleet from the French torpedos. With their water-tube or locomotive boilers, the torpedo boat destroyers reached 27 knots in 1893 and 30 knots in 1899, while the turbine driven *Viper* did 37 knots in 1900.[7]

The early years of the twentieth century saw perhaps the most rapid changes in naval design. The détente with France and the Japanese alliance allowed the Admiralty to withdraw the capital ships from far eastern waters and the Mediterranean and concentrate them in home waters, against Germany, the new potential enemy. At the same time, the destruction of the Russian Navy in 1904 and the rapid growth of German warship tonnage upset the building target, which had been a naval strength of 10 percent above that of the next two European powers combined. Within a very short time, therefore, new decisions on the type of capital ships and on their numbers had to be made. The Russo-Japanese War, and the Battle of Tsushima in particular, had also taught important tactical and technical lessons to warship designers. The most far-reaching of them was the realization that the fear of torpedos would, in future, keep hostile warships at least 3,000 yards apart, and since heavy guns were fairly accurate at distances up to 6,000-7,000 yards, the lighter guns with their shorter range became useless. The argument of the few officers who had been advocating the "all big-gun ship" thus found very strong support in practice. It was not long before the energetic First Sea Lord, Admiral Sir John Fisher, had the first ship of the new type, the *Dreadnought*, on the stocks.

The design of the *Dreadnought* was revolutionary enough. The vessel had ten heavily protected 12-inch guns, no secondary armament, a large number of light, widely dispersed antitorpedo boat guns, and turbines for higher speeds. Even more revolutionary, however, was the building program of the dreadnoughts, which was drawn up on the assumption that the new ships would, in fact, render all existing vessels out of date. No new capital ships were to be laid down after 1904, and those under construction were to be hurriedly completed, while a prototype of the new vessels was to be built for tests and rapidly copied afterward. In future, there were to be fewer new vessels on the stocks at any one time, but these were to be completed more rapidly.[8]

Fisher had succeeded in forcing all other navies to follow him, at a good distance behind, in building dreadnoughts and in discounting their earlier battleships as obsolete. The design itself was subject to no major changes before the war, though each new program increased the size, weight, armament, and cost. In 1910 the 12-inch gun gave way to the 13.5-inch gun, the armor was thickened from 11 inches to 12 inches, and all five turrets were placed on the center line. Also, the two end turrets forward and aft were built on the Michigan plan, that is, placed above, not behind, each other. In 1913 the 13.5-inch gun was replaced by the 15-inch gun, but the number of turrets was reduced to four, and the space of the central turret was taken up by boilers in order to raise the speed from 21 to 25 knots. The total displacement tonnage of a dreadnought rose from 18,000 d.t. in 1906 to 28,000 d.t. in 1913, and the cost from £1.7 million to £2.5 million, excluding the armament.[9]

The committee that had approved the dreadnought type also laid down new directives for the building of cruisers. These were also designed for big guns only. They were not only to carry out the duties of the old frigates, acting as scouts, protecting commerce, and destroying enemy raiders and cruisers, but also to take their place in the line of battle. As a result they had to be designed with armor, armament, and a displacement tonnage almost as heavy as the battleships themselves, while the speed and power were greater. The battle cruiser was perhaps the least successful of Fisher's ideas. His ambitious plan of building vessels to be used as cruisers *and* in the line of battle, and his motto that "speed is armour," "literally exploded with the magazines of three of his battlecruisers" in the Battle of Jutland, when the British vessels failed to stand up to the more heavily armed German battleships.[10] Battle cruisers had proved too costly to be used in routine commerce-protection duties even before 1916, and light, fast cruisers were built from 1908 on, armed with 4-inch or 8-inch guns and unprotected.[11]

As we have shown in chapter 7, as late as 1914 the majority of scientifically educated naval architects in the United Kingdom had, at one time or another, been in the Admiralty service. Because of their superior training, these men were better equipped than most private naval architects to design both naval and merchant vessels. It is significant that of the five occasions when the government invited private designs for warships, four elicited results so inadequate in their calculation of displacement, buoyancy, and stability (not to mention fighting qualities), that they had to be discarded; while the one vessel that was designed and built entirely by private means, the *Captain*, capsized in the open sea.[12]

The Shipbuilding Department of the Navy

Throughout most of the nineteenth century the Royal Naval Dockyards, free as they were from competition and the incentives of profit, were considered to be by definition inefficient. Backed by the current economic and political ideology, and opposed only by the feeble voices of government officials, themselves under the influence of the same ideas, the defenders of private shipbuilding could allege not only a lack of initiative and inventiveness in the Dockyards, but also extravagance, waste, and a level of costs higher than that of any private competitive yard. The attack was strongest in the early 1860s, when the rapid conversion to steam, iron, and armor had encouraged the foundation of a number of speculative private yards that could not possibly hope for sufficient commercial orders and strove instead to supplant the Dockyards and their work. After that period, the inferiority of the Admiralty was taken for granted in many quarters.[13]

Nevertheless, in a number of areas of technology, the Dockyards were in the vanguard. In addition to the scientific training provided in the Dockyard schools and to Froude's research with his experimental tank, investigations in many other areas were financed by the Admiralty. The Committee on Iron of 1860, the Report on Coals Suitable for the Steam Navy of 1848-1849, the Boiler Committee of 1874, and the Water Tube Boiler Committee of 1901 all worked under naval auspices. Original research on stability and on oil fuel was also carried out by Admiralty constructors. Perhaps the greatest service was the introduction of steel by Barnaby in 1873 and its general adoption in the navy before the mercantile community had begun to be alive to its constructional value. Later on, it was the Admiralty officers who ensured that a high quality of steel be used, as they had been equally instrumental in enforcing bulkheads, cellular double bottoms, and other improvements on the liners they subsidized as auxiliary cruisers. Even in marine engineering, though none of the engines was built in the Dockyards, the introduction of twin screws, forced draft, and the watertube boiler and marine steam turbine, to name only a few improvements, were encouraged by expenditures from public funds. "A large part of the positive and accurate knowledge which now exists," wrote one observer, "has come from sources more or less supported or connected with Government institutions."[14]

The charge of inefficiency and high costs in the Dockyards is not easy to disprove because the inadequate system of accounting used by the navy does not allow for comparisons with ships built under contract in private yards. Throughout the period there was no complete valuation of all capital items used in the Dockyards. Moreover, the varied nature of activities performed by the navy did not make it possible to assign overheads to any one ship under construction. To omit all overhead charges would show Dockyard vessels as cheaper, but would be clearly unfair in view of overheads in the private yards. To include them all would show privately built vessels to be cheaper, but would be unfair to the Dockyards, which had a large expenditure on maintaining facilities for repairs and victualing the fleet, and which designed both the ships built in the Dockyards and those built on contract. To include an arbitrary percentage of the "establishment" charges would be to beg the question. Even Admiral Robinson's suggestion of 1868, to compare the labor and material costs only, did not lead to any valid results. Changes in accounting practices in the 1880s and 1890s, under which "the balance sheet of the Admiralty dockyard and shipbuilding transactions became for the first time clearly intelligible,"[15] did not make direct comparisons feasible.

Making due allowances for these difficulties, the Dockyard costs can be assumed to have been somewhat above the private costs of shipbuilding, even including profits. This difference tended to disappear after the reequipment of the Dockyards in the 1890s. The Select Committee on Estimates of 1913 came to the conclusion that Dockyard costs were no greater than in private yards. The costs of ship repairs were always less in the Dockyards, with the result that almost no repair work was performed under outside contracts. Even in the late 1890s, when more than 60 percent of new construction was in outside yards, a little more than 80 percent of the expenditures for ship repairs was in the Dockyards.[16]

Before 1870 much of the difference in costs was accounted for by the large stores of seasoned timber the Admiralty had to keep, and by the higher quality of work performed in the Dockyards. After 1870 the difference was caused mainly by the long delays in the Dockyards while vessels were under construction. In 1885 it was alleged that no ironclad warship was ever commissioned in under five years after being laid down and that some vessels took eight or nine years to build. To some extent, this was because of delays while the armament, armor, and so on were changed in accordance with the latest research, because of temporary transfers of craftsmen to emergency repairs, and because of problems in getting armor of the proper quality from the makers, all of whom after 1900 also built ships themselves. There is no doubt that it was wasteful, however, and after the reforms of the 1890s the Dockyards worked much more quickly. The *Royal Sovereign*, the first battleship built after the reforms, took only two years and eight months to complete.[17]

The British Admiralty has also been accused of never initiating a change, but merely following the innovations of other navies. There have always been two naval building policies: one that tries to introduce innovations to make all present fleets obsolete, and another that tries to match or exceed in quality the navies of potential enemies. The latter had always been the British tradition until the days of Admiral Fisher. "Among naval powers the weaker have a more obvious interest in revolutionising naval warfare than have their stronger rivals . . . when the dominant power can build ships faster than its weaker rivals, and can hope to overtake them rapidly if a lead be momentarily conceded, the argument for conservatism in constructive policy becomes all the stronger."[18] Of the two policies of building ships as rapidly as possible to a set plan, or improving them as they progress on the stocks, so that the completed vessel was more up-to-date than the original design, Britain commonly chose the latter. Up to 1904, at least, Britain's warship building capacity, Dockyard and private, was more than twice that of all other nations put together.[19]

The Dockyards were only in part shipbuilding establishments. They also served the navy in other ways, and a naval officer was always in command. They were intended as repairing and fitting-out yards, primarily in times of war, when casualties from actions in the North Sea could withdraw to the Thames, and casualties from the Channel to the South Coast yards. Pembroke was the only yard intended purely for building, Portsmouth became the largest, and Devonport (with Keyham) was the third important shipbuilding yard. When Chatham was modernized at the beginning of our period, Woolwich and Deptford were closed and Sheerness used chiefly as a fitting-out yard for Chatham. After 1900, neither of the Thames yards did much building.[20]

The control of the Dockyards, and also the internal organization of the Admiralty Construction Corps, underwent many changes after 1860. As a result of upheavals in 1858-1861, a treasury committee was appointed in 1858, and another committee to inquire into Dockyard management sat from 1858 to 1859. They were followed by a further committee in 1860 and by a royal commission, which reported in 1861. The report of the commission recommended the end of the piece work system with its 95,000 different prices, a proper system of cost accounting instead of the ludicrous "Doomsday Book," and the centralization of shipbuilding and dockyard control under one officer. These recommendations were acted upon, and Admiral Spencer Robinson was appointed Controller even before the report was published.[21]

Until the 1860s the Admiralty Board consisted of the First Lord who had four Naval Lords and one Civil Lord under him, each of whom was in charge of a department under a Principal Officer. The Controller was responsible to the First Naval Lord, and there were also two Secretaries. An Order in Council of January 14, 1869, changed the composition of the Board, dispensing with two Naval Lords and elevating the Controller—who was in charge of material, which included shipbuilding—to the Board as the Third Naval Lord. The resulting absence of contact among the officers was soon felt to be unsatisfactory, but other reforms introduced at the same time, including a separate purchasing department and the placing of ordnance, Dockyard building, and stores under the Controller, were more successful. The Controller now had under him the Director of Naval Construction, the Engineer-in-Chief, the Director of Ordnance, the Director of Dockyards, the Director of Stores, and the Inspector of the Dockyard Expense Account, among whom the Director of Naval Construction was, at best, only a first among equals.[22]

The sinking of the *Captain* and the resignation of E. J. Reed as Director of Naval Construction in 1870 left the department in a state of flux. It was placed under N. Barnaby, who was finally appointed Chief Naval Architect in 1872. Dissatisfactions with the reform of 1869 led to the appointment of a House of Lords Select Committee in 1871, and as a result of their report the Admiralty Board reverted to its former position and the Controller lost his seat on it. The Chief Naval Architect was put in charge of contracts as well as Dockyard work, and between 1873 and 1885 he was, in fact, in effective charge of Dockyard work and of naval construction. As "adviser" to the Controller and to the Board he was responsible for most of the planning. His office, with its assistant constructors and draftsmen, produced all the final designs and calculations. His inspectors at the private contractors' yards and the other constructors, foremen, and junior officers in the Dockyards ensured the carrying out of his plans. There was also close cooperation with the department of the Engineer-in-Chief. Outside his control remained only the ordnance department, a branch of the army.[23]

This organization continued, with minor changes, until the revolution in the Admiralty preceding and following the appointment of W. H. White. In 1882 the Controller was readmitted to the Admiralty Board, and a Civil Lord, with technical knowledge, was added. By an Order in Council dated August 23, 1883, a Corps of Naval Constructors was formed to give the naval architects in the service a regularized employment, salary scale, and route of promotion. This important new reform was a direct result of the favorable report of Lord Brassey's committee and was said to be advantageous in two respects. First, it offered an opportunity for persons of "good social position" to enter the corps near the summit and thus bypass the laborious climb through apprenticeship, work as a shipwright or study at the Naval College, and gradual promotion. Secondly, it gave a stimulus to members of corps by creating new opportunities for circulating between the Admiralty, the Dockyards, and the private yards.[24]

The first few years after White's appointment in 1885 were spent in reorganizing the Dockyard work, contract work, and naval construction, following a series of detailed departmental inquiries. The main result was to strengthen the Director of Naval Construction's position by making him Assistant Controller and establishing a clear line of command to the staff at the Dockyards and elsewhere. On the resignation of Lord Salisbury, Lord Ripon became First Lord in the Gladstone administration, and E. J. Reed, a frequent and bitter critic of White, was rewarded with a Treasury lordship for his long parliamentary services. It would be too much to believe that the appointment of Francis Elgar, a friend of Reed, to the newly created position of Director of Dockyards on almost an equality with White, was prompted by regard for the benefit of the service alone. White certainly protested strongly against the newly divided authority in the Dockyards.[25] The chain of responsibility after the reforms is shown in figure 12.

Not until Fisher's appointment did the greatly expanded naval program require a further reorganization in the Admiralty. By a series of

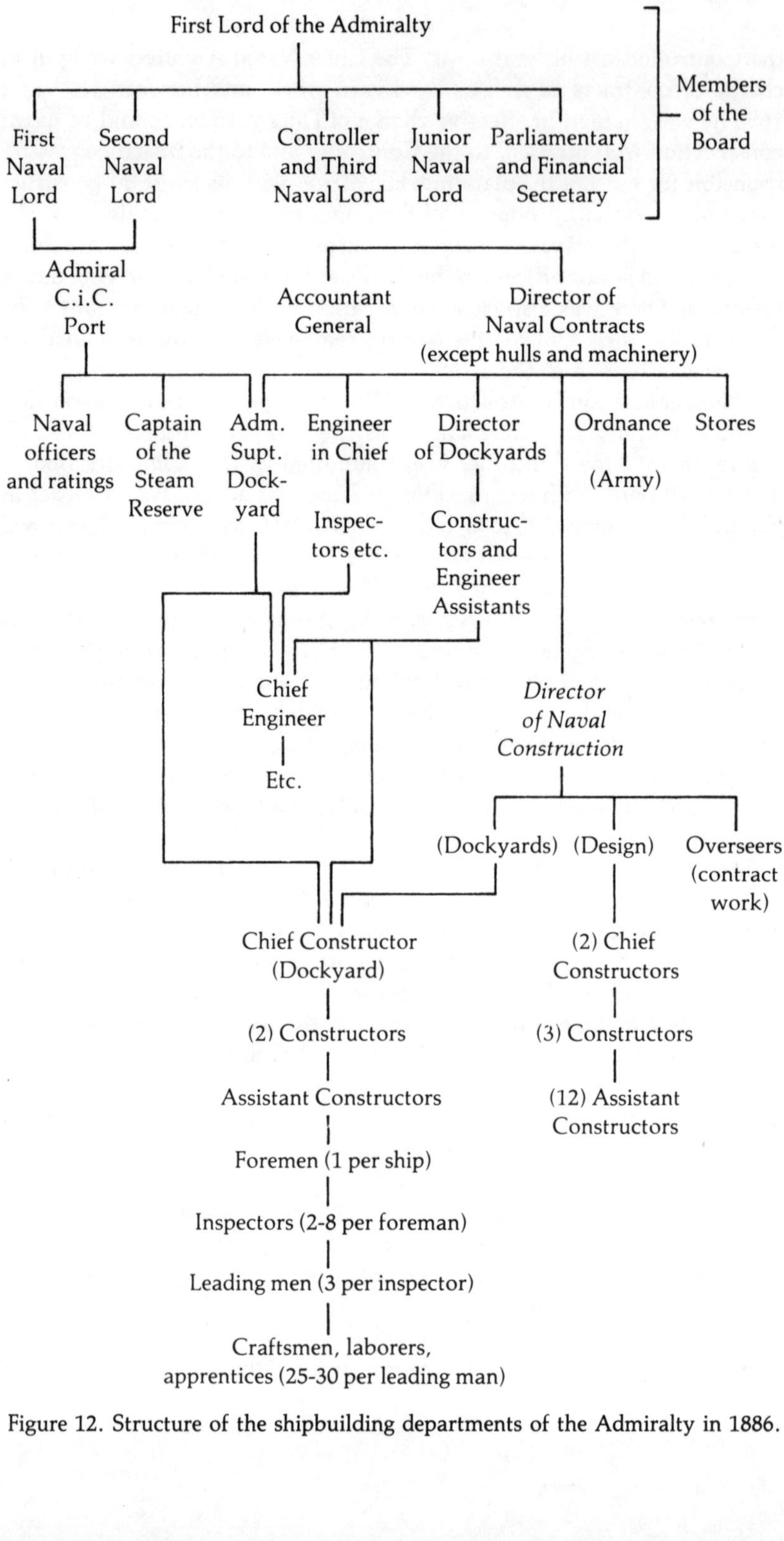

Figure 12. Structure of the shipbuilding departments of the Admiralty in 1886.

Orders in Council between 1902 and 1904 the Controller and the Admirals supervising the Dockyards were freed from a great deal of detailed work; naval ordnance became a separate branch; the Director of Naval Construction was made responsible for designs only and the Director of Dockyards became a traveling inspector instead of an official at the Admiralty. Finally, the Chief Constructors and Chief Engineers in the Dockyards were given much greater power and independence, on a basis similar to that of managers in private yards.[26]

Admiralty Contracts for Private Builders

Warships that were not built in the Dockyards were built by contract with private firms. With the introduction of iron and steam into the navy, contract work became of increasing importance, especially in periods of war, emergencies, or rapid technical change, when the Dockyard building capacity was manifestly deficient.

There was a fundamental distinction between the hulls and the rigging on the one hand, which were designed at the Admiralty and could be built in the Dockyards, and, if they were tendered for by private yards, were built from the plans and under the close supervision of Admiralty officials; and the engines and later torpedo boats on the other hand, for which there were never any building or repair facilities in the Dockyards and whose private builders were given much greater latitude in design. The other fittings and most raw materials, anchors, chains, armor plates, steel plates (the last of which were rolled in the Dockyards in 1883), gun mountings, and so on were also bought from outside suppliers, and guns were obtained through the army ordnance department.[27]

Hulls, armor, and engines were bought by the Admiralty on the advice of the Chief Constructor and Chief Engineer, and other stores by the Director of Contracts after 1869.[28] All purchases were made, in theory, by inviting competitive tenders from British firms. In practice, the range of tenderers was closely circumscribed and their choice not entirely haphazard.

Tenderers were chosen only from the "Admiralty List," to which firms were admitted on application after inspection of their technical equipment. Lists existed for the building of large engines, small engines, armorclads, cruisers, and torpedo vessels. On satisfactory completion of some orders, firms could be promoted from the building of smaller to the building of larger vessels or engines. The early lists, especially those for marine engines, were dominated by Thames firms, but in time the northern firms were promoted to them also. When an entirely new departure in shipbuilding was in question, even firms outside the lists were invited to tender, as happened in the building of the *Warrior* in 1858 and of the fast torpedo vessels of 1900. Once the contract was given out, faulty execution never led to a rejection of the work, but firms might be struck off the list and readmitted only after a thorough change in their control.[29]

When the selected firms were invited to tender on the skeleton of first specifications, the lowest tenders were generally, but by no means always, accepted. On one occasion, Laird Brothers got a contract without competition for having sent in the best of the private designs submitted for a cruiser, even though their design was not used. Exceptionally low tenders by inexperienced firms were nearly always rejected—or a higher price suggested—since it was known that they had miscalculated and were likely to skimp the work at the end, when they saw their error, or demand a higher price. In any case, quality was often also taken into account. On occasions, an attempt would be made to spread the work among the regions, especially to depressed districts, as in the case of the gunboats in 1867, and there were also attempts to secure a steady flow of orders to the largest contractors in order to prevent overexpansion and unemployment. After the Fair Wages Resolution of 1891 the Admiralty for a time also took local wage levels into consideration, though normally neither the price of raw materials nor the level of wages could account for the differences in some of the tenders. In the case of the tenders for the engines of the *Nile* and *Trafalgar,* which had been severely criticized, a different design had been arbitrarily overvalued as a pretext to give the contracts to two higher tenderers in which the Admiralty officials had greater confidence. Subject to the "advice" of the Chief Constructor and Chief Engineer on these considerations, the tenders accepted were the lowest ones.[30]

During the whole building period in private yards, Admiralty "overseers"—senior members of the Constructors' Corps—were in attendance to ensure Admiralty standards in the quality and the methods of building and to supervise any changes of design introduced after the keel was laid. Complaints from builders about the interference of these overseers were, of course, plentiful. Equally strong were the complaints about the vague specifications and the wide terms of the contracts, which allowed the Admiralty to order many changes in detail after construction had begun, although any major alteration from the original contract was paid for separately. It could not be denied, however, that the Admiralty standards of building and seaworthiness were much higher than those of foreign governments or those for merchant ships. As long as private yards were profit-making firms and most of the work was done by the piece or by subcontract, close control was necessary. Well-run firms, such as Laird's, had no complaints about the overseers.[31]

In the case of marine engines, specifications were kept very general and the designs were left to a small number of makers, who were credited with great ability, experience, and reputation, and who were keen to improve the designs competitively. The reports of the committee of 1859, strongly critical of the monopoly held by the Thames marine engineering firms, were opposed by the surveyor and ignored by his department. In the 1850s torpedo boats and destroyers were similarly left entirely to the design of a handful of specialist firms.[32]

Although the supervision was closer, the Admiralty contracts had points of similarity with the builders' mercantile contracts. Payment was in installments of which a last portion was retained until after the steam trials. The system of premiums for early delivery and penalties for delay, however, was seldom implemented. Delivery on time was unknown—the delays were due partly to changes in the design and partly to more profitable work for the builders elsewhere. While delay in delivery to private firms might result in an action for damages, delay in delivery to the Admiralty earned merely a letter of rebuke. Vague attempts were made to keep contractors supplied with a steady flow of orders, but no allowance could be made by the Admiralty for cyclical fluctuations or for foreign government contracts. The building completed year by year, and the proportion completed in the Dockyards and privately, depended upon the naval vote and on the amount of Dockyard labor needed for repair. This was a naval, not an economic, decision.[33]

Not all tenders were genuine. Firms would tender occasionally at an absurdly high price without any intention of accepting the contract, in order to see the latest Admiralty design, which they could then use for foreign navies. Most firms were keen to erect all fittings, mountings, and riggings, for a similar reason—experience for foreign contracts. Work for the British Admiralty was often accepted at a loss, which could be recouped either by "extras"—alterations during building for which the firm made its own price—or by building for foreign navies on the basis of British prestige. The important contractors, such as Armstrong, Earle, Clydebank (J. Brown), Fairfield, Palmer, Vickers, the Thames Ironworks, Robert Napier, Laird, and Scott and Company, usually had large foreign contracts on hand. Most of them had agents in foreign capitals, and if their interests demanded it, new inventions were made available to other navies before the Royal Navy, or work on foreign ships might be speeded by retarding progress on a British vessel.[34]

The repair of ships was not, in peacetime, carried out in private yards, since no price could be fixed until the vessel was opened out and the full extent of the damage known, and payment by list prices on items which could not be easily checked by the overseers proved much more expensive than Dockyard work. Engines, however, were commonly repaired by the private makers while the vessel was being overhauled in the Dockyards.[35]

Between 1869, when the separate department for contracts was set up, and 1885, when W. H. White returned to the Admiralty from Armstrong's, bringing with him the system of close personal and financial links between private firms and public servants, the contracts for ships and engines seem to have been given out on a fairly unbiased system of competitive tenders. Armor plates, however, a growing item of expenditure on all major warships, were always supplied by a cartel, and hydraulic gun mountings by a monopoly. An attempt was made to sell Armstrong's monopoly at a very high price at one point when a member of the firm was on the Board as an "expert" Civil Lord. Armstrong's offered to make its patents available to others on an ad valorem royalty or for a lump sum of £100,000, but the scheme fell through. Many of the firm's patents had been evolved in experiments on Admiralty property, at Admiralty cost, and with the aid of the Admiralty's scientific establishment. Nevertheless, Armstrong's position as patentee of previous inventions was so strong that it could take out its patents before the Admiralty could even stipulate for reasonable contract conditions for itself.[36]

Compound armorplates were made by Charles Cammell and John Brown only, who divided the contracts between them before tendering. The report of the Forwood Committee (1887), describing the cartel, led to an attempt to encourage competition, but the entry of Vickers and Beardmore (see chapter 5) and later Armstrong into the field of production was no improvement, and the Admiralty, closely connected as it was with the armaments giants, refused to go into armor plate production itself, although this was admitted to form the worst bottleneck in shipbuilding in 1900. It was in armor plates that the high costs, and high profits, of warship building accumulated. By the end of the century the five producers had bought their own shipyards, and not only could they drive out outsiders, such as the Thames Ironworks, but even the dockyards themselves were neglected by the "armaments ring" in favor of their own shipbuilding for the royal and other navies. Although the dockyards had to order their armor plates 8 to 12 months in advance, they could not be sure of punctual delivery.[37]

Between 1870 and 1914 warship building grew from a minor specialist activity to one cornerstone of prosperity in the shipbuilding industry. After the inglorious collapse of 1866, it had been treated as a sideline in many ports, perhaps, at best, credited with a slight stabilizing influence on employment. From about 1880 on, however, its growing importance for the prosperity of the industry was recognized (see tables 10.1-10.3). After the turn of the century warship building was decisive. The Naval Defence Act of 1889 was perhaps the first major building program involving large long-term contracts for private firms.[38]

According to J. H. Biles, the total displacement tonnage built in Britain

Table 10.1. Naval expenditure on contracts, selected major firms (£000).

	Naval defense acts						
Firm	Total 1889-1893	Total 1893-1897	1899-1900	1900-1901[a]	1902-1903	1903-1904	1904-1905
Fairfield	—	529	835	903	823	657	829
Clydebank (John Brown)[b]	1426	1404	581	873	340	950	679
London and Glasgow Company	478	423	122	200	524	532	183
Vickers[b]	457	1170	729	1017	876	1303	1575
Thames Ironworks	396	16	381	760	228	534	400
Cammell-Laird[b]	941	923	360	529	334	892	495
Palmer	1400	115	152	340	225	514	260

Source: *Economist*, March 3, 1900; *Statist*, March 26, 1904; British Parliamentary Papers, 1890, XLIV(17.), p. 641, and 1895, LXV(260.), p. 723.

a. Naval vote.

b. Including predecessors and constituent firms.

Table 10.2. Shipbuilding for the Royal Navy, thousands of displacement tons, 1866-1885.

Year	Dockyards (ton W.H.CLE.)[a]	Contract (ton W.H.)[a]	Total (ton W.H.)[a]
1866-67	12.2	1.3	13.5
1867-68	19.1	8.4	27.5
1868-69	12.6	11.6	24.2
1869-70	13.6	11.5	25.1
1870-71	13.5	7.0	20.5
1871-72	13.5	8.1	21.6
1872-73	12.4[b]	2.1	14.5
1873-74	11.3	3.4	14.7
1874-75	11.3	5.1	16.4
1875-76	14.1	5.5	19.6
1876-77	13.4	10.8	24.2
1877-78	12.0	4.5	16.5
1878-79	11.7	6.1	17.8
1879-80	12.4	2.8	15.2
1880-81	13.4	3.4	16.8
1881-82	15.4	3.2	18.6
1882-83	16.2	3.8	20.0
1883-84	16.2	3.0[c]	19.2
1884-85	16.2	4.2	20.4

Source: British Parliamentary Papers, 1884-85, XLVIII(330.), pp. 566-567.

a. Ton W.H. = ton weight of hull built in that financial year. Ton W.H.CLE. = ton W.H. calculated on the expenditure of labor—the actual weight cannot be calculated until the ship has been completed.

b. Ton W.H. calculated from the dockyard labor employed.

c. Including two purchased vessels.

rose from 119,000 in the decade 1858-1868 to 236,000 in 1868-1878, 256,000 in 1878-1888, 875,000 in 1888-1898, and 1,063,000 in 1898-1908; at the same time the cost of a capital ship grew from £270,000 for a screw threedecker in 1859 to the £375,000 of the *Warrior*, £620,000 of the first dreadnought, and well over £1,700,000 of the later dreadnoughts. Tonnage statistics are not reliable indicators of the degree of activity, since displacement tons of warships are in no way comparable to merchant gross tons, but the share of labor, or the value of output of private warship construction in the six or eight years before 1914 has been estimated at between 20 and 25 percent of total shipbuilding, merchant and naval (see table 10.4).[39]

At the beginning of our period a complaisant attitude at the Admiralty

Table 10.3. Shipbuilding for the Royal Navy, thousands of displacement tons, 1885-1913.

	Dockyards		Contract				Total
Year of launch	Chatham, Sheerness	Portsmouth, Devonport, Pembroke	Clyde	North East coast	Barrow, Liverpool	Others	
1885	14.3	25.0	3.1	2.8	—	10.0	55.2
1886	—	14.3	13.1	16.3	1.9	1.5	47.3
1887	7.3	22.1	5.6	10.5	—	10.5	55.9
1888	8.0	20.5	5.9	—	—	—	34.5
1889	15.5	20.7	6.8	6.6	—	—	49.5
1890	4.9	18.2	6.8	11.4	13.8	9.5	64.5
1891	28.5	39.6	14.2	18.1	—	6.8	107.2
1892	13.7	36.7	21.8	36.0	18.2	14.8	141.3
1893	9.8	22.7	—	—	0.3	3.6	36.3
1894	16.8	9.9	0.9	—	1.4	1.8	30.9
1895	20.5	49.9	41.8	3.4	21.8	1.1	138.5
1896	19.2	47.0	23.8	3.9	22.9	1.2	118.1
1897	7.9	23.9	12.7	8.5	13.2	1.4	67.6
1898	29.8	41.1	41.2	3.1	12.1	14.0	141.4
1899	17.2	49.7	28.1	1.0	26.6	0.3	122.9
1900	3.0	2.2	14.0	2.7	15.1	0.8	37.8
1901	17.2	47.7	62.2	14.7	40.1	28.8	210.7
1902	20.8	30.6	31.6	9.8	—	1.6	94.4
1903	2.1	26.1	54.4	29.2	33.2	5.0	150.2
1904	10.8	46.2	28.4	11.1	13.0	16.1	125.6
1905	16.3	29.9	13.9	14.6	17.0	2.8	94.6
1906	14.6	32.5	16.5	16.5	2.9	1.8	84.8
1907	—	51.8	38.2	37.9	5.2	4.4	137.5
1908	1.9	41.8	1.5	0.6	2.5	3.9	52.3
1909	1.9	45.3	14.4	8.1	28.2	3.0	101.0
1910	1.9	52.2	41.3	31.0	1.8	8.0	136.2
1911	6.6	52.9	76.9	27.1	37.7	26.9	228.1
1912	1.6	53.4	52.4	31.7	31.6	2.3	173.0
1913	12.3	69.8	66.6	10.3	29.2	5.6	193.8

Source: *Annual Dockyard Expense Accounts; Annual Abstract of Labour;* British Parliamentary Papers, 1890-91, LII(219.), p. 31.

gave support to foreign building by private firms. At least from 1869 onward it had been the practice of the Admiralty Board to allow and order their constructors to design warships to be built by private yards for foreign governments. Such a contract for the Argentine government

Table 10.4. Shipbuilding expenditures for the Royal Navy (£000).

Year	Dockyard building and repair[a]	Contract building	Total[a]
1869-70	1,864	591	2,455
1870-71	1,992	508	2,500
1871-72	1,715	390	2,105
1872-73	1,615	157	1,772
1873-74	2,292	297	2,589
1874-75	2,565	446	3,011
1875-76	2,533	564	3,097
1876-77	2,420	1,020	3,440
1877-78	2,421	2,092	4,513
1878-79	2,457	782	3,139
1879-80	2,244	496	2,740
1880-81	2,322	469	2,791
1881-82	2,485	402	2,887
1882-83	2,394	776	3,170
1883-84	3,040	366	3,406
1884-85	3,298	531	3,829
1885-86	3,989	2,033	6,022
1886-87	3,011	1,947	4,958
1887-88	3,076	1,088	4,164
1888-89	2,983	797	3,780[b]
1889-90	3,436	1,446	4,882[c]
1890-91	4,025	2,885	6,910[c]
1891-92	4,029	2,932	6,961[c]
1892-93	3,635	2,041	5,676[c]
1893-94	3,380	1,268	4,648[c]
1894-95	3,972	2,093	6,065[c]
1895-96	4,757	2,824	7,581
1896-97	4,735	4,471	9,206
1897-98	4,364	2,852	7,216
1898-99	5,295	3,699	8,994
1899-1900	6,031	4,057	10,088
1900-01	6,240	5,658	11,898
1901-02	7,013	4,999	12,012
1902-03	6,737	5,694	12,431
1903-04	7,992	8,088	16,080
1904-05	7,513	7,770	15,283
1905-06	7,598	5,315	12,913
1906-07	6,995	5,535	12,530
1907-08	7,321	4,665	11,986
1908-09	8,094	3,847	11,941
1909-10	7,527	6,253	13,780

Table 10.4. (cont.)

Year	Dockyard building and repair[a]	Contract building	Total[a]
1910-11	8,101	9,243	17,344
1911-12	8,374	8,704	17,078
1912-13	9,462	8,874	18,336
1913-14	10,219	9,073	19,292

Source: *Annual Dockyard Expense Accounts;* British Parliamentary Papers, 1884-95, XLVIII(330.), pp. 566-567; and 1890-91, LII(219.), p. 31.

a. These figures include building, repair, and "incidental" (overall) charges in the dockyards. Statistics generally quoted include only annual expenditures on building and omit the overhead charges, which amount to up to 30 percent of the total. These cannot be divided between repairing and building, and it has been thought advisable here to include them, as well as the expenditure on repair. The total is therefore higher than the actual building expenditure only, but it is more representative than the statistics commonly quoted.

b. Including expenditure under the Imperial Defence Act of 1888.

c. Including expenditure under the Naval and Imperial Defence Act of 1888-1889.

in 1879, the first for which the designer was to be paid, was granted on Barnaby's advice to W. H. White. The Admiralty Board protested at not having been informed first, but Barnaby's explanation was much the same as the Controller's, two years later, when he defended the continuation of such arrangements: it would be of advantage to the country to have foreign ships built in its private yards with the knowledge of the Chief Constructor; the yards were kept busy when there was little work on hand from the British Admiralty; and the public standing of the constructors in the department was raised and they were less likely to leave for private employment.[40]

When White took up employment with Armstrong's in 1883, he designed the vessels for the largest supplier of warships abroad. By 1912 the firm had completed 128 warships of a total displacement tonnage of 494,000 t. for 16 different nations, the chief of which were: United Kingdom, 168,000 t.; Japan, 101,000 t.; Brazil, 82,000 t.; Chile, 61,000 t; and China, 26,000 t.[41] White, as his admiring biographer wrote, "was not unwilling to play the part of *honnête courtier*, by pointing out the growth of the Japanese Navy to [his] Chinese clients, or that of the Chinese to their indomitable rivals. In doing so he was always careful to insist upon the confidential nature of his designs, and the daily progress of our scientific knowledge. By such means he was able to increase the profits of the great Company [Armstrong's] which employed him; to extend and perfect the resources of what is perhaps the most important of our national industries; and to kindle in the hearts of two Asiatic peoples the flames of an enlightened and sacred patriotism."[42]

A number of contemporaries were not so enthusiastic about the practice of stirring up international hatred and suspicion for the sake of profits for the armament combines, and the practice was always officially denied. The classic answer given by the chairman of Armstrong, Whitworth at its annual meeting in 1914 does not sound as convincing to posterity as it appears to have done to the shareholder who asked for a denial of the allegation that the firm was inciting unrest in order to sell armaments: "the question, as put in the public Press, is such a silly one that it is scarcely worth answering. No firm of our standing can possibly lend itself to playing such pranks, and I think there is no foundation for the idea that any of the armament firms could do such a thing. It is really against our interests. We do not do such things. These suggestions you see in the papers are incredibly wicked and excessively mischievous. In fact, the only thing we can do, to quote Mr. Winston Churchill's words, is to treat them as 'hellish suggestions.' "[43]

The large naval programs of the 1890s caused an expansion in the shipbuilding and armament works, which in turn used all their considerable lobbying power to clamor for more armaments to keep them going. This happened particularly in the building lull of 1900 and after the reduction in repair work and its confinement to the Dockyards in 1905. In 1913 Churchill, the First Lord, asked for an extra £1 million for supplementary estimates for contracts, for the specific reason that the private yards had completed their building program earlier than had been anticipated, and had to have profitable contracts to keep them going. In 1914 part of the increase of Churchill's program was for the purpose of allowing for heavy earnings for the contractors, and an extra £750,000 was asked for "increased earnings under existing contracts upon New Construction."[44]

Among the private firms Armstrong's maintained the closest contact with official patronage. Since the Crimean War, when Armstrong was admitted to Woolwich to see the latest British improvements, which were later used to his own profit at Elswick, the relation of his firm to the government was, in Reed's word, "remarkable." Armstrong himself gained the confidence of the Under Secretary for War, Sir Benjamin Hawes, who married a sister of I. K. Brunel. Brunel's son was then made a partner at Elswick. In 1867 Armstrong's arranged to build ships at Mitchell's yard, later amalgamated, under George Rendel, who became a Civil Lord of the Admiralty from 1881 to 1885 and later returned to the firm to organize its Italian works at Pozzuli. Sir Andrew Noble, later to rise to the chairmanship of the firm, started as an assistant inspector of artillery. In 1883 W. H. White was seconded from his Admiralty appointment to Elswick, where his salary was trebled. He had been a successful protege of N. Barnaby, the Chief Constructor, himself a brother-in-law of Reed, his predecessor. In Elswick, White laid out the new shipyard and designed the powerful armored cruisers, the export of which was to become most profitable in the following decade. He also secured the firm the contract for the *Renown* and the *Sans Pareil*, which would otherwise have gone to Palmer's. Only two years later, in 1885, the naval scare brought him back to the Admiralty, still bound to advise Armstrong's for a further two years, and only released by the firm on the condition that Philip Watts, another able constructor, be sent up from London to take his place. White retired from the Admiralty in 1901 and became a director of Swan and Hunter and of the Parsons Turbine Company, among others, before his death in 1913.[45]

In the 1880s, Armstrong's chief designer was Perret, another old Admiralty constructor. Sir Philip Watts rose to the position of managing director at Elswick before he succeeded White, in 1902, as Chief Constructor at the Admiralty. The offer to succeed A. Noble, managing director of Armstrong's, was made to Fisher, the "strong man" at the Admiralty, but fell through in 1901. One of Armstrong's premium apprentices, and a family friend of the leading officials of the Admiralty, Tennyson D'Eyncourt, succeeded Watts as Chief Constructor of the navy in 1912, after having worked for Armstrong's from 1891 to 1898 and from 1902 to 1912. Watt's nephew, Sir George Carter, meanwhile became manager of the Elswick works. In 1912-1913 four more of Armstrong's directors were drawn from the public service: Sir George H. Murray, Permanent Secretary to the Treasury; Rear Admiral Sir Charles Ottley, Secretary of the Imperial Committee of Defence; Sir William E. Smith, Naval Construction; and Sir Percy Girouard, Governor of the East Africa Protectorate. Perhaps the large government contracts of the firm, and their rapid expansion, might not be entirely unconnected with the fact that from 1883 to 1914 all Chief Constructors were drawn directly from its staff.[46]

Vickers Ltd. was also not without contacts when it bought its way into shipbuilding. James Dunn, a Dockyard-trained constructor who had risen to become Assistant Director of Naval Construction by 1894, became a director of the firm in 1897 and extended the Barrow yard for naval shipbuilding. Sir A. Trevor Dawson, experimental officer at Woolwich, was later the firm's managing director.[47]

Francis Elgar, at first the principal assistant of Reed at the Admiralty, left to become general manager of Earle's, at Hull. After some private practice in London and in Japan, and some teaching at the University of

Glasgow, he returned to the Admiralty as Director of Dockyards from 1886 to 1892. In the latter year he reverted to private industry, rising to the position of managing director and later chairman of the Fairfield Company, and also chairman of Cammell-Laird after their reorganization. Reed himself joined Whitworth's after his retirement, and later became chairman of Earle's, designing warships for foreign flags. Colonel Boxer joined the Whitworth firm from Woolwich, where the firm had been allowed to make heavy guns, including those sold to foreign powers. Other prominent former Admiralty employees associated with contractors were Admiral Sir Digby Morant, a director of Fairfield, J. W. Shepherd, a Dockyard-trained constructor, who was in turn managing director of Robert Napier and Sons, the Fairfield Company, and finally the London and Glasgow Company, Rear-Admiral Sir Colin R. Keppel, a director of Hawthorn-Leslie, Engineering-Captain W. Onyon, of Beardmore, and Rear-Admiral Bacon and Commander Crease, both of the Coventry Ordnance Works. In 1914 the following numbers of directors and important shareholders were connected with the government, the House of Commons, the Civil Service, and the Service Ministries: Armstrong's, 10; Vickers-Beardmore, 11; Cammell-Laird-Brown-Firth-Fairfield-Coventry Group, 18.[48]

Subsidies

Even in the years between 1860 and 1914, when British shipping and shipbuilding enjoyed a unique position of world leadership, when the theory of laissez-faire was still on the ascendant and the shipping community was bitterly hostile to the actions of the Civil Service, there was never a time when the position of the British merchant marine was left entirely to competitive ability. British subsidies were perhaps not as openly granted as, for example, the French: government support was hidden in postal subsidies, in payments by colonial administrations, even in tonnage laws; but it was never in doubt.

In the early postal subsidies, which began with the contract of the P. and O. in 1830, the British ocean steamship companies were supported to aid the navy, as well as to beat the lines' foreign competitors in a field in which neither the success of the regular packet service nor the superiority of the British had been proved. Not until 1853 was the clause in the subsidy contracts that specified that ships must be able to carry the heaviest guns repealed on the advice of a departmental committee, and until 1861 it was the Admiralty that paid the subsidies.[49] As in the Navigation Laws, commercial and imperial motives were mixed.

After American competition in the Atlantic crossing had finally collapsed during the Civil War, the lucrative subsidies granted to the transatlantic steamship companies lost most of their justification and formed the subject of much envy and reproach. The contracts, accordingly, were nominally thrown open to public bidding in 1868, and were distributed by genuine competitive tender in 1874, though the British companies, the Inman, Cunard, and White Star lines, were still paid at double rates. When the West Indian mail contracts were also thrown open in 1874, and the East Indian and Australian contracts in 1879, they were accepted by their former holders at much lower rates. These were the years of undisputed British supremacy in steam shipping, and subsidies reached their lowest point; but the interval did not last long.

The naval scare of 1885 ushered in the first new series of naval subventions. The idea of converting ocean liners to fast auxiliary cruisers for the navy in times of war was a natural one in an era of rapid improvements in marine engineering. It was broached to the Admiralty in 1877 by Lord Brassey and by the leading shipowners Donald Currie, Lord Inverclyde, John Burns, and David McIver, and received some notice in a paper by Barnaby to the Institution of Naval Architects in 1878. The Admiralty confidently turned it down at the time, in view of the adequate reserve of merchant shipping under the British flag that might be hired, but the scare brought rapid disillusion: for the hire of 16 vessels as cruisers for an average of six months, from the most respectable companies, the Admiralty had to pay no less than £476,000, while the transports needed for the Egyptian campaign of 1883-1885 cost the Admiralty nearly £1.5 million.[50]

To bring down the cost, contracts were entered into with the Cunard and White Star companies in 1887, and with other companies later. The companies undertook to allow the Admiralty to hire a number of their steamers as transports at predetermined rates in case of need. In addition, they agreed to build a number of very fast liners as cruisers to Admiralty specifications, in particular with regard to the proper provision of bulkheads and gun mountings, and having their engines and steering apparatus well below the water line. A proportion of their crews and officers was to belong to the Royal Naval Reserve, and there were safeguards to prevent the sale of the vessels to foreign flags, in return for an annual retention fee.[51] The vessels in question were among the fastest in the British merchant marine; the terms on which they were subsidized and on which they were to be hired in case of need were extremely generous.

With the turn of the century the attitude toward subsidies changed. Hitherto they had been given rather shamefacedly, and few outside the "strong men" in the Admiralty and perhaps their immediate beneficiaries would have cared to defend them in public. About the year 1901 a number of developments combined to alter the position. The Boer War had conjured up a picture of an isolated Britain, depending upon the navy for the defense of the empire against a number of enemies; the sudden increase in shipbuilding output in Germany and America on the basis of more up-to-date equipment and backed by adequate industrial resources seemed to imperil the position of the British merchant marine; the naval programs of Germany and Russia grew almost as rapidly as the British; the German supremacy in the top class of fast "greyhound" liners became unmistakable; and, above all, the projected International Mercantile Marine seemed to transfer, at one blow, the majority of the British transatlantic liners, including some subsidized ones, to American ownership and control.

From the moment the Cunard Company had decided to remain outside this American-controlled combine, Lord Inverclyde, supported by all shades of chauvinist propaganda, set about to convince the British public that the British maritime supremacy was threatened, that increased subsidies were called for, and that the Cunard Company, the only loyal transatlantic line, was the proper agency to receive them.

In May 1901 a Select Committee on Steamship Subsidies was appointed, under Evelyn Cecil. After a preliminary report in August it was reappointed in 1902 and issued its final report in December 1902. The evidence was by no means all in favor of larger subsidies or increased payments for auxiliary cruisers. Renwick, an M.P., and Admiral Colomb objected to the advantage they gave to the large, monopolistic lines; Lord Pirrie, head of Harland and Wolff, pointed out that during the Boer War shipowners had been falling over themselves in their eagerness to hire their vessels out to the government, so that a retention fee was wasteful. Lords Brassey and Beresford, and the shipowners, on the other hand, urged increased subsidies. In the final report, the Select Committee admitted that the progress of some countries had been materially assisted by their subsidies, yet noted that Germany had forged ahead very fast with hardly any subventions and that Britain had not lost ground. The committee preferred competition to subsidies, which led to the formation of shipping rings, and was in favor of granting them, if at all, for specific purposes only (such as the preservation of British imperial interests in East Africa) and on condition of a guarantee from the recipients as to the speeds of the vessels and the continued British character of their boards, officers, and crews. Subsidies for auxiliary cruisers should be given to a very small number of the fastest vessels only.[52]

The formation of the International Mercantile Marine also caused the appointment of an Admiralty committee in April 1902 under Lord Camperdown to inquire into the means of securing faster and safer mercantile ships, and to prevent their transfer to foreign flags. The committee recognized that vessels became uneconomical at high speeds, and suggested a number of different ways of subsidizing their construction: (*a*) the contribution of a lump sum to their cost; (*b*) a loan for first cost at only 3 percent; (*c*) annual payments over a minimum of 10 years.[53] To prevent their subsequent transfer to foreign owners, the committee proposed that the Admiralty should own 33/64 of the vessels, leaving the management and all the profits to the owners.

These reports and the agitation that accompanied them were quickly followed by an agreement between the Admiralty and the Cunard Company. The Admiralty undertook to provide up to £2.6 million at 2.75 percent for the building of two new "greyhounds" of at least 24-25 knots speed, and to pay the company £150,000 annually as an auxiliary cruiser retention fee (compared with the £15,000 paid hitherto) in addition to the mail subsidies. In return, the Admiralty was granted the right to determine the construction of the two steamers with a view to war service, and to amend the design of all other new ships of the company of over 17 knots. The company also agreed to employ a minimum proportion of British sailors and officers in all vessels, and to hand over a sufficient number of shares to a government nominee to prevent the transfer of ownership abroad.[54]

Opinion was virtually unanimous on the good bargain driven by Lord Inverclyde.[55] Even more impressive was the extent to which the British government had been committed in support of a private firm. In the sequel, American control of the International Mercantile Marine was loosened with the collapse of many other of Pierpont Morgan's schemes. The Admiralty was able to extend its hold over those British lines which had originally joined the combine, by a new agreement in 1903, providing for subsidies in return for the usual safeguards of continued British ownership and crews.

British spokesmen were generally reticent about the subsidies paid by their own country. These were frequently described as mere payment for postal services or as "subventions," not subsidies; they were "assumed not to be bounties in excess of the actual service performed . . . still, notwithstanding this assumption, such has been their practical effect."[56] It is true that Britain had no limitation on her coastal trade, though her colonies had, and she paid no direct construction or mileage bounties, as did France, Italy, Japan, and many of the minor maritime nations, nor did her government have quite as close relations with her mercantile marine as the Russian. But neither the United States, whose subsidies were limited to postal payments (many of them to British lines), nor Germany, which paid bounties only to a few colonial lines, paid more than Britain. It must be admitted, on the other hand, that the frequent foreign allegations of a preference shown in various ways to home vessels in British ports were greatly exaggerated.

In the last resort, the British mail lines, like their foreign competitors, were dependent on subsidies for their existence.[57] In 1901, before the Cunard contract, subsidies compared as follows: United Kingdom, £1,003,000; colonies, £189,000; total empire, £1,192,000; France, £1,787,000; Russia, £375,000; Germany, £440,000; United States, £346,000 (the figures for Germany and the United States include minor amounts paid to foreign vessels).

The State and the Armaments Industry

The state also took an active interest in other sectors that affected national defense. A study of the cordite industry reveals that the government maintained a "special relationship" with the largest producers of cordite in which the government first helped to create private sources of supply (as well as beginning production itself), "nursed" the infant firms to maturity, and then, through a selective purchasing policy, promoted co-operation among the leading producers in order to ensure the availability of sufficient supplies at short notice in case of emergency demands.[58] Government involvement in the petroleum industry is even more striking, for in this instance the state purchased a share in an industry that was primarily involved in supplying civilian consumers through normal marketing channels. On the eve of the First World War, the government bought a 51 percent share of the Anglo-Persian Oil Company—the forerunner of British Petroleum. The owners of the company initially proposed that the government take a majority holding because they wanted to obtain help in securing access to rich new oil fields in Iraq. They argued that without government sponsorship they might lose control of the fields to the "foreign" Royal Dutch Shell Company, thus making it possible for supplies to fall into enemy hands in wartime. The Foreign Office accepted these arguments, but the Admiralty, the principal government user of petroleum, was unconvinced, as it was obvious that, regardless of ownership, no source so far removed from effective political control could ever be regarded as safe. At length, however, Admiralty officials were brought around by the possibility of gaining good oil at a very cheap price, thus beginning an involvement that has continued to the present day.[59]

Despite irregularities, the government never completely abdicated its responsibilities in regulating the industries with which it became involved. In order to be well prepared for conflicts, the War Office and the Admiralty wanted to guarantee the rapid delivery of ample stocks of armaments and munitions, and manipulated their purchases in such a way as to encourage private firms to maintain their ability to produce arms despite frequent slack periods; purchase price was not always the primary consideration. In Trebilcock's words, "the War Office planned to create capacity [for cordite production], not necessarily to use what it created."[60] And if the cordite producers were guilty of collusion in their dealings with the government, the War Office was not loath to use its position as a monopsonist to fight back.

It could at least be argued that the government and the nation profited in other ways from their connection with the armaments and munitions firms. As anticipated, the Admiralty was able to obtain petroleum supplies from the Anglo-Persian Oil Company at far below market prices. Even the manufacture of armor had beneficial, if largely unintentional, side effects. The technology developed for the production of warships and other armaments was profitably employed in producing, among other things, merchant vessels and machine tools.[61]

It is unreasonable, of course, to expect that laissez-faire would have prevailed in the armaments and munitions industries, because the market conditions for perfect competition never existed. Not only was the state traditionally an active, and on occasion a monopolistic, producer itself, but it was always the dominant buyer that could, if it saw fit, regulate the market to suit its own purposes. Moreover, the production of many types of arms required a large capital investment in areas in which the volume of demand was strictly limited. Under the circumstances officials may well have thought it necessary to provide special enticements to induce private investments when it was clear that demand might fluctuate tremendously, depending on the vicissitudes of the world situation.

At the same time, this created opportunities for abuse by manufacturers or others if government officials were corrupt, deluded, or unvigilant. Large steamship subsidies in the prewar period are a good example of the extent to which officialdom could be deceived into a belief that private profit coincided with national security. What gave armaments and munitions producers the ability to gain subsidies or super-profits—and to use lobbying and publicity to influence government policy in ways that were closed to other manufacturers—was the powerful emotive quality of the defense movement. If an external threat seemed strong enough, no government could resist for long a clamor for new and better arms. Even the most parsimonious Liberal ministries were faced with demands for greater defense spending. As "security was valued above cheese-paring,"[62] expenditure was not begrudged the services if the need seemed at all justified. This in turn created ample openings for abuse by oligopolistic producers, who saw it in their interests both to allege Britain's weakness and her enemies' strength and to take advantage of an atmosphere of fear to gain greater profits. Again in Trebilcock's words, "the armaments industry was closely connected with the 'national interest,' and, since this was an even more potent concept than *laissez-faire*, modifications in the weaker idea are to be looked for."[63]

A comparison of British, German, and American governmental policies toward the shipping and shipbuilding industries between 1870 and 1914 provides a neat illustration of the varying effectiveness of state intervention over the period. In general, although at certain times naval contracts were sizable and the government did sponsor several important innovations, the effects of British policy on the shipbuilding industry as a whole were slight. A few shipbuilders certainly benefited from naval contracts, and a few major shipping lines were helped by subsidies. It is possible that this led to some misallocation of resources, but the overall costs to the British consumer and taxpayer were not great in comparison to the total outlay on shipbuilding and shipping. Most shipbuilders did not concentrate on naval or giant passenger vessels, and subsidized mail liners made up but a small proportion of the British merchant fleet. The majority of builders and shipowners depended for their prosperity on the construction and operation of merchant cargo vessels or small passenger ships. In these areas, in which Britain was clearly competitive internationally, the state interfered very little with the free market.

State intervention in Germany was both greater and more beneficial than in Britain. Shipping subsidies were less significant, but through legislation the German government was able to promote the rapid maturation of an infant shipbuilding industry. By pressing shipping lines to buy domestically built vessels and by further aiding the shipyards with orders for naval vessels, the government encouraged the development of a shipbuilding industry that in time became internationally competitive.[64]

State intervention was far less efficacious in the United States, perhaps because the infant was stillborn. American policy was a maze of contradictions that was ultimately more harmful than helpful. The shipbuilding industry was protected by legislation requiring that (with few and decreasing exceptions) American-registered vessels be domestically produced. Duties on foreign iron and steel and the high cost of American shipyard labor, however, raised the price of finished ships to uncompetitive levels. Foreign-owned and foreign-built vessels were excluded from coastal shipping, but the American oceangoing fleet was totally unprotected. The net effect of government regulation was to stifle the shipping industry while not offering any substantial advantages to shipbuilders whose largest potential market had been suffocated. As coastal shippers were hampered by their high capital costs, much of the trade fell to the competing railways, which were able to charge lower rates. Overseas trade fell almost entirely into the hands of foreigners. American naval building programs, which periodically boosted the shipbuilding industry, had further adverse effects on merchant shipping and shipbuilding, for they led to even higher labor costs in the yards, which made merchant vessels that much less profitable to construct and own. Large-scale subsidies, to help compensate shipowners for the high costs of their fleets, were proposed but never enacted. None of this would have been of avail since, if Hutchins's evaluation is accepted, unlike the situation in Germany, differentials in wage and material costs were so great between the United States and her competitors that no amount of government encouragement could have led to the establishment of a self-sufficient American shipbuilding industry before 1914.[65]

To sum up, although many aspects of British shipbuilding were unregulated, the state did to some extent help to ensure the continuing viability of the industry. The government was able to compensate in part for the reluctance of private builders to invest in research and education. From the 1890s naval building was an important specialty of some of the largest British yards, most of which were adjuncts of armaments and armor manufacturers. These firms were able to benefit from high profits and reduced fluctuations in output, but they also had to conform to the higher standards of quality of construction that the Admiralty required. Many of these same firms built large passenger vessels and benefited as well from government subsidies to mail lines. Taken together, the measures of the government represented an important exception to the principles of laissez-faire that British shipbuilders of the period so enthusiastically espoused. This was to expected, however, in a situation in which the government was the largest, and in some cases the only, customer.

11. Conclusion

WE HAVE studied a variety of aspects of the British shipbuilding industry between 1870 and 1914 to determine the forces that were acting on the industry over the period and to provide a basis for evaluating the effectiveness of shipbuilding entrepreneurs in dealing with rapid change.

One of the most important influences on the industry was the changing nature of the end product, ships. Although most of the trends described here had begun in earlier decades, it was in the years 1870-1914 that the industry came to be dominated by the construction of metal-hulled steam vessels. The techniques that had evolved in the era of the wooden sailing ship were of only limited applicability to the design and construction of the new types of vessels. Moreover, the improving technologies of metallurgy and steam power made it possible, and in many cases desirable, to build ever-larger and more powerful ships. A wider selection of shipboard appliances became available and ships were increasingly built to serve specialized functions.

A second important influence was the highly cyclical nature of the demand for ships. Although the average tonnage produced did increase from decade to decade, annual output fluctuated greatly as each period of rapid improvement was followed by one of almost equally sharp decline. At times, a decade or more passed before a previous record in output was surpassed.

Thirdly, shipbuilding differed from other branches of heavy industry in that it remained primarily a construction activity. Most ships were built to designs specified individually by their purchasers, and the scope for mass production was limited. Handicraft skills were of great significance and mechanization proceeded slowly.

In combination these factors were vital in determining the nature of the development of the shipbuilding industry. The growing size and complexity of vessels forced yard owners to increase the size and capitalization of their establishments. It was simply impossible to build modern vessels in the cramped and poorly equipped yards suitable for wooden shipbuilding. On the other hand, the severe fluctuations in output made it expedient for builders to minimize their investments, as their fixed capital would be underutilized for extended periods. The third factor, the necessary reliance of shipbuilding as a constructive industry on handicraft skills, provided a solution to the dilemma. British shipbuilders, who were able to rely on a large force of highly skilled men, were in a position to economize on capital investment to some extent. This helped to strengthen the international competitiveness of the British in comparison to builders in other countries, who were forced to mechanize heavily because they could not draw on a properly trained and reasonably priced labor force.

This did not mean that British shipbuilders rejected new types of equipment out of hand. New machines were tested carefully, if cautiously, before decisions were reached as to whether they should be adopted or rejected. Although the rate of adoption may have been somewhat slower than in Germany or the United States, new equipment was added steadily, and by 1914 the shipbuilding industry in Britain was highly mechanized.

Nevertheless, the reliance upon skilled labor and the evolution of the industry from a traditional craft set the patterns of actually constructing ships. The workforce was differentiated along craft lines, and as new techniques developed additional trades found employment in the yards. Most of these, especially among the metal workers, brought skills that were already utilized in other industries. While the tendency to replace skilled workers by semiskilled machine operatives grew after the turn of the century, most of the new equipment that was adopted was intended to be used within the established context of the skilled trades, for example, to improve the efficiency of traditionally designated platers or shipwrights.

The continued organization of the workforce along craft lines was the most efficient way of building ships in the short run, but it did have disadvantages. For one thing, it led to a multitude of demarcation disputes among the various crafts over different categories of work. Increasing mechanization made matters worse because new types of equipment often provided new areas of conflict by blurring the distinctions between trades. From the standpoint of the workers, attempts to defend and extend a trade's domain were sensible in an industry in which unemployment was commonplace, but demarcation disputes definitely complicated the tasks of the yard owners in organizing their operations.

Secondly, the emphasis upon skilled crafts engendered attitudes among the employers that probably contributed to the decline of British shipbuilding in the interwar years. Because a high degree of manual dexterity was deemed essential, yard owners were encouraged to lay stress on apprenticeship programs and to skimp on the provision of technical and scientific education. A belief in the virtues of the "practical man" who could work with his hands seemed to render superfluous or even dangerous scientific research or advice tendered by men whose training had been theoretical. These deficiencies were only partly compensated for by government-sponsored education and research.

In general, if the effects on the subsequent interwar period are discounted, British shipbuilders had an exemplary record in matters of labor force organization and mechanization. The rate of increase of labor productivity was high and probably exceeded that for British industry as a whole between 1900 and 1910. Before 1900 this increasing productivity was reflected in higher real earnings among the workforce, but after that date, as foreign competition increased, it allowed for reductions in the price of finished vessels.

British shipbuilders tried other means of conquering the challenges imposed by severe fluctuations in demand. Many firms saved on capital by concentrating on the production of a limited number of types of vessels. Equally important was the attempt to even out fluctuations by establishing strong relationships with customers who might then be induced to place orders in periods of low demand. This was done most successfully by the handful of builders who built giant passenger liners. Unlike the majority of owners of cargo ships, the large mail lines had ample cash reserves and could afford to order in slack periods. Such relationships were mutually beneficial; the lines were able to purchase ships more cheaply, and yard owners could make more continuous use of the expensive capital equipment necessary to build large vessels. From the 1890s, some yards also came to rely increasingly on orders for naval vessels. These offered high profits and, while the builders could not control the rhythm of demand, the orders were frequently placed at times when the demand for merchant vessels was low.

Shipbuilders also established strong relations with suppliers. As ships became more complex builders had to rely increasingly on outside firms to provide the myriad necessary components. Competition among suppliers helped to reduce the price of components and to speed up the rate of technological change.

Finally, there was a growing degree of vertical and horizontal integration in the industry. Integration with marine engine builders was especially common, and many of the yards that concentrated on the construction of warships were controlled by armaments and armor manufacturers. Competition among builders remained high for certain classes of work, particularly in the construction of cargo ships, but over the period the warship and armaments and armor producers became closely connected through such devices as interlocking directorates and cartels.

When judged by the criteria we set out in our Introduction, British shipbuilders come off very well. Their efforts to deal with matters of marketing and supply, mechanization, and labor force organization were generally successful. The only areas in which they may be severely faulted are research and education. As it took several decades for the adverse effects of their neglect of the latter to be fully felt, Britain was able to maintain her international competitive superiority until 1914.

The broader implications of our findings for any general study of the quality of British entrepreneurship may be less encouraging. It could be argued that British shipbuilders were successful precisely because their behavior conformed in many respects to the pattern so roundly condemned by many critics of the economy during this period. Not only were they reluctant to support basic research and education; their labor relations policies, a mixture of harshness and paternalism, left a record that was at best uninspiring; most shipyard layouts were outdated and inefficient; and although limited companies were becoming more common, in 1914 the industry was still dominated by family firms. Even some aspects of the shipbuilders' behavior that have been shown to have been most astute, such as the organization of the productive process along craft lines left over from the preindustrial era and the caution with which machinery was adopted, could be interpreted as being more backward-looking than progressive.

Coleman's distinction between "players" and "gentlemen" cannot be easily applied to the shipbuilding industry because it contained almost no gentlemen in his sense of the word. Second- and third-generation shipbuilders were generally educated in the same way as their contemporaries who entered the industry from other backgrounds. Few were exposed to an educational bias against industry or the acquired sense of guilt about making profits that Coleman finds in industrialists who attended public schools. Most shipbuilders' sons who were going into the business left school early and served rigorous apprenticeships in the yards, and in the process they were made constantly aware of the importance of practice as opposed to or in addition to theory. Thus they acquired many of the cautious, empirical attitudes Coleman attributes to "players."

There can be no doubt that caution was one of the more important characteristics of British shipbuilders. The "steady state hypothesis" of behavior that is outlined in the Introduction was one manifestation of this. As they were not sanguine about their industry's growth potential and as they clearly recognized the perils that resulted from the extreme fluctuations in output, British shipbuilders took care to avoid overinvestment. In practice this was one of their strengths and bolstered the position of British shipbuilding relative to that of some other countries where builders were less realistic and overcapitalized their concerns. If this extreme caution was a general characteristic of British entrepreneurs, however, it could have had an adverse effect on the overall rate of economic growth. In some industries, which were not as suited as shipbuilding to traditional methods of production or as beset by cyclical demand, the British may have refused to modernize or to enter new lines of activity because they did not perceive that the opportunities existed. In such cases, the cost of taking a "safe" course of action could have been great indeed.

It would be wrong, however, to close on a negative note. Before passing judgment we should recall that this is the history of an industry that was highly successful in world competition during our period, whatever problems may have appeared later. Most of the "dubious" practices listed above can be easily justified in the case of shipbuilding, where the quality of entrepreneurship before 1914 was quite good. As several recent studies have demonstrated, this may have been true as well of other important sectors of the British economy.[1]

APPENDIXES

NOTES

ABBREVIATIONS

SELECTED BIBLIOGRAPHY

INDEX

Appendix A: A Note on Tonnage Measurements

From the middle of the nineteenth century, four principal measures of a vessel's size have been used in Britain. Gross tonnage refers to the total volume of a vessel, one ton being equivalent to 100 cubic feet. Net tonnage is calculated by subtracting from gross tonnage the volume of areas, such as the quarters of the crew, engine rooms, and bunkers, which cannot be used to carry cargo or passengers. Net tonnage is generally used in computing harbor tolls and other charges.

Deadweight tonnage is the weight of fuel, stores, cargo, and passengers carried when a ship is fully laden. It varies depending on the density of the cargo. Displacement tonnage, however, is a constant, the weight of the water displaced when a ship is fueled up but empty of cargo and passengers. Deadweight tonnage is usually applied only to cargo vessels, and the size of warships is expressed in displacement tons. The size of all merchant cargo and passenger vessels on the official register is given in net tons, but many shipbuilding statistics are published in gross tons.

Unfortunately, there is no good way of converting from one set of measurements to another. The *Queen Mary* measured 81,235 gross tons or 77,500 displacement tons. A cargo ship of 8,000 deadweight tons might have a displacement of 11,500 tons and measure 5,200 tons gross or 3,200 tons net.[1] The problem of comparison is exacerbated by anomalies in methods of calculating and collecting tonnage statistics that can lead to inconsistencies within the same series.[2] From 1881, for instance, erections above deck that could not be "efficiently closed" were not included in gross or net tonnage calculations, leading builders to leave open hatchways (which in reality could be closed if necessary) in permanent parts of the superstructure in order to reduce taxes and harbor and canal charges.

Conversions from gross to net tonnages also vary according to the arrangement of the engine rooms. For screw steamers, if between 13 and 20 percent of the gross tonnage is occupied by machinery, 32 percent is deducted from gross to obtain net tonnage. If less than 13 or more than 20 percent is taken up by machinery, the deduction is 1.75 times the volume occupied. Therefore, for a vessel of 3,000 g.t. with an engine room capacity of 389 tons, 680.75 tons can be deducted. If the engine room capacity were 390.10 tons, however, the deduction would be 960 tons.[3] Furthermore, the proportion deducted in computing net tonnages varies according to the type of vessel. Liners, tugs, and other ships with large engine rooms in proportion to their total volume have larger percentages deducted than do cargo vessels.

The method of collecting statistics also contributed to inconsistencies. Newspapers gathered data on annual launchings from returns sent in by shipbuilders. Some builders included open erections that were not included in calculating gross register tonnage, while others did not. As a result, the returns of these builders were inflated both in comparison to those of their competitors and to the figures that eventually appeared on the official register.[4]

Finally, tonnage figures did not accurately reflect the value of a ship. A vessel with extra fittings cost more to build than one of equal tonnage but without trimmings. Similarly, passenger liners and warships, which required more work per ton, appeared in the statistics on a par with cargo steamers.

Appendix B: Statistical Tables

Table B.1. Money earnings of all workers and of joiners and joiners machinemen, Leven Shipyard, 1852-1914 (£ per week).

Year	Earnings of all workers	Earnings of joiners and joiners machinemen
1852	0.66	0.98
1853	0.71	0.94
1854	0.75	0.97
1855	0.79	1.01
1856	0.76	1.26
1857	0.76	1.04
1858	0.72	1.03
1859	0.72	0.83
1860	0.76	0.95
1861	0.77	1.01
1862	0.76	1.07
1863	0.78	1.13
1864	0.83	1.22
1865	0.85	1.22
1866	0.87	1.27
1867	0.84	1.17
1868	0.82	1.07
1869	0.86	1.16
1870	1.18	1.17
1871	1.16	1.29
1872	1.20	1.27
1873	1.12	1.28
1874	1.13	1.41

(Cont.)

Table B.1. (cont.)

Year	Earnings of all workers	Earnings of joiners and joiners machinemen
1875	1.11	1.12
1876	1.12	1.20
1877	0.99	1.90
1878	1.19	1.55
1879	1.14	1.52
1880	1.14	1.62
1881	1.19	1.77
1882	1.31	1.68
1883	1.34	—
1884	1.26	1.87
1885	1.13	1.46
1886	1.16	1.49
1887	1.11	1.35
1888	1.24	1.59
1889	1.27	1.58
1890	1.39	1.84
1891	1.37	1.81
1892	1.34	1.66
1893	1.37	1.81
1894	1.37	1.95
1895	1.36	1.80
1896	1.37	1.78
1897	1.35	1.79
1898	1.42	1.89
1899	1.58	2.03
1900	1.55	2.05
1901	1.59	2.03
1902	1.64	2.23
1903	1.46	—
1904	1.45	1.92
1905	1.42	1.91
1906	1.49	—
1907	1.59	2.00
1908	1.51	—
1909	1.51	—
1910	1.52	—
1911	1.75	—
1912	1.79	—
1913	1.94	—
1914	1.99	—

Source: Unpublished book of "wages notes" of William Denny and Brothers.

Table B.2. Real earnings of all workers, Leven Shipyard, 1852-1913 (£ per week).

Year	Real earnings	Year	Real earnings
1852	0.70	1883	1.33
1853	0.63	1884	1.33
1854	0.60	1885	1.28
1855	0.63	1886	1.40
1856	0.61	1887	1.37
1857	0.60	1888	1.48
1858	0.64	1889	1.51
1859	0.63	1890	1.60
1860	0.63	1891	1.59
1861	0.67	1892	1.63
1862	0.63	1893	1.67
1863	0.64	1894	1.85
1864	0.70	1895	1.89
1865	0.73	1896	1.88
1866	0.73	1897	1.82
1867	0.71	1898	1.82
1868	0.71	1899	1.88
1869	0.80	1900	1.70
1870	1.07	1901	1.85
1871	1.01	1902	1.91
1872	0.94	1903	1.70
1873	0.88	1904	1.75
1874	0.94	1905	1.65
1875	0.95	1906	1.60
1876	0.97	1907	1.64
1877	0.90	1908	1.74
1878	1.18	1909	1.66
1879	1.16	1910	1.57
1880	1.12	1911	1.72
1881	1.20	1912	1.72
1882	1.30	1913	1.83

Source: Table B.1 deflated by the Rousseaux overall price index given in B. R. Mitchell and Phyllis Deane, *Abstract of British Historical Statistics* (Cambridge: Cambridge University Press, 1962), pp. 471-473. The average of the prices of 1865 and 1885 is taken to equal 100.

Table B.3. Unemployment rate of members of engineering, metal, and shipbuilding unions in the United Kingdom, 1851-1913 (percent).

Year	Rate	Year	Rate	Year	Rate
1851	3.9	1872	0.9	1893	11.4
1852	6.0	1873	1.4	1894	11.2
1853	1.7	1874	2.3	1895	8.2
1854	2.9	1875	3.5	1896	4.2
1855	5.4	1876	5.2	1897	4.8
1856	4.9	1877	6.3	1898	4.0
1857	6.1	1878	9.0	1899	2.4
1858	12.2	1879	15.3	1900	2.6
1859	3.9	1880	6.7	1901	3.8
1860	1.9	1881	3.8	1902	5.5
1861	5.5	1882	2.3	1903	6.6
1862	9.0	1883	2.7	1904	8.4
1863	6.7	1884	10.8	1905	6.6
1864	3.0	1885	12.9	1906	4.1
1865	2.4	1886	13.5	1907	4.9
1866	3.9	1887	10.4	1908	12.5
1867	9.1	1888	6.0	1909	13.0
1868	10.0	1889	2.3	1910	6.8
1869	8.9	1890	2.2	1911	3.4
1870	4.4	1891	4.1	1912	3.6
1871	1.3	1892	7.7	1913	2.2

Source: B. R. Mitchell and Phyllis Deane, *Abstract of British Historical Statistics* (Cambridge: Cambridge University Press, 1962), pp. 64-65.

Table B.4a. Wage rates of blacksmiths, carpenters, and joiners machinemen, 1880-1912.

Date	Blacksmiths	Joiners machinemen	Carpenters
Dec. 1880	28/2	33/2	29/3
Jan. 1881	31/6	36/6	31/6
Jan. 1882	31/6	36/7	31/6
Mar. 1882	—	—	32/9
Apr. 1882	32/7	—	—
Mar. 1883	—	—	34/10½
May 1883	—	37/8	—
June 1883	—	—	36/-
Feb. 1884	31/6	36/7	33/9
July 1884	29/3	34/4	31/6
Jan. 1885	29/3	34/4	31/6
Mar. 1885	28/2	33/9	30/4½
Oct. 1888	29/3	34/11	32/1
Apr. 1889	30/5	36/7	33/9
Sept. 1889	—	—	36/-
Feb. 1890	32/8	—	—
Nov. 1892	31/6	35/5	34/10½
June 1893	—	36/-	—
Apr. 1896	32/7½	37/2	—
Aug. 1896	—	38/3	36/-
Sept. 1896	33/9	—	—
Apr. 1897	34/10½	39/5	37/1½
Sept. 1898	36/-	41/8	38/3
Mar. 1903	34/10½	36/7	37/1½
Jan. 1908	33/9	36/2	—
Feb. 1908	—	—	36/-
May 1910	33/9	35/1	36/-

Source: Unpublished book of "wages notes" of William Denny and Brothers.
Note: All rates are for a standard week's work.

Table B.4b. Wage rates of engineers, blacksmiths hammermen, and laborers, 1880-1912.

Date	Engineers	Blacksmiths hammermen	Laborers
Dec. 1880	29/3	17.5	15/9
Jan. 1881	31/6	—	16/10½
Jan. 1882	31/6	—	16/11
May 1882	33/9	—	—
July 1883	34/10½	—	—
Feb. 1884	—	18/-	—
July 1884	—	17/5	16/4
Jan. 1885	34/10½	17/5	16/4
Mar. 1885	33/9	15/9	15/9
Oct. 1888	33/9	17/5	16/4
Apr. 1889	—	18/7	16/11
Aug. 1889	34/8	—	—
Jan. 1890	34/11	—	—
Nov. 1892	33/9	18/-	16/3
Feb. 1896	34/10½	—	16/10½
Apr. 1896	—	18/7	—
June 1896	36/-	—	—
Sept. 1896	—	19/1½	—
Dec. 1896	37/1½	—	17/5½
Apr. 1897	—	19/8	—
May 1897	—	20/3	—
Sept. 1898	36/- & 38/3	21/11	—
Oct. 1898	—	—	18/-
Mar. 1903	—	23/7½	18/-
June 1903	36/-	—	—
Dec. 1905	36/6 & 38/3	—	
Jan. 1908	—	23/7½	18/-
Jan. 1909	35/5	—	—
May 1910	36/-	21/9½	20/-
Dec. 1911	37/1½	—	20/3
Nov. 1912	38/3	—	21/4½

Source: Unpublished book of "wages notes" of William Denny and Brothers.
Note: All rates are for a standard week's work.

Table B.5a. Wages in shipbuilding and engineering, 1866-1914: overall indices (1900=100).

Year	Earnings	Wage rates, eight maritime districts	Year	Earnings	Wage rates, eight maritime districts
1866	79	74	1891	93	94
1867	77	70	1892	92	91
1868	77	70	1893	91	89
1869	77	69	1894	91	89
1870	78	72	1895	91	89
1871	79	75	1896	94	94
1872	83	80	1897	96	96
1873	86	82	1898	98	100
1874	87	84	1899	100	100
1875	87	83	1900	100	100
1876	87	82	1901	100	100
1877	88	82	1902	100	99
1878	87	82	1903	99	98
1879	83	79	1904	99	98
1880	84	80	1905	99	95
1881	86	83	1906	—	96
1882	88	86	1907	—	96
1883	88	89	1908	—	94
1884	87	84	1909	—	94
1885	86	83	1910	—	95
1886	84	79	1911	—	100
1887	85	80	1912	—	103
1888	88	86	1913	—	104
1889	91	91	1914	—	104
1890	93	94			

Sources: Earnings index: Bowley and Wood, *Journal of the Royal Statistical Society*, 1906, p. 185. Wage rates index, 1866-1904: based on table II, ibid. The wage rates index is an average of the eight districts for which data are given in table B.5b. Wage rates index, 1905-1914: based on the wage changes given in the Board of Trade, *Labour Gazette* (monthly), for shipbuilding only, and on other sources. Indices have been fitted on the earlier series by overlapping for the years 1902-1904, and should therefore be treated with caution. For a full explanation of the Bowley and Wood indices, see ibid., pp. 154ff.

Table B.5b. Wages in shipbuilding and engineering, 1866-1914: indices for various districts (weighted average for all districts = 100).

Year	Barrow	Dundee	Hull	Mersey	Belfast	Clyde	Tyne	Wear
1866	78	70	90	96	74	79	90	89
1867	76	67	85	97	—	(75)	84	—
1868	76	67	85	95	—	(75)	83	—
1869	76	—	—	93	—	(75)	83	—
1870	79	—	—	—	—	77	85	—
1871	83	71	(92)	96	78	(85)	91	93
1872	89	—	(96)	101	83	92	98	98
1873	91	—	100	106	87	95	100	99
1874	93	82	100	106	86	95	105	104
1875	91	—	99	106	86	93	104	103
1876	92	—	99	106	86	92	102	103
1877	92	84	99	106	86	91	102	103
1878	89	—	98	106	85	90	101	105
1879	86	—	95	99	82	88	98	100
1880	84	82	96	—	—	(88)	100	94
1881	90	—	99	—	99	(90)	102	(98)
1882	95	—	100	—	95	(93)	106	(102)
1883	95	91	100	(108)	95	97	107	106
1884	89	—	96	103	90	93	102	102
1885	87	—	96	102	87	90	102	102
1886	85	81	96	93	85	86	99	97
1887	87	(85)	96	93	87	(86)	99	97
1888	92	(88)	99	104	94	(94)	104	102
1889	98	(92)	101	110	101	101	111	110
1890	101	95	102	115	101	104	113	113
1891	101	95	103	115	101	103	113	113
1892	97	93	102	112	99	100	111	108
1893	95	92	100	109	96	99	105	102
1894	95	92	100	109	96	99	105	102
1895	95	92	100	109	96	99	105	102
1896	100	95	104	113	102	106	111	108
1897	103	98	109	115	105	108	115	113
1898	107	100	111	119	109	113	119	117
1899	107	101	111	119	111	113	119	117
1900	107	101	112	119	111	113	119	117
1901	107	101	112	119	111	113	119	117
1902	107	101	112	119	111	113	117	114
1903	104	99	112	119	109	110	115	114
1904	104	99	112	119	109	110	115	113
1905	101	97	103	120	105	109	111	109
1906	101	97	106	120	105	109	114	113

(Cont.)

Table B.5b. (cont.)

Year	Barrow	Dundee	Hull	Mersey	Belfast	Clyde	Tyne	Wear
1907	101	97	106	120	105	109	114	113
1908	100	95	105	118	107	107	112	111
1909	100	95	105	118	107	107	112	111
1910	100	95	105	118	107	108	112	111
1911	107	102	112	121	116	114	118	117
1912	109	105	114	124	117	117	121	120
1913	110	106	115	125	118	119	123	122
1914	110	106	115	126	118	119	123	122

Sources: 1866-1904: based on table II in Bowley and Wood, *Journal of the Royal Statistical Society*, 1906, p. 185. The data have been calculated with the aid of other information in the papers by Bowley and Wood to a common base to facilitate comparisons of the wage levels in the various ports. 1905-1914: based on the wage changes given in the Board of Trade, *Labour Gazette* (monthly), for shipbuilding only, and on other sources. Indices have been fitted on the earlier series by overlapping for the years 1902-1904, and should therefore be treated with caution.

Note: Figures in parentheses denote incomplete data.

Table B.6. Total labor costs as a percentage of total hull costs, ships built by William Denny and Brothers, 1881-1913.

Year	%	Year	%
1881	27.23	1898	34.42
1882	29.55	1899	35.28
1883	29.43	1900	35.55
1884	30.23	1901	31.40
1885	29.40	1902	31.45
1886	29.88	1903	33.44
1887	29.69	1904	32.37
1888	30.10	1905	32.72
1889	30.85	1906	32.53
1890	30.73	1907	34.96
1891	31.31	1908	30.97
1892	34.26	1909	31.34
1893	34.16	1910	34.92
1894	36.63	1911	37.85
1895	33.50	1912	33.21
1896	34.51	1913	36.48
1897	34.85		

Source: Derived from data in William Denny and Brothers, Cost Books 1-10.

Table B.7. Mercantile shipbuilding output of the chief shipbuilding countries, 1892-1914 (in thousands of gross tons).

Year	U. K.	France	Germany	Holland	Italy	Japan	Norway	U. S. A.	World
1892	1110	17	65	14	14	—	25	63	1358
1893	836	20	60	1	11	1	17	27	1027
1894	1047	20	120	15	5	3	17	67	1324
1895	951	28	88	8	6	2	13	85	1218
1896	1160	45	103	12	7	8	12	184	1568
1897	925	49	140	20	13	7	17	87	1332
1898	1368	67	153	19	27	11	23	173	1893
1899	1417	90	212	34	49	7	28	224	2122
1900	1442	117	205	45	68	5	33	334	2304
1901	1525	178	218	30	61	37	37	433	2618
1902	1428	192	214	69	46	27	38	379	2503
1903	1191	93	184	59	50	35	42	382	2146
1904	1205	81	202	56	30	33	50	239	1988
1905	1623	73	255	44	62	32	53	303	2515
1906	1828	35	318	67	31	42	61	441	2920
1907	1608	62	275	69	45	66	58	475	2778
1908	930	83	208	59	27	60	53	305	1833
1909	991	42	129	59	31	52	29	210	1602
1910	1143	81	159	71	23	30	37	331	1958
1911	1804	125	256	93	17	44	35	172	2650
1912	1739	111	375	99	25	58	50	284	2902
1913	1932	176	465	104	50	65	51	276	3333
1914	1684	114	387	118	43	86	54	201	2853[a]

Source: Lloyd's Register, Annual Returns.
a. Data for several countries unobtainable.

Table B.8. Merchant tonnage built in the United Kingdom, 1860-1913 (net tons).

Year	Sailing	Steam	Total	% built for foreigners
1860	152,178	67,699	225,871	6.2
1861	130,202	78,124	208,326	3.6
1862	164,061	97,871	261,932	7.8
1863	254,120	124,187	378,307	4.6
1864	274,778	186,055	460,833	6.3
1865	236,504	211,665	448,169	7.4
1866	211,319	168,220	379,539	10.1
1867	179,501	126,478	305,979	12.1
1868	246,358	115,970	362,328	12.7
1869	242,562	145,530	388,092	8.7
1870	126,461	267,896	394,357	13.1
1871	60,260	330,798	391,058	9.4
1872	58,757	415,961	474,718	17.2
1873	89,626	363,917	453,543	18.2
1874	189,094	414,773	603,867	13.7
1875	245,357	226,701	472,058	10.9
1876	241,088	136,932	378,020	4.7
1877	214,514	236,405	450,919	3.8
1878	148,273	322,446	470,719	9.0
1879	61,579	344,412	405,991	12.1
1880	58,065	414,831	472,896	14.6
1881	93,768	515,110	608,878	17.7
1882	147,839	635,212	783,051	14.8
1883	148,090	744,126	892,216	13.9
1884	173,179	415,095	588,274	15.4
1885	219,094	221,918	441,012	8.1
1886	145,249	186,279	331,528	11.6
1887	87,127	290,071	377,198	18.9
1888	96,393	477,554	573,947	15.8
1889	137,147	717,582	854,729	21.4
1890	148,692	663,946	812,638	19.8
1891	229,610	579,883	809,493	17.2
1892	287,072	514,476	801,548	13.6
1893	123,874	460,800	584,674	15.3
1894	99,628	569,864	669,492	14.2
1895	64,717	582,917	647,634	19.8
1896	73,291	663,523	736,814	29.4
1897	89,700	554,997	664,697	25.2
1898	44,890	825,718	870,608	20.1
1899	50,483	898,527	949,010	21.0

Table B.8 (cont.)

Year	Sailing	Steam	Total	% built for foreigners
1900	46,122	898,145	944,267	22.0
1901	61,663	921,470	983,133	21.1
1902	81,064	869,361	950,425	15.8
1903	56,408	701,969	758,377	17.0
1904	51,728	832,531	884,259	16.8
1905	38,090	1,010,358	1,048,448	18.8
1906	49,083	1,107,688	1,156,771	20.3
1907	39,790	997,589	1,037,379	28.6
1908	37,392	564,309	601,701	31.4
1909	33,915	597,610	631,535	19.1
1910	28,250	670,457	698,707	14.0
1911	38,044	1,073,576	1,111,620	17.7
1912	52,816	1,053,313	1,106,129	18.6
1913	30,382	1,201,539	1,231,921	20.8

Sources: 1860-1908: "British and Foreign Trade and Industry," Cd. 4954 (1909), pp. 104-105; 1909-1913: *Statistical Abstract of the United Kingdom.*

Table B.9. Regional shipbuilding output in Britain, 1870-1913 (gross tons).

Year	(1) Clyde[a]	(2) Newcastle	(3) Sunderland	(4 North East Coast[b]	(5) United Kingdom[a]
1870	189,800	86,000	70,084	—	—
1871	196,200	77,628	81,903	—	—
1872	224,000	83,672	134,825	—	—
1873	251,846	87,913	99,371	—	—
1874	266,838	110,000	88,022	—	—
1875	221,854	98,000	79,904	—	—
1876	188,323	—	55,041	—	—
1877	169,710	87,968	87,587	—	—
1878	222,353	126,307	112,602	—	637,302
1879	174,750	139,843	87,432	—	469,024
1880	236,579	149,082	108,626	—	660,000
1881	341,022	177,165	130,862	—	1,000,000
1882	391,934	208,406	183,350	—	1,200,000
1883	419,664	216,573	212,313	—	1,250,000
1884	296,854	124,221	99,597	—	750,000
1885	193,458	106,447	61,761	—	540,000
1886	172,440	82,760	56,699	—	473,600
1887	185,362	104,296	84,254	—	578,600
1888	280,037	213,203	142,508	—	903,687
1889	335,201	281,710	217,383	—	1,300,933
1890	362,168	235,062	194,307	—	1,272,263
1891	325,477	185,367	188,715	—	1,268,470
1892	336,414	181,508	186,440	570,296	1,109,950
1893	280,160	144,261	118,317	431,405	836,383
1894	340,885	190,601	168,257	544,768	1,046,508
1895	360,152	161,476	125,266	497,564	950,967
1896	420,841	200,746	215,956	611,727	1,159,751
1897	340,037	169,585	174,496	498,594	952,486
1898	466,832	238,551	258,754	763,825	1,367,570
1899	491,074	249,038	242,611	766,282	1,416,791
1900	486,337	265,142	244,371	794,300	1,442,471
1901	511,990	292,989	268,069	872,723	1,524,739
1902	516,977	280,860	230,091	701,005	1,427,558
1903	446,869	219,360	189,500	581,343	1,190,618
1904	417,870	236,055	229,135	671,580	1,205,162
1905	539,850	310,391	305,169	872,314	1,623,168
1906	598,841	385,987	326,701	1,005,148	1,828,343
1907	619,919	292,814	291,606	817,510	1,607,890
1908	355,586	174,259	86,547	355,859	929,669

(Cont.)

Table B.9 (cont.)

Year	(1) Clyde[a]	(2) Newcastle	(3) Sunderland	(4) North East Coast[b]	(5) United Kingdom[a]
1909	403,187	182,235	132,371	434,810	991,066
1910	392,392	203,831	179,435	578,315	1,143,169
1911	630,583	412,959	186,828	977,278	1,803,844
1912	640,529	317,654	305,734	888,683	1,738,514
1913	756,976	366,331	299,964	974,109	1,921,153

Sources: Column 1 and Column 5, 1878-1891: annual shipbuilding reports of the *Glasgow Herald* as reproduced in W. S. Cormack, "An Economic History of Shipbuilding and Marine Engineering, with Special Reference to the West of Scotland" (Ph.D. diss., University of Glasgow, 1930), appendixes D1 and H1. Columns 2 and 3, 1870-1891: David Dougan, *The History of North East Shipbuilding* (London: George Allen and Unwin, 1968), pp. 230-231. Columns 2-5, 1892-1913: *Lloyd's Register of Shipping* as given in Dougan, pp. 221-223.

a. All figures for the Clyde and those for the United Kingdom from 1878 through 1891 include warships as well as merchant vessels. One displacement ton is taken to equal one gross ton.

b. Figures for the North East Coast include Hartlepool, Whitby, Middlesbrough, and Stockton as well as Newcastle and Sunderland.

Table B.10. Assets of the Clydebank Shipyard, 1884-1913, shipbuilding department only (in £).

Year	Machinery	Other	Property	Ground	Railway	Wetdock	Total
1884	37,181	—	39,663	20,740	2,737	15,763	116,084
1885	37,971	—	43,859	20,740	3,956	16,020	122,546
1886	38,026	—	44,476	20,765	4,072	16,084	123,423
1887	41,236	—	44,263	20,765	4,456	28,560	139,280
1888	43,302	—	44,533	20,765	4,514	30,524	143,638
1889	44,948	—	45,784	21,257	4,526	30,601	147,116
1890[a]	44,912	11,253	34,416	21,572	4,540	31,334	148,027
1890[a]	45,046	10,800	33,040	53,016	4,515	34,000	180,417
1891	43,559	9,991	35,636	53,016	4,384	34,000	180,586
1892	48,836	9,991	35,342	53,016	4,308	41,909	193,402
1893	59,787	10,448	38,003	53,016	5,856	39,843	206,953
1894	62,406	10,611	40,376	53,016	6,880	39,843	213,132
1895	68,553	11,889	43,078	53,016	7,393	39,843	223,772
1896	66,183	13,813	41,971	53,016	6,839	36,855	218,677
1897	74,130	14,688	44,983	53,016	6,332	34,000	227,149
1898	78,337	18,166	57,470	75,703	7,619	34,000	271,295
1899[b]	99,602	17,197	83,235	82,218	13,629	34,000	329,881
1899[b]	131,220	19,710	101,250	112,790	15,500	55,000	435,470
1900	145,473	24,865	101,316	123,504	16,000	58,884	470,042
1901	145,090	18,518	102,323	121,150	16,030	57,706	460,817
1902	141,438	18,310	99,302	121,029	16,997	57,335	454,411
1903	140,364	26,999	96,409	121,078	16,669	61,118	462,637
1904	141,922	28,639	94,594	121,078	16,335	58,276	460,845
1905	181,453	33,044	93,181	123,644	16,110	57,259	504,691
1906	198,782	38,266	94,071	123,644	15,784	54,140	524,687
1907	223,211	38,035	92,189	123,644	15,468	70,856	563,403
1908	213,939	37,949	90,345	123,644	15,159	74,621	555,657
1909	207,913	37,191	88,539	123,644	14,856	80,933	553,076
1910	200,695	36,521	87,297	123,644	14,745	80,352	543,254
1911	223,111	42,190	106,226	126,991	20,105	78,923	597,546
1912	231,462	44,947	119,681	127,747	25,375	105,687	654,899
1913	230,958	43,559	117,288	127,182	24,916	83,573	627,476

Source: J. and G. Thomson and John Brown and Company, Property Valuation Books, UCS1, 25/1-8.

a. The first set of figures for 1890 is as of March 31; the second is from a new valuation as of April 1.

b. The second set of figures for 1899 resulted from a new inventory taken at the time of the sale of the yard to John Brown and Company.

Notes

Introduction

1. See the bibliography in Saul, *The Myth of the Great Depression 1873-1896.* More recent contributions include McCloskey, ed., *Essays on a Mature Economy: Britain after 1840,* and *Economic Maturity and Entrepreneurial Decline: British Iron and Steel, 1870-1913.*

2. See, for example, the following exchange: McCloskey, "Did Victorian Britain Fail?"; Aldcroft, "McCloskey on Victorian Growth"; and McCloskey, "Victorian Growth: A Rejoinder."

3. Lewis, "International Competition in Manufactures," pp. 583-584.

4. Kindleberger, *Economic Growth in France and Britain: 1851-1950,* pp. 148-149, and "Germany's Overtaking of England, 1806-1914," pp. 497-498; Landes, "Entrepreneurship in Advanced Industrial Countries: The Anglo-German Rivalry," pt. 2, pp. 5-7; Saul, "The Engineering Industry," pp. 193, 202.

5. Landes, "Entrepreneurship in Advanced Industrial Countries," pt. 2, p. 7.

6. Landes, "The Structure of Enterprise in the Nineteenth Century: The Cases of Britain and Germany," pp. 120-121.

7. Landes, *The Unbound Prometheus,* p. 247.

8. Aldcroft, "Technical Progress and British Enterprise, 1875-1914," in Aldcroft and Richardson, *The British Economy 1870-1939,* pp. 178-179.

9. Aldcroft, "The Entrepreneur and the British Economy, 1870-1914," pp. 118-119.

10. Coleman, "Gentlemen and Players."

11. Saul, *Myth of the Great Depression,* p. 46.

12. Landes, "Entrepreneurship in Advanced Industrial Countries," pt. 2, pp. 21-22; Sawyer, "Entrepreneurship in Periods of Rapid Growth: The United States in the 19th Century."

13. Landes, "The Structure of Enterprise," p. 121.

1. The Evolution of the Steel Steamship

1. Fry, *History of the North Atlantic Steam Navigation,* p. 24; Basset-Lowke and Holland, *Ships and Men,* p. 139; Lindsay, *History of Merchant Shipping,* III, 291.

2. Lindsay, *History of Merchant Shipping*, III, 291, 362, 364; *Nautical Magazine*, 1859, pp. 48-50; Glover, "Statistics of Tonnage during the First Decade under the Navigation Law of 1849," p. 17; British Parliamentary Papers, 1870 LX (288.). The most comprehensive treatment of the Passenger Acts is given in MacDonagh, *A Pattern of Government Growth: The Passenger Acts and their Enforcement, 1800-1860.*

3. Bates, *American Marine*, pp. 141-145, 151-156, 160. Of the vessels in cargo or ballast entering British ports in 1850, 66.19 percent of the tonnage was British, as was 63.51 percent of the tonnage clearing. By 1860 these percentages had dropped to 56.19 and 56.00 respectively. Glover, "Statistics of Tonnage," p. 13. In 1870, of the tonnage entered and cleared in British ports, 70.18 percent was British. Glover, "Tonnage Statistics for the Decade 1860-70," p. 220.

4. See, for example, Tregelles, *Iron and Wood in Shipbuilding*, p. 1; and Grantham, *Iron as a Material for Shipbuilding*, p. 71; S.C. on the Navigation Acts (Commons), 1847, ev. C. Graham, q. 3366.

5. Thurston, "Iron Manufactures in Great Britain," p. 29.

6. Barry, *Dockyards of France*, pp. 52, 171, 173-174; Clyde Shipbuilders and Clyde Shipwrights, *Arbitration Proceedings*, 1877, p. 35.

7. Wallace, *Wooden Ships and Iron Men*, esp. pp. 42, 85, 94, 192, 230, 277; Ross, Newton, and Benians, eds., *Cambridge History of the British Empire*, VI, *Canada*, pp. 367, 569-570, 609, 663.

8. Öhlrich, *Deutschland zur See*, p. 36; Lehmann-Felskowski, ed., *The Shipbuilding Industry of Germany*, p. 4; S. C. on Navigation Acts (Commons), 1847, ev. G. F. Young, q. 5767; 1848, ev. C. Allhusen, qs. 6714-6715; Godeffroy, q. 7013. A table of international cost comparisons is given in Pollard, "The Economic History of British Shipbuilding," p. 9, n. 2.

9. Grantham, *Iron as a Material*, p. 118, quoting Chatfield. The *Hindostan*, a 78-gun ship, needed 70 acres growing for 80 years. Tregelles estimated the need for a 500-ton merchantman as 12 acres for 75 years. *Iron and Wood*, p. 1. Rawson's estimate for a three-decker was 80 acres.

10. Barry, *Dockyards and Private Shipyards*, pp. 26-27. He considered the timber dockyards so inefficient as to amount to a "disguised form of outdoor relief" (p. 29).

11. "Hogging" and "sagging" are the raising and lowering, respectively, of the middle section of the ship in relation to the stem and stern ends. The largest timber ships ever built were launched in America after 1900—the *Wyoming* (3,730 g.t.) and the *Roanoke* (3,347 g.t.)—but these were exceptional. See also Usher, "Growth of English Shipping," p. 477.

12. Harley, "On the Persistence of Old Techniques: The Case of North American Wooden Shipbuilding."

13. Ward, "Presidential Address," p. 9; Robb, "Ship-building," p. 352; Clapham, *An Economic History of Modern Britain*, II, *Free Trade and Steel*, pp. 62-63.

14. Zimmermann, *Ocean Shipping*, pp. 140-141; Kirkaldy, *British Shipping*, p. 52; "British Foreign Trade and Industry," Cd. 4954 (1909), p. 104; Graham, "The Ascendency of the Sailing Ship 1850-1885," pp. 74-88.

15. Harley, "The Shift from Sailing Ships to Steamships, 1850-1890," pp. 215-

234; North, "Ocean Freight Rates and Economic Development 1750-1913," pp. 542-543, and "Sources of Productivity Change in Ocean Shipping, 1660-1850"; Walton, "Productivity Change in Ocean Shipping after 1870: A Comment"; Hughes and Reiter, "The First 1,945 British Steamships," pp. 376-378.

16. Harley, "The Shift from Sailing Ships to Steamships," table 3, p. 228; Kirkaldy, *British Shipping*, pp. 130-131; Hyde, *Blue Funnel*, pp. 15, 22; Fayle, *A Short History of the World's Shipping Industry*, pp. 240-241; Ward, "Presidential Address," pp. 11-13; Talbot, *Steamship Conquest of the World*, p. 41; Graham, "Ascendency of the Sailing Ship," pp. 86-87.

17. Kirkaldy, *British Shipping*, pp. 131-138; Talbot, *Steamship Conquest*, p. 55.

18. Parsons, "The Marine Steam Turbine from 1894 to 1910," pp. 80, 95; Ward, "Presidential Address," p. 20; Kirkaldy, *British Shipping*, pp. 142-146.

19. Biles, "Fifty Years of Warship-Building on the Clyde," pp. 368-369; Zimmermann, *Ocean Shipping*, p. 18.

20. Kirkaldy, *British Shipping*, pp. 136-137.

21. Talbot, *Steamship Conquest*, pp. 72-85.

22. Rolt, *Isambard Kingdom Brunel*, passim.

23. Fayle, *Short History*, p. 240; Course, *The Deep Sea Tramp*, pp. 24, 36-41; Kirkaldy, *British Shipping*, p. 112 and appendix XXII. Between 1892 and 1911, William Doxford and Sons, Ltd., of Sunderland built about 200 "turret" ships under patent. Designed to reduce Suez Canal tolls, which were calculated using the breadth of the upper deck as a measure of the breadth of the ship, their sides rounded concavely from just above the water line and then rose vertically, leaving a very narrow upper deck. For differences in tonnage see Appendix A.

24. The German-built *Imperator* was larger at 52,000 g.t. The 56,000 g.t. *Bismarck*, although begun before the war, was not completed until 1922 when she became the White Star Line's *Majestic*. Bonsor, *North Atlantic Seaway*, p. 46.

25. Changes in naval technology are discussed in Chapter 10.

26. Fayle, *Short History*, p. 253; Haji, *Economics of Shipping*, p. 51.

27. Rawson, *Ships and Seamen*, p. 193.

28. Pollock, *The Shipbuilding Industry*, p. 80.

29. Huldermann, *Albert Ballin*, p. 63.

30. Corson, *The Atlantic Ferry in the Twentieth Century*, pp. 6-7; *Syren and Shipping*, February 3, 1897; March 10, 1897; November 25, 1904; Rawson, *Ships and Seamen*, pp. 183-192; Michon, *Les grandes compagnies anglaises de navigation*, p. 89; Brassey, *Steamship Subsidies*, p. 5; *Economist*, July 1, 1911; Kirkaldy, *British Shipping*, p. 99.

31. The largest White Star Liners cost £200,000 in 1874 (15 knots), £350,000-400,000 in 1889 (20 knots), £500,000-550,000 in 1894 (22 knots). The Merchant Cruiser Committee estimated the cost in 1902 of a 24-knot liner at £850,000 and of a 25-knot liner at £1 million. White, *Proceedings of the Institution of Civil Engineers*, 1903-04, pp. 92-99; *Shipping*, January 1, 1895.

32. Fry, *North Atlantic Steam Navigation*, pp. 81, 136, 268; Corson, *The Atlantic Ferry*, pp. 21, 24, 81, 135.

33. According to the *Economist*, of 160 shipping companies in the United Kingdom 73 declared that their ships operated as tramps. See Michon, *Les*

grandes compagnies anglaises, p. 19; R. C. on Shipping Rings, 1909, *Report*, p. 34; Associated Shipwrights' Society, *Annual Report*, 1903; Gregg, "Decline in Tramp Shipping," p. 339; Crammond, *British Shipping Industry*, p. 16.

34. *Economist*, July 1, 1911.

35. *TINA*, 1907, pp. 94-95; Holmes, *Ancient and Modern Ships*, II, 109.

36. *Transactions of the North-East Coast Institution of Engineers and Shipbuilders* (hereafter *TNECI*), 1888-89, pp. 151-152.

37. Ibid., pp. 153-155; Pollock, *Shipbuilding Industry*, pp. 124-125; *Syren and Shipping*, January 6, 1904; Holmes, *Ancient and Modern Ships*, II, 112-114.

38. Holmes, *Ancient and Modern Ships*, II, 117; *Proceedings of the Institution of Civil Engineers*, 1903-04, p. 157; Biles, *Design and Construction of Ships*, I, 72-73; Board of Trade Committee on Tonnage Measurement, 1906, ev. A. W. Bibby, q. 1244.

39. Holmes, *Ancient and Modern Ships*, II, 104-106, 115-116; *TNECI*, 1892-93, pp. 277-279.

40. Lisle, *Tanker Technique, 1700-1936*, pp. 16-23; *TINA*, 1934, p. 202; *Society of Naval Architects and Marine Engineers* (hereafter *SNAME*), 1945 (Historical transactions), p. 135; *Annals of Lloyd's Register*, pp. 144-145; *TINA*, 1887, pp. 1-2; Kirkaldy, *British Shipping*, p. 122.

41. *TINA*, 1934, pp. 204-207; *SNAME*, 1945 (Historical), p. 135; Lisle, *Tanker Technique*, p. 27.

42. *TINA*, 1893, pp. 13-14; 1934, p. 208; *SNAME*, 1945 (Historical), pp. 137-138; *Annals of Lloyd's Register*, p. 146; Lisle, *Tanker Technique*, pp. 34-46.

43. Haldane, *Steamships and Their Machinery*, pp. 451-457; *Transactions of the Institute of Marine Engineers*, 1903-04, p. 5; *Syren and Shipping*, December 28, 1904; January 4, 1911; July 21, 1926; Fry, *North Atlantic Steam Navigation*, p. 193; *The Steamship*, May 15, 1883; Kirkaldy, *British Shipping*, p. 118; Smith, "Some Pioneers of Refrigeration"; Critchell and Raymond, *History of the Frozen Meat Trade*, pp. 13-41, 128ff.; Waters, *Clipper Ship to Motor Liner*, p. 54.

44. *Annals of Lloyd's Register*, pp. 152-156; Holmes, *Ancient and Modern Ships*, II, 130; *TINA*, 1903, p. 134.

45. Committee on Industry and Trade ("Balfour Committee"), pt. IV, 367; Hesketh, *J. and E. Hall Ltd.*, pp. 43-45; Stephen, *A Shipbuilding History*, p. 109; Blake, *Down to the Sea*, p. 154.

2. Fluctuations in Shipbuilding Output

1. Thornton, *British Shipping*, p. 75; Fayle, *The War and the Shipping Industry*, pp. 2-6.

2. An annual index of shipbuilding output from 1855 is given in Feinstein, *National Income, Expenditure and Output of the United Kingdom, 1855-1965*, table 52, pp. T114-115. As can be seen from figure 1, the fluctuations in this index, which is based on inputs, do not differ appreciably from those of output given in gross tons for the entire United Kingdom. As most other series are given in tons, for purposes of comparability we shall rely most heavily here on the traditional tonnage data.

3. See table 3.3.

4. Robertson, *A Study of Industrial Fluctuation*; Ballard, "The Relation Between Shipbuilding Production, Prices, and the Freight Market"; Robertson and Morison, "Shipping and Shipbuilding"; Sanders, "Foreign Trade and Shipbuilding"; James, *Cyclical Fluctuations in the Shipping and Shipbuilding Industries*; Tinbergen, "Ein Schiffbauzyklus?"; Einarsen, *Reinvestment Cycles*; Koopmans, *Tanker Freight Rates and Tankship Building*.

5. Before World War I, a cargo steamer of 7500 g.t. without "trimmings" of any kind required a minimum of seven months to be constructed, and the average building period for all ships was probably about one year. James, *Cyclical Fluctuations*, p. 41.

6. Given a fleet of 2,000 vessels of equal size with a replacement rate of 5 percent per year, a 5 percent increase in the demand for shipping would create a demand for 200 new vessels, 100 to replace older ones and 100 to meet the new demand. A further increase of 5 percent in the demand for shipping in the second year would raise the demand for new ships to 210, but an increase in shipping demand of only 3 percent in the third year would reduce the demand for new ships to only 177 vessels since, while 110 would be needed as replacements, only 67 ships would be necessary to fulfill the new demand.

7. James, *Cyclical Fluctuations*, p. 53.

8. For a description of the poor earnings of shipbuilders in 1912, a good year for tonnage launched, see Kirkaldy, *British Shipping*, pp. 590-591.

9. The variations in the number of men employed at the Dumbarton yard of William Denny and Brothers illustrate the severe annual fluctuations in activity which shipbuilders faced. See tables 9.2, 9.3, and 9.7.

10. Variations in the prices of ship plates are given in McCloskey, *Economic Maturity and Entrepreneurial Decline*, appendix B, pp. 135-136.

11. *Engineering*, December 29, 1871, p. 417; August 5, 1881, p. 137. The development of the shipbuilding industry was affected in several ways by the availability of steel. Steel was adopted more quickly on the Clyde than on the North East Coast because there were several manufacturers of steel plates in the Glasgow region but none in the north of England before the early 1880s. Until then it was unprofitable for the North East Coast builders to import steel from Wales or Scotland, and they continued to use iron. *Engineering*, September 1, 1882, p. 218.

12. To retain some uniformity, the "shipbuilding industry" will be taken to mean all firms engaged in ship and boat building and repair, including their marine engineering departments, but excluding firms engaged in marine engineering alone, as well as the Royal Naval Dockyards.

13. Maywald, "The Construction Costs and Value of the British Merchant Fleet," p. 57; Feinstein, *National Income*, p. 188.

14. Maywald, "Construction Costs," pp. 53-55.

15. See also the criticism of Cormack, "An Economic History of Shipbuilding," p. 388.

16. McCloskey, *Economic Maturity*, pp. 46-53 and table II, pp. 135-136.

17. *Statist*, August 16, 1884; *Economist*, "Commercial History," various years; *Statistical Abstract of the United Kingdom*.

18. In 1890, only 10.5 percent of the tonnage sold from the Register was of steel, as against 45 percent of the tonnage on the Register and 92.5 percent of the

tonnage launched. In the same year, only 48 percent of the ships sold were steamers, compared with 63.2 percent of those on the Register and 81.8 percent of those launched. These proportions are similar to those of other years in the period. See also Cormack, "An Economic History of Shipbuilding," p. 146.

19. Appendix 25, p. 256. Because of rounding, totals may not be exact.

20. Wallace, *Wooden Ships*, p. 313; Fitger, *Schiffbau und Seeschiffahrt*, p. 16; *Statist*, April 7, 1900; R. C. on the Depression of Trade, 1886, ev. J. Burke, qs. 10601-10602; J. Scott, qs. 11893-11896, 11938-11940; S.C. on Foreign Ships, 1904-05, appendix 16, p. 112.

21. R.C. on the Depression of Trade, 1886, ev. Steam Shipowners Association, qs. 10916-10920; S.C. on Steamship Subsidies, 1901, ev. A.N. Hill, qs. 2033-2036; 1902, ev. Pirrie, qs. 1805-1806, 2012-2013; ev. R. Giffen, q. 11; *Report*, p. XVIII; S. C. on Foreign Ships, 1904-05, appendix 12, p. 104.

22. For data on British sales to foreigners, see Cormack, "An Economic History of Shipbuilding," appendix 6, p. 371; and Schwarz and Halle, *Die Schiffbauindustrie*, p. 88. Stephen, *A Shipbuilding History*, pp. 68, 103-105; Corson, *Atlantic Ferry*, pp. 121, 140; Fry, *North Atlantic Steam Navigation*, passim; *Syren and Shipping*, January 13, March 10, 1897; Pollock, *Shipbuilding Industry*, p. 180. Total sales of new tonnage in 1899-1907 were, to Japan, £12.75 million; Germany, £9.5 million; Austria-Hungary, £5 million; Norway, £4.5 million; and British Possessions, £8.75 million. Tariff Commission, *Fourth Report*, paragraph 95.

23. Lehmann-Felskowski, *Shipbuilding Industry of Germany*, p. 3; Schroedter, "The Ironmaking and Shipbuilding Industries in Germany," p. 79; Halle, *Schiffbau in Deutschland*, p. 199; Neumann, *Schiffbauindustrie*, p. 73; Leckebusch, "Der Beginn des deutschen Eisenschiffbaues 1850-1890," p. 182.

24. *The Steamship*, June 15, 1883; Ross, *Produktionsbedingungen der Seewerke*, pp. 38-40; Schroedter, "Ironmaking and Shipbuilding in Germany," pp. 26-28; Lehmann-Felskowski, *Shipbuilding Industry of Germany*, pp. 6-7, 107; Haack, "Development of German Shipbuilding," p. 740; Lange, "Stettiner Vulkan," pp. 469-470; Halle, *Schiffbau in Deutschland*, pp. 208-209. Details of the German shipbuilding and iron and steel industries are given in Leckebusch, *Die Beziehungen der deutschen Seeschiffswerften zur Eisenindustrie an der Rhein in der Zeit von 1850 bis 1930.*

25. Halle, *Schiffbau in Deutschland*, pp. 210-212.

26. Schroedter, "Ironmaking and Shipbuilding in Germany," pp. 10-25; Halle, *Schiffbau in Deutschland*, pp. 212-214.

27. Schwarz and Halle, *Die Schiffbauindustrie*, pp. 178-179; Tariff Commission, *Fourth Report*, paragraph 1312; Pollock, *Shipbuilding Industry*, p. 187; *Statist*, January 6, April 29, 1899; April 16, 1904; *Economist*, December 1, 1900; January 16, 1904; *Syren and Shipping*, February 24, 1897; November 6, 1904. For statistics on the output of the chief countries, see appendix table B.7.

28. Hutchins, *The American Maritime Industries and Public Policy, 1789-1914*, pp. 448, 451; Tyler, *The American Clyde*, p. 16; Swann, *John Roach, Maritime Entrepreneur*, p. 52.

29. Tyler, *American Clyde*, pp. 26, 35, 43, 77, 110-112; Swann, *John Roach*, ch. 3; Hutchins, *American Maritime Industries*, p. 452.

30. *SNAME*, 1894, p. 39; Tyler, *American Clyde*, pp. 49, 109-111.

31. *Statist*, July 12, 1902; Schwarz, "Der Amerikanische Schiffbau im letzten Jahrzehnt," pp. 226-227; Roberts, "Formation of the Society of Naval Architects and Marine Engineers," pp. 539-543; *The Shipbuilder*, no. 8. Shipyards on the Great Lakes had developed before 1893 with the needs of trade, sheltered by law and geography from the competition of foreign yards.

32. Kenny, "Steel for Shipbuilding," p. 446; Schwarz, "Amerikanischer Schiffbau," pp. 226-229.

33. Schwarz, "Amerikanischer Schiffbau," pp. 226-229; *Statist*, March 1, July 12, 1902; *Engineering Magazine*, April 1899; "Histories of Private Shipyards," *SNAME*, 1945 (Historical); Jordan, "Shipping on the Great Lakes," pp. 330-332; Wheeler, "Commerce of the Great Lakes," pp. 82-84; Renner, "Schiffbau und Schiffahrt auf den Grossen Seen," pp. 247-248; *Steamship*, September 1899.

34. *Economist*, October 15, 1898; December 8, 1900; June 21, 1902; May 26, 1906; Dunmore, *Ship Subsidies*, p. 28; Schwarz and Halle, *Die Schiffbauindustrie*, pp. 178-179; *Syren and Shipping*, January 4, 1899; *Statist*, October 7, 1899.

35. Crouzet, "Essai de construction d'un indice annuel de la production industrielle française au XIXe siècle," p. 80.

36. *Capital and Labour*, June 17, 1874; March 17, 1875; *Economist*, May 28, 1881; April 29, 1882; Board of Trade, Departmental Committee on Shipping and Shipbuilding, *Second Report*, 1918, p. 22; Meeker, *History of Shipping Subsidies*, p. 45.

37. S. C. on Steamship Subsidies, 1902, *Second Report*, appendix 4; Colin, *La navigation commerciale au XIXe siècle*, pp. 185-186; Schwarz and Halle, *Die Schiffbauindustrie*, pp. 178-179.

38. Russo, "Fifty Years Progress of Shipbuilding in Italy," pp. 255-257, 261-270; Meeker, *History of Shipping Subsidies*, p. 100; *Syren and Shipping*, December 16, 1896; March 10, 1897; August 31, 1904; Terano and Yukawa, "Development of Merchant Shipbuilding in Japan," pp. 138-145; Orchard, *Japan's Economic Position*, p. 86.

39. A general account of the competitiveness of the industry is given in Pollard, "British and World Shipbuilding, 1890-1914: A Study in Comparative Costs."

40. R. C. on the Depression of Trade, 1886, esp. J. Price, qs. 10978-10985, 11013-11118; J. Scott, qs. 11878-11879, 11902. 11976-11977, 12005-12016, 12051; appendix A. 1, p. 342, written by Lowthian Bell; also *Statist*, November 25, 1885; and Bell, *Principles of the Manufacture of Iron and Steel*, pp. 503, 706.

41. R.C. on the Depression of Trade, 1886, qs. 14761-14763, 14774, 14785-14810. J. Scott's experience of French shipyard workers was that they still worked their own plots and had to absent themselves three times a year for sowing, reaping, and cutting peat. Q. 12013.

42. R. C. on the Depression of Trade, 1886, ev. J. Price, q. 10991; J. Scott, qs. 11946-11947.

43. Meeker, *History of Shipping Subsidies*, pp. 46-47, 179; Colin, *La navigation commerciale*, pp. 178-179; Jones, *Government Aid*, p. 47; Dunmore, *Ship Subsidies*, p. 49; Fayle, *War and Shipping*, p. 17; Halle, *Schiffbau in Deutschland*, p. 219; *SNAME*, 1894, p. 48; 1900, pp. 193, 195.

44. See also Jeans, *Iron Trade*, p. 222. The figures are from Board of Trade, Departmental Committee, *Second Report*, 1918, p. 41; Jeans, *Iron Trade*, pp. 131-132; Jones, *Is Unemployment Inevitable?*, p. 286; Neumann, *Schiffbauindustrie*, p. 74; R. C. on the Depression of Trade, 1886, appendix A.1, p. 331; Halle, *Schiffbau in Deutschland*, p. 217; *Journal of the Iron and Steel Institute*, 1883, p. 835.

45. Francis Elgar told the Tariff Commission that he had never found American or German ship plates to be priced below the British. *Fourth Report*, paragraph 502. In fact the high price of plates forced many American shipyards to integrate vertically with the steelmakers—on unfavorable terms. Schwarz, "Amerikanischer Schiffbau," p. 230; *Economist*, August 8, 1900; January 8, 1910; *Statist*, March 19, 1898; January 28, 1899; May 4, December 28, 1901; September 28, 1912; Associated Shipwrights' Society, *Quarterly Report*, 1905, second quarter; Jeans, *Iron Trade*, p. 160; Neumann, *Schiffbauindustrie*, pp. 40, 78; Dunmore, *Ship Subsidies*, pp. 42-43; *Palmer Record*, no. 6; *Journal of the Iron and Steel Institute*, 1883, p. 835; Laeisz, "Shipbuilding in Germany," p. 17; Board of Trade, Departmental Committee, *Second Report*, 1918, appendices 43, 45, pp. 24-25; *Fairplay*, January 28, 1910; Furness, *The American Invasion*, p. 14.

46. Pollard, "British and World Shipbuilding," table 3, p. 438; *SNAME*, 1900, pp. 176-178; Schwarz, "Amerikanischer Schiffbau," p. 231; Grantham, *Ocean Steam Navigation*, p. 129. Some estimates put American wages much higher, for example Dunmore, *Ship Subsidies*, p. 48; but see *SNAME*, 1900, p. 190 for examples of the higher output of the American worker. Also, Schwarz and Halle, *Die Schiffbauindustrie*, p. 158. For German wages see *The Shipbuilder*, no. 3, and the open letter by the Board of Trade to the Shipbuilding Employers' Federation, in *Shipbuilding Trades Agreement*, June 3, 1910.

47. Schwarz and Halle put them at 10 percent above the British in 1901. *Die Schiffbauindustrie*, p. 184. See also, Cormack, "An Economic History of Shipbuilding," p. 239; Tariff Commission, *Fourth Report*, paragraphs 92, 500, 529.

48. Schwarz and Halle, *Die Schiffbauindustrie*, pp. 184-185; *Statist*, July 12, 1902; also White's Presidential Address, *Proceedings of the Institution of Mechanical Engineers*, p. 156; Colin, *La navigation commerciale*, pp. 178, 183-184; Associated Shipwrights' Society, *Quarterly Report*, 1891, first quarter; Fry, *North Atlantic Steam Navigation*, p. 242; Meeker, *History of Shipping Subsidies*, p. 46.

3. Regional Shipbuilding

1. Although in general wages were lower in Scotland than in England.

2. Aberconway, *The Basic Industries of Great Britain*, p. 163; Dillon, *Palmer's Shipbuilding and Iron Company Limited*, p. 12; Wilkinson, *The Town that Was Murdered*, pp. 72ff.

3. See for example the report of the protest meeting on Barrow Island held by the local slum-dwellers, reported in the *Barrow Pilot*, January 31, 1874; and the letter by Robert Loxham, an opposition councillor, to the same weekly, October 25, 1873. Also, compare the *Enquiry as to the Division in Dalton-in-Furness*,

1871, pp. 9ff. For a fuller account, see Pollard, "Barrow-in-Furness and the Seventh Duke of Devonshire," pp. 215-227.

4. *Barrow News*, September, 1881.

5. Wilkinson, *Town that Was Murdered*, pp. 106-107.

6. *The Steamship*, January, 1890.

7. Richardson, *Vickers, Son and Maxim*, pp. 191-193; *Vickerstown Souvenir*, 1903; *The Shipbuilder*, nos. 30, 34.

8. Burnett, *History of the Engineers' Strike*, p. 5; Jefferys, *The Story of the Engineers*, p. 85; Dougan, *The History of North East Shipbuilding*, p. 125.

9. Smith and Holden, *Where Ships Are Born, Sunderland, 1346-1946*, p. 49; Stephen, *A Shipbuilding History*, p. 62; *Denny, Dumbarton*, p. 12.

10. Blake, *Down to the Sea*, p. 299.

11. Tralau, *Über den Standort der Schiffbauindustrie*, p. 9; Schwarz and Halle, *Die Schiffbauindustrie*, p. 140.

12. For example, R. C. on Labour, 1892, Group A, qs. 26520-26525.

13. For differences among the Mersey, the Clyde, the Thames, the Tyne, and the Royal Dockyards, see Reed, *Shipbuilding in Iron and Steel*, pp. 428-463.

14. Of the total of 138 vessels of 4000 g.t. and over built between 1858 and 1884, 80 were built on the Clyde, 12 in Barrow, 11 each on the Mersey and the Tyne, 7 in Belfast, 7 in the rest of the United Kingdom, and 10 in the rest of the world. Pollock, *Modern Shipbuilding*, p. 199.

15. Ibid., pp. 185-186.

16. *Economist*, Commercial Histories of 1885 and 1896; December 27, 1890; November 8, 1902; *Statist*, January 15, 1910; *Clyde Arbitration, 1877*, p. 19; *Syren and Shipping*, January 2, 1907; *TNECI*, 1909-10, p. 102.

17. Napier, *Life of Robert Napier*, p. 135.

18. Lindsay, *History of Merchant Shipping*, IV, 70, 593; *TIESS*, 1876, p. 4.

19. *Clyde Arbitration, 1877*, p. 110; R. C. on Trade Unions, 1867-68, ev. J. P. Smith, qs. 17454-17456; Strang, "The Improvement of Tidal Rivers," p. 395; Lindsay, *History of Merchant Shipping*, IV, 594; Cormack, "An Economic History of Shipbuilding," p. 384; *Capital and Labour*, September 15, 1875; Pollock, *Shipbuilding Industry*, p. 175.

20. *Economist*, October 13, 1906; Levy, *Trust und Kartellentwicklung in Grossbritannien*, p. 156; Pearce stated in 1877 that the freight charges for iron from the East Coast were 7/6 per ton to the Thames and between 12/5 and 15/- per ton to the Clyde, but it is doubtful if, as alleged, iron was shipped to the Clyde to any extent. *Clyde Arbitration, 1877*, p. 23. McCloskey contends, however, that cartels in ship plates were rarely, if ever, effective. *Economic Maturity*, pp. 28-34.

21. *Nautical Magazine*, "Sunderland Ships," 1892, p. 585; Smith and Holden, *Where Ships are Born*, pp. 2, 36, 51, 56; *Victoria County History of Durham*, II, 303-304.

22. *Syren and Shipping*, January 4, 1899; December 26, 1900; January 2, 1907; Harcourt, *Achievements in Engineering in the Last Half Century*, pp. 210-213; *Industrial Rivers of the United Kingdom*, pp. 54-57; Pollock, *Modern Shipbuilding*, p. 191, and *Shipbuilding Industry*, p. 176; British Association, *Trade and In-*

dustry, p. 123; S. C. on Combination Laws, 1825, ev. Temple, p. 806; Associated Shipwrights' Society, *Statements on Burt's Award*, p. 11, and *Annual Reports*, esp. 1907.

23. *Syren and Shipping*, January 23, 1901; R. C. on Labour, 1892, Group A, Employers' Association Replies (311); *Victoria County History of Yorkshire*, II, 307.

24. Pollock, *Modern Shipbuilding*, p. 194; Rutherford, *The Man Who Built the "Mauretania,"* p. 20; Board of Trade, *Industrial Survey of the North-East Coast*, p. 232; *Clyde Arbitration, 1877*, pp. 103, 107, in which Pearce also refers to lower rents on the Tyne.

25. R. C. on Labour, 1892, Group A, ev. Doxford, q. 26604; Smith, *History of Marine Engineering*, p. 246; *TNECI*, 1901-02, p. 239.

26. D'Eyncourt, *Shipbuilder's Yarn*, p. 45; *Clyde Arbitration, 1877*, p. 136.

27. Barry, *Dockyards of France*, 228n.; Rawson, *Ships and Seamen*, pp. 35-40.

28. *Engineering Magazine*, 1897, pp. 365, 776; Cormack, "An Economic History of Shipbuilding," p. 260; S. C. on Government Contracts, 1897, ev. Hills, qs. 4108, 4191, 4600; Maudslay, qs. 3385-3423, 3518; Langton, qs. 3377-3378; *Syren and Shipping*, November 9, 1904. For a more detailed account as well as estimates of employment, etc., see Pollard, "The Decline of Shipbuilding on the Thames."

29. *Liverpool Daily Courier*, April 3, 1925; Webb MSS., I, 399; Rathbone, "Shipwrights' Trade Societies," p. 496.

30. *Manchester Guardian Commercial*, July 14, 1927; *Liverpool Echo*, March 2, 1934; Rathbone, "Shipwrights' Trade Societies," p. 497.

31. Burstall, *Shipbuilding in Liverpool*; Rees, *Some Merseyside Shipbuilders*, pp. 3, 9. Liverpool was the only port where the main mercantile output continued to be sailing vessels right down to 1893.

32. Thurston, "Iron Manufactures in Great Britain," p. 31; *Syren and Shipping*, January 3, 1900.

33. Sulley, *History of Ancient and Modern Birkenhead*, pp. 58ff.; Mott, *Reminiscences of Birkenhead*, pp. 8-12; *TINA*, 1874, p. 268.

34. Sulley, *Birkenhead*, p. 231; *Nautical Magazine*, 1853, p. 167; *Syren and Shipping*, January 7, 1903.

35. Associated Shipwrights' Society, *Quarterly Report*, 1901, second quarter; *Economist*, November 19, 1916; R. C. on Labour, 1892, Group A, ev. MacIlwaine, q. 26526; Jefferson, *Viscount Pirrie of Belfast*, p. 203; R. C. on Trade Unions, 1886, ev. Burke, q. 10577.

36. Jefferson, *Pirrie*, pp. 24-27, 49, 69, 72, 200-201; *Economist*, June 22, 1907; November 10, December 3, 1910; *Engineer*, July 10, 1886; *The Shipbuilder*, nos. 37, 44, 48.

37. As calculated from the Censuses of Population for Ulster, 1861-1911. The proportion of Roman Catholics declined from 28 percent for all shipbuilding workers in the province in 1861 to only 8 percent in 1911, while the proportion of Episcopalians increased from 23 to 36 percent. Over the same period the proportion of Presbyterians decreased slightly from 49 to 44 percent. In 1911, 44 percent

of all employed males in Ulster were Roman Catholic, 23 percent Episcopalian, and 27 percent Presbyterian.

38. Bevan, "The Industrial Resources of Ireland," pp. 707-708; S. C. on Trade Unions, *Second Report*, 1838, ev. Morton, qs. 5907-5914, 5940-5944, 5960, 5967-5989; Fagan, qs. 4143, 4188-4190, 3926-3943; Rathbone, "Shipwrights' Trade Societies," p. 517; Webb MSS., I, 32, 411; Barry, *History of the Port of Cork Steam Navigation*, p. 3; Cliffe Leslie to the Manchester Statistical Society, 1853, Webb MSS., I, 241; R. C. on Labour, 1892, Group C, ev. O'Reilly-Simmons, qs. 16411-16415, and Group A, Replies to Trade Union Questionnaires (132); Smellie, *Shipbuilding and Repairing in Dublin*, pp. 51-60.

39. *Victoria County History of Lancaster*, VIII, 306ff.; Fisher, *Popular History of Barrow*, pp. 78, 84, 108ff.; Leach, *Barrow-in-Furness*, pp. 27-28; Casson, *A Few Furness Worthies*, pp. 80ff.; Richardson, *History and Antiquities of Furness*, II, 244, 271-276; Gradon, *Furness Railway*, p. 27; Melville and Hobbs, *Some Early Railway History in Furness*, pp. 14ff.; Leach, *Duke of Devonshire*, passim.

40. The formation of the Naval Armament Company in 1888 meant only a minor change. The chairman, Lord Hartington, was a nephew of the Duke of Devonshire, and there was also a Cavendish on the board. The only significant addition was Sir Thomas Brassey, the new vice-chairman. Fisher, *Popular History*, p. 163; *Barrow Herald*, March 31, 1888; Borough of Barrow, *Minutes of Council Meetings*. Extended accounts of this phase in Barrow's history are given in Pollard, "Barrow-in-Furness," and in Marshall, *Furness and the Industrial Revolution*.

4. The Firms

1. See the evidence of a foreman in de Rousiers, *The Labour Question in Britain*, p. 277.

2. S. C. on Trade Unions, 1838, q. 5982; Barry, *Dockyards of France*, p. 177. Abroad the yards were even smaller. In Germany 15,000-20,000 Rthl. (£2,000-£3,000) and in France as little as £100- £200 was considered a sufficient investment. In America many large builders had no more than $500 invested, although yards with steam gear were worth $15,000-$20,000. Halle, "Die Volkswirtschaftliche Entwicklung des Schiffbaues in Deutschland und in der Hauptländern," p. 199; Hall, *American Marine*, p. 199; Stephen, *A Shipbuilding History*, p. 15. In 1824 William Stephen employed 18 journeymen, 8 apprentices, 2 sawyers, and 2 blacksmiths, a total of 30 men. Russell, a well-known Clyde firm, employed only 60-70 men in 1846. R. C. on Labour, 1892, Group A, ev. Law, qs. 21664-21665.

3. Chapman, "The New Shipwright Building Co. of Dundee."

4. The dockyards and the leading Thames yards were among the largest of their day. Among the provincial yards, Laing's at Sunderland employed up to 150 men in 1825, Straker at South Shields, 60-100; Scott's of Greenock employed up to 220 in 1828. S. C. on the Combination Laws, 1825, pp. 915, 917. Chains and cordage were usually made by specialist firms, who also made all other accessories for the smaller yards. For a description of the large Blackwall Yard in

1841, see Penny Magazine's *Days at the Factory*, pp. 36ff.; for William Pile's yard on the Wear in 1850, see Rutherford, *The Man Who Built the "Mauretania"*, p. 34.

5. *Days at the Factory*, p. 37; *TIESS*, 1858-59, p. 36; *Proceedings of the Institution of Civil Engineers*, 1882-83, pp. 136ff. Early power for the hauling was manual, by horses or by steam. A hydraulic ram with a spur-geared winch was invented by Daniel Miller in 1849 and erected at Kelvinhaugh in 1850.

6. Large sums were tied up in the work in progress, although they were usually advanced by the owner. The damage done to iron ships of Tod and M'gregor by the gale of 1856 was £18,000, and the damage by fire to Stephen's Dundee yard in 1867 was £20,000. Stephen, *A Shipbuilding History*, pp. 34, 43. R. C. into the Control and Management of the Naval Yards, 1861, ev. Lloyd, qs. 10978-10982. Also the reminiscences of the Dockyards in 1859 by White, *Proceedings of the Institution of Mechanical Engineers*, 1899, p. 158.

7. Thompson, *Hundred Years of Shipbuilding*, p. 10.

8. Of the 80-90 Sunderland firms in existence in 1838-1840, more than 40 had collapsed by 1848. S. C. on Navigation Laws, 1848, qs. 4072-4075, 4080, 8013. But the list contained such names as W. Doxford, R. Dixon, and W. Thompson, all of whom later built up successful businesses.

9. *Directory of Shipowners*, 1914.

10. One should not take too seriously the contention of T. J. Byres that Scottish shipbuilders were more successful than other entrepreneurs in heavy industry over the period because they were more devoted to their work and did not enter politics or other outside activities. Byres, "Entrepreneurship in the Scottish Heavy Industries, 1870-1900." Byres seems to have missed much information on the political and social backgrounds of the shipbuilders.

11. For a list of firms and their successful apprentices, see Pollard, "The Economic History of British Shipbuilding," pp. 59-60.

12. A careful analysis is given in Schwarz and Halle, *Die Schiffbauindustrie*, p. 24.

13. Investment in shipbuilding was not very different in this respect from investment in other industries. Compare Todd, "Joint Stock Companies," pp. 69-71. Of all the limited companies formed between 1844 and 1868, banking and finance companies were the least stable and shipbuilding companies came next. Only 26 percent of the latter survived. Hunt, *Development of the Business Corporation in England*, p. 157.

14. Calculated from data given in Shannon, "The Limited Companies of 1866-83," pp. 308-313.

15. Marwick, "The Limited Company in Scottish Economic Development," pp. 415-429.

16. Shannon, "The First Five Thousand Limited Companies and Their Duration," p. 408, and "The Limited Companies of 1866-83," p. 292; Aberconway, *Basic Industries*, p. 232; *Denny Dumbarton*, p. 19; Stephen, *A Shipbuilding History*, pp. 82, 112; and Carvel, *Stephen of Linthouse*, p. 5.

17. See Shannon, "Limited Companies of 1866-83," p. 300, for information on the changing share values of all limited concerns. Also, Jefferys, *Engineers*, pp. 159-160, 165, 454.

18. In Germany the average nominal capital of all shipbuilding companies rose from M690,000 in 1870 to M850,000 in 1880, M1,440,000 in 1890, and M1,500,000 in 1900. The nominal capital of 13 of the larger firms grew from M42.7 million in 1896 to M102.4 million in 1908. Neumann, *Technische Entwicklung*, p. 34. Also, Pollard, "The Economic History of British Shipbuilding," table A.4.

19. See Neumann, *Technische Entwicklung*, for data on capital per worker employed in Germany (pp. 621-624), for the increase in the number of workers employed by the leading firms (pp. 57-58), and for the development of limited liability in shipbuilding (p. 66). Also, Halle, "Volkswirtschaftliche Entwicklung," p. 215.

20. Cowens, *TNECI*, vol. XV; R. C. on the Depression of Trade, 1886, *Third Report*, qs. 11058-11060; Irving, "New Industries for Old? Some Investment Decisions of Sir W. G. Armstrong, Whitworth & Co. Ltd., 1900-14," p. 152.

21. S. C. on Admiralty Money and Accounts, 1868. Laird Brothers in a letter dated June 18, 1868, put it at 18-20 percent of annual output, p. 276; Samuda, at 13-17 percent of the annual expenditure on materials and labor, qs. 4637-4638; Peter Rolt, Chairman of the Thames Ironworks, thought an armor-clad costing £200,000 would need "floating stores" of £10,000-£15,000, qs. 1059-1060. Assuming a building period of two years, this would mean 10-15 percent of annual output.

22. *Capital and Labour*, June 2, 1875, quoting the *North American* of Philadelphia; *Syren and Shipping*, January 1, 1902; Cormack, "An Economic History of Shipbuilding," p. 185.

23. For example, at Palmer's incorporation in 1865, 80.6 percent of the £2 million subscribed came from 73 people who each subscribed £5,000 or more. Of these, 11 bankers subscribed a total of £266,250; there were also 19 "gentlemen," 14 merchants, 5 manufacturers, and 8 engineers. Of these, 22 gave a Manchester address (£296,000) and 9 were from Newcastle (£90,000).

24. The connections and social standing of shareholders were also of importance. At the formation of the Parsons Marine Steam Turbine Company Ltd. in 1897, Parsons (himself a son of the Earl of Rosse) intended to refuse shares to outside buyers in the hope of attracting Lord Brassey and the Duke of Devonshire to his company. Appleyard, *Charles Parsons*, p. 95. Stephen, *A Shipbuilding History*, p. 81; Rutherford, *Mauretania*, p. 40; Yarrow, *Alfred Yarrow*, p. 25.

25. Clapham, *An Economic History*, II, 114-115; Pollard, "The Economic History of British Shipbuilding," table A.5.

26. Irving, "New Industries for Old?" p. 152.

27. In an interesting letter to *The Shipbuilder*, no. 5, 1907, "An English Shipbuilder" complained that too many different types were being produced in Britain and that still more specialization would improve British chances against foreign competition. He also thought that there was too much competition among builders in England. An editorial in the same issue significantly suggested a cartel of shipbuilders to regulate output and pay compensation for leaving berths idle and reducing orders.

28. *Syren and Shipping*, January 3, 1900; January 2, 1901; July 3, 1912, etc.

29. Lubbock, *Last of the Windjammers*, p. 245.

30. The low output of Harland and Wolff in 1904 was due to a changeover to electricity and other yard modernization schemes. The sudden drop in 1909 was caused by the temporary clearing of three berths in order to turn them into two giant berths for the White Star Liners *Titanic* and *Olympic*.

31. *Fairplay*, December 25, 1913, p. 1084.

32. In 1904 the 10 most productive firms launched 40.1 percent of total British output; in 1905, 38.9 percent; and in 1913, 41.1 percent. Firm output figures for 1883 are given in Pollock, *Modern Shipbuilding*, p. 183; for 1904 and 1905, in the *Glasgow Herald* "Engineering and Shipbuilding Supplement" for those years; and for 1913 in *Fairplay*, December 25, 1913, p. 1084.

33. Pollock, *Modern Shipbuilding*, p. 183; *Fairplay*, December 25, 1913, p. 1084.

5. **Industrial Organization**

1. Bowen, *A Century of Atlantic Travel*, p. 124; D. Napier, *Autobiographical Sketch*, p. 51; J. Napier, *Life of Robert Napier*, pp. 94, 149. The marine engineer was not necessarily an inferior partner. In the case of Robert Napier, the reverse was certainly true.

2. On a slightly different basis, the *Directory of Shipowners* listed, in the years 1903-1914, 76 shipbuilding firms, 56 firms engaged in both shipbuilding and marine engineering, and 38 in marine engineering alone. In addition, a number of firms had closed down by 1914: 3 in shipbuilding, 2 of both trades, and 7 marine engineers.

3. Committee on Admiralty Contracts, 1884, ev. Laird, q. 1598. In 1868 Grantham complained that British builders neglected their building for repairs, which were more profitable. *Iron Shipbuilding*, p. 117.

4. R. C. on Trade Unions, 1867-1868. According to Wigram, q. 16611, the journey to the Clyde cost more than any difference in the prices charged.

5. Some firms, for example Clover, Clayton and Company of Birkenhead, bought up damaged ships, repaired them, and sold them on their own account. Rousiers, *Labour Question*, p. 271; Jefferson, *Pirrie*, p. 201; Smellie, *Shipbuilding in Dublin*, p. 103.

6. R. C. on Trade Unions, 1867-68, ev. Palmer, qs. 17698-17702; Pollock, *Modern Shipbuilding*, p. 171; Jeans, *Pioneers of the Cleveland Iron Trade*, pp. 241-244; Dillon, *Palmer's*, pp. 48-68; Hichborn, *Report on European Dock Yards*, p. 57.

7. *Days at the Factory*, p. 246; Pollock, *Shipbuilding Industry*, p. 89, and *Modern Shipbuilding*, p. 131; *Journal of the Iron and Steel Institute*, 1884, pp. 151-152. The Thames Ironworks and Millwall Iron Works were said to be the exceptions before 1865, making and rolling their own plates. Barry, *Dockyard Economy*, pp. 216, 224.

8. *Shipping*, December 1, 1896.

9. League of Nations, *Memorandum on Shipbuilding*, 1927, p. 29; Scott, *Industrial Survey*, p. 39; *Engineer*, July 10, 1896. When the *Renown* was built at the Royal Dockyard at Pembroke, and the contract for the machinery went to Maudslay, the following items were contracted for with other firms: steering en-

gines, electric dynamos, evaporators and distilling condensers, hydraulic machinery, boat hoists, and indicators. Maudslay's themselves subcontracted to specialist firms for the crank shafts, intermediate and stern shafts, pistons and connecting rods, cross heads, cast steel crank bearing frames, cast steel pistons, cylinder covers, steel springs, brass condenser tubes, copper steam pipes, boiler plates and furnishings, and so on. Hamilton, *Naval Administration*, 1896, p. 178. For the accessories bought from outside for the *Great Eastern*, see the letter of Brunel to his directors, dated May 18, 1853, in Brunel, *Life of Isambard Kingdom Brunel*, p. 303. Also, Westgarth, *TNECI*, 1898-99, p. 87. Hunter preferred his own foundry and forge. Ibid., p. 93.

10. Jefferys, *Engineers*. The driving shafts, for instance, of the largest liners were made by Jessop and Sons, Sheffield. Rousiers, *Labour Question*, p. 267.

11. Hichborn, *European Dock Yards*, p. 65; *Shipbuilder*, Special Number, June, 1911; *Vickers, Son and Maxim, Their Works and Manufactures*, p. 138.

12. The development of the Parsons steam turbine was a typical example. Appleyard, *Charles Parsons*, pp. 88, 132, 145, 267.

13. Pollock, *Modern Shipbuilding*, p. 179; *Statist*, January 19, 1884.

14. Barry, *Dockyard Economy*, pp. 206-207. See also, Schwarz and Halle, *Die Schiffbauindustrie*, p. 143; Neumann, *Technische Entwicklung*, pp. 67-70; Halle, "Volkswirtschaftliche Entwicklung," p. 212.

15. Kindleberger, *Economic Growth*, pp. 146-147.

16. *Shipbuilder*, Summer 1910, pp. 13-14; Peirson, *Great Shipbuilders, The Rise of Harland and Wolff*, pp. 43, 207, 277; *Fairplay*, May 15, 1902, pp. 780-781; William Todd Lithgow, Inventory of Estate, Renfrewshire, vol. 71, 1908, pp. 418-426; John Scott, C.B., Inventory of Estate, Ayrshire, vol. 1, 1903, pp. 325-327, and vol. 2, 1903, pp. 885-886; Robert Sinclair Scott, Inventory of Estate, Ayrshire, vol. 2, 1905, pp. 285-287, and vol. 2, 1906, pp. 935-936; Davies, "The African Steam Ship Company."

17. Connections between shippers and shipbuilders could turn out very badly. See, for example, Robertson, "Shipping and Shipbuilding: The Case of William Denny and Brothers," p. 43, n. 1.

18. A more detailed account is given in Robertson, "Shipping and Shipbuilding." The compilation of the holdings of the Denny partners is taken from information in *Denny, Dumbarton;* from an obituary of Col. J. M. Denny, *Glasgow Herald*, December 11, 1922, p. 5; and from the inventories of the estates of Peter Denny, Sr., Walter Brock, and John Ward, Scottish Record Office, Edinburgh. See also, William Denny and Brothers, journal entry for June 1, 1859.

19. Peter Denny, Sr., Inventory of Estate.

20. These percentages are calculated from data on more than 770 vessels given in William Denny and Brothers, Cost Books 1-10. *Profit* as used here means simply the difference between the sale price on the one hand and the costs of materials, labor, and overhead on the other.

21. Robertson, "Shipping and Shipbuilding," pp. 45-47.

22. R. C. on Shipping Rings, 1909, *Report*, pp. 11, 48, 58-79; *Minority Report*, pp. 100, 105-107, 114; R. C. on the Depression of Trade, 1886, qs. 10075-10080, 10165-10184, 10299-10346, 10663-10678, 11259-11260, 11343ff.; Bates, *American Navigation*, p. 331; Kirkaldy, *British Shipping*, pp. 179-187; Eucken,

Verbandsbildung in der Seeschiffahrt, pp. 18ff., 137ff.; Corson, *The Atlantic Ferry*, pp. 18-19; Kreuter, *Preisbildung in der Linienreederei*, p. 29; *Economist*, February 18, 1882; October 13, 1883; December 25, 1905; December 8, 1906; June 5, June 19, 1909; *Statist*, September 6, 1879; December 1, 1883; February 25, 1888; *Syren and Shipping*, January 27, 1897; January 1, 1902; Fayle, *Short History*, p. 280.

23. The combine also hoped for a monopoly of the Chicago meat shipments. *Statist*, March 22, April 26, 1902; *Economist*, November 23, 1901; Moody, *Truth about Trusts*, p. 105; Bates, *American Navigation*, p. 334.

24. Fayle, *Short History*, p. 278; *Economist*, December 23, 1911; *Syren and Shipping*, January 2, 1901; Jefferson, *Pirrie*, p. 142; Pollard, "The Economic History of British Shipbuilding," diagram A.3, p. 602.

25. Michon, *Les grandes compagnies*, p. 16; Zimmermann, *Ocean Shipping*, pp. 440-441.

26. Harland and Wolff also desired to obtain a secure supply of armor from John Brown as part of Harland and Wolff's new policy of bidding for Admiralty contracts. The merger between the two firms came as a shock to British public opinion, which had firmly believed John Brown to be associated with the Cunard Line only, a preserve of the British Admiralty, and to be immune from the American and German connections of Lord Pirrie. *Economist*, June 22, 1907; *Statist*, July 20, 1907; *Shipbuilder*, no. 5.

27. Chandler, *Strategy and Structure*, pp. 290-291.

28. According to the *Statist*, Armstrong's plan at the end of the century to develop its own armor plate works was what induced the armor plate makers to integrate forward. July 1, 1899.

29. Dougan, *History of North East Shipbuilding*, pp. 89-91; Scott, *Vickers*, p. 34; *Engineering*, November 22, 1889, p. 593; Macrosty, *The Trust Movement in British Industry*, p. 40.

30. Scott, *Vickers*, pp. 40-45; *Fairplay*, June 24, 1897, p. 1065; July 29, 1887, p. 194.

31. Grant, *Steel and Ships*, pp. 20-21, 38; Scott, *Vickers*, pp. 42-43; *Fairplay*, August 10, 1899, pp. 244-245. At the time of the sale, however, the total physical plant at Clydebank, including all machinery, land, buildings, tenements, and so forth of both the shipyard and the engine works, came to only £728,664 10s. John Brown and Co., UCS1, 25/6.

32. Macrosty, *The Trust Movement*, pp. 44-45; Scott, *Vickers*, pp. 49, 92; *Fairplay*, January 23, 1902, p. 133; *Engineering*, October 7, 1904, pp. 455-458.

33. Fairfield's acquired shares in Cammell-Laird in 1906, and together with John Brown the two companies formed the Coventry Ordnance Works in 1904-1905. Similarly, in 1906-1907 Armstrong-Whitworth and Vickers obtained control of the Whitehead Torpedo Company both in Britain and in Austria-Hungary. *The Fairfield Shipbuilding and Engineering Works*, p. 21; Noel-Baker, *The Private Manufacture of Armaments*, pp. 38-39, 459; *Shipbuilder*, Summer 1907, p. 39; Pollard, "The Economic History of British Shipbuilding," diagram A.3, p. 602.

34. Irving, "New Industries for Old?" p. 159.

35. *Statist*, May 3, 1903; October 6, 1906.

36. *Statist*, January 25, 1914; *Economist*, March 1, 1914.

37. For a detailed account of the activities of the steelmakers' associations, see Macrosty, *The Trust Movement*, pp. 66-72. Also, Board of Trade, Departmental Committee on Shipping and Shipbuilding, *Second Report*, pp. 24-25. McCloskey has recently shown, however, that these associations were rarely able to enforce compliance with their policies. *Economic Maturity*, pp. 28-34.

38. Pollard, "Barrow-in-Furness," pp. 213-221; Dougan, *North East Shipbuilding*, pp. 41, 48; *Engineering*, December 29, 1871, p. 417; August 5, 1881, p. 137; Burn, *The Economic History of Steelmaking 1867-1939*, pp. 225-226.

39. Aberconway, *Basic Industries*, pp. 228-231; Carvel, *Stephen of Linthouse*, p. 116; Burn, *Steelmaking*, pp. 373-374; Jones, *Shipbuilding in Britain*.

40. Levy, *Monopolies, Cartels, and Trusts in British Industry*, 2nd English ed., p. 224.

41. Pollock, *Shipbuilding Industry*, p. 87; R. C. on Tonnage Measurement, 1881, qs. 2435-2440; Admiralty Committee on Contracts, 1887, q. 3792.

42. In exceptional cases builders might approach owners directly and offer to build a certain type of ship. Committee on Industry and Trade, Metal Industries, 1928, p. 370; Pollock, *Modern Shipbuilding*, p. 104.

43. Pollock, *Shipbuilding Industry*, p. 87; Mitchell, *Shipbuilding and the Shipbuilding Industry*, p. 32; Barry, *Dockyard Economy*, p. 150.

44. Haldane, *Steamships and Their Machinery*, p. 112.

45. Bruce, *William Denny*, pp. 165-166.

46. Committee on Contracts, 1884, qs. 1581-1582; R. C. on Unseaworthy Ships, 1873, *Final Report*, qs. 9251-9255; Committee on Contracts, 1887, qs. 3803-3805; Rutherford, *Man Who Built the "Mauretania"*, p. 39.

47. Reid, *James Lithgow, Master of Work*, p. 33; Shannon, "Limited Companies of 1866-83," pp. 306-307. Vessels built "on spec" were usually fitted as sparsely as possible to allow the prospective buyer full freedom of choice.

48. Napier, *Robert Napier*, appendix I, pp. 251-258.

49. Board of Trade, Departmental Committee on Shipping and Shipbuilding, 1918, p. 22; Mitchell, *Shipbuilding*, p. 32; Stephen, *A Shipbuilding History*, p. 79; Committee on Contracts, 1884, q. 1656; Robertson, "Shipping and Shipbuilding," p. 40.

50. Jefferson, *Pirrie*, pp. 43, 203; Schwarz and Halle, *Die Schiffbauindustrie*, p. 182; Huldermann, *Albert Ballin*, p. 177.

51. *TINA*, 1913, pp. 2-6.

52. *Clyde Arbitration, 1877*, p. 12; *Statist*, July 15, 1899; R. C. on Unseaworthy Ships, 1873, *Final Report*, q. 16468. The delays in delivery might be serious. Barnes, an Admiralty overseer, in his evidence to the Committee on Contracts, 1884, stated that Palmer's could deliver torpedo vessels in 4 months because they made and rolled their own plates; other firms would take 12 months. Qs. 2021-2024. The smaller yards were especially dependent upon their "iron merchants." Schwarz and Halle, *Die Schiffbauindustrie*, p. 182.

6. The Development of the Shipyard

1. Brief descriptions of the processes involved in shipbuilding, including ordering and design, are given in Talbot, *Steamship Conquest*, pp. 17-38, and in Pollock, *Shipbuilding Industry*, pp. 81-119.

2. Shields, *Clyde Built*, pp. 114-118; *Engineering*, February 7, 1868, p. 111.

3. *Engineering*, October 7, 1904, p. 455.

4. *TIESS*, 1880-81, p. 10; *Engineering*, February 7, 1868, p. 111. Tod and M'gregor also owned a marine engine works, the Clyde Foundry, in Glasgow.

5. *Engineering*, April 18, 1913, pp. 516-518. Rapid increases in the capital value of the yard during the periods when the *Lusitania* and *Aquitania* were being built are shown clearly in table 4.3.

6. *Engineering*, August 3, 1888, p. 103; June 12, 1908, p. 79; July 5, 1912, pp. 3-6.

7. Marshall, *Furness*, p. 330; *Engineering*, October 7, 1904, p. 455.

8. *Engineering*, September 25, 1874, p. 237; August 6, 1880, p. 108.

9. Pollard, "Barrow-in-Furness"; Marshall, *Furness*, pp. 389, 398; Scott, *Vickers*, p. 44; *Engineering*, August 7, 1896, p. 163; October 7, 1904, p. 455.

10. Dunn, "The Arrangement and Equipment of Shipbuilding Works," p. 169.

11. *Engineering*, August 28, 1896, p. 269; Dunn, "Arrangement and Equipment," p. 169. See the map in *Engineering*, July 12, 1895, p. 32.

12. For a discussion of the proper way of locating the various shipyard departments in relation to the berths, see Dunn, "Arrangement and Equipment," p. 169.

13. *Engineering*, December 11, 1908, pp. 775-776.

14. *Engineering*, October 18, 1889, p. 451; July 3, 1914, pp. 1-3.

15. John, "Shipbuilding Practice of the Present and Future," p. 68; *SNAME*, 1900, pp. 192-194.

16. See the list of machinery at Barrow in Dunn, "Arrangement and Equipment," pp. 216-218.

17. *TIESS*, 1902-03, pp. 248-264; 1910-11, p. 343; *TINA*, 1899, p. 121; *TNECI*, 1905-06, p. 54; *SNAME*, 1898, p. 49; R. C. on the Poor Law, 1910, appendix vol. VIII, ev. Browne, q. 86,339.

18. *TINA*, 1899, p. 121; *TNECI*, 1905-06, p. 29; *TIESS*, 1902-03, pp. 249-250; R. C. on the Poor Law, 1910, appendix vol. VIII, ev. Browne, qs. 86,332-86,334.

19. Dunn, "Arrangement and Equipment," p. 169.

20. *TINA*, 1899, p. 353; John, "Shipbuilding Practice," p. 68; *Shipbuilder*, no. 2; *Steamship*, June 1894; *Engineering*, October 3, 1879, p. 257; August 8, 1902, p. 172.

21. *SNAME*, 1900, p. 168; *TIESS*, 1894-95, pp. 186-187; *TINA*, 1897, pp. 305-306; *TNECI*, 1894-95, pp. 14, 27, 35, 125-126.

22. *TIESS*, 1880-81, pp. 17-27; *SNAME*, 1900, p. 170; *Proceedings of the Institution of Civil Engineers*, 1876-77, p. 68; *Engineering*, June 12, 1908; Schwarz and Halle, *Die Schiffbauindustrie*, p. 150.

23. *TNECI*, 1905-06, pp. 28-30, 53; *TIESS*, 1893-94, pp. 25-34.

24. *Factory Inspector's Report*, 1902, Wilson, p. 325; *Engineering*, August 8, 1902; June 12, 1908; *The Engineer*, July 10, 1869; *Shipbuilder*, nos. 44, 46; *Swan, Hunter and Wigham Richardson*, p. 21; Richardson, *Vickers*, pp. 166-168; *TNECI*, 1900-01 and 1901-02, visits to Furness-Withy and Swan and Hunter; Scott's, *Two Centuries of Shipbuilding*, pp. 96, 104; *The Palmer Record*, nos. 2, 9; Thames Iron Works, *Quarterly Gazette*, no. 12, p. 169; no. 21, p. 38; *Shipping*, July 1, 1896; Stephen, *A Shipbuilding History*, p. 86; *SNAME*, 1900, p. 169.

25. *TIESS*, 1894-95, p. 279; *TNECI*, 1895-96, p. 106; 1905-06, pp. 28, 53, 59; *TINA*, 1897, pp. 297-304.

26. *TINA*, 1897, p. 309; Dunn, "Arrangement and Equipment," p. 169.

27. *TINA*, 1899, p. 121.

28. *TINA*, 1879, pp. 232, 236-237, 241-242; *TIESS*, 1877-78, p. 5; 1880-81, pp. 20-21; *TNECI*, 1895-96, p. 101; *Engineering*, August 6, 1880, p. 108; September 11, 1880, p. 245.

29. *TIESS*, 1896-97, p. 174; *TINA*, 1898, pp. 16-18.

30. *TNECI*, 1905-06, pp. 54-55; *TINA*, 1899, p. 123; 1902, p. 230; *SNAME*, 1898, p. 30.

31. *TINA*, 1907, pp. 308-310; John, "Shipbuilding Practice," pp. 69-70.

32. *TNECI*, 1918-19, pp. 49-50; John, "Shipbuilding Practice," p. 71.

33. *Engineering*, February 7, 1868, p. 111; December 29, 1871, p. 417; September 25, 1874, p. 237; October 7, 1914, p. 458; John, "Shipbuilding Practice," p. 69; *TINA*, 1903, p. 154; *TIESS*, 1910-11, pp. 403-404.

34. *TIESS*, 1910-11, pp. 345-357; *Shipbuilder*, Summer 1902, p. 12; Spring 1908, p. 226; *Engineering*, December 11, 1908, pp. 778-779.

35. John, "Shipbuilding Practice," p. 68.

36. *SNAME*, 1900, pp. 167-168.

37. *TINA*, 1902, pp. 243-246; John, "Shipbuilding Practice," p. 68; *Engineering*, November 7, 1913, p. 611.

38. *TINA*, 1902, pp. 246-249; Dunn, "Arrangement and Equipment," p. 183; *Engineering*, August 26, 1910, pp. 291-292; July 5, 1912, pp. 11-12.

39. *TIESS*, 1894-95, p. 206; 1896-97, pp. 159-160; *TINA*, 1897, pp. 297-304; 1902, p. 230.

40. *SNAME*, 1900, p. 176; *TINA*, 1902, p. 270; 1903, p. 153.

41. For a discussion of the implications of the "head start" theories, see Ames and Rosenberg, "Changing Technological Leadership and Industrial Growth."

42. Wright, *Freshwater Whales*, pp. 11-16; *SNAME*, 1900, p. 195.

7. Scientific Shipbuilding and the Evolution of Technical Education

1. For example, the works of Tenaud (1689 and 1695), Huygens (1693), James Bernoulli (1696), John Bernoulli (1714), Paul Hoste (1697), Beuguer (1746), Euler (1749), Duhamel (1752), and Daniel Bernoulli (1757).

2. Slade, *Naval Construction and Naval Promotion*, pp. 1-2; *TINA*, 1860, p. 186; Abell, *Shipwright's Trade*, p. 149; *Nautical Magazine*, 1852, p. 590; Smith and Holden, *Where Ships Are Born*, p. 20; Rutherford, *Man Who Built the "Mauretania"*, p. 33.

3. Rolt, *Isambard Kingdom Brunel*, chs. 12-14; *Quarterly Review*, March 1856; Brunel, *Life of Isambard Kingdom Brunel*, pp. 255ff.; Pollock, *Shipbuilding Industry*, pp. 50-52; Noble, *The Brunels*, p. 216; Russell, *Very Large Ships*, p. 6; *TINA*, 1872, pp. 173-180.

4. Cormack, "An Economic History of Shipbuilding," p. 125; Russell, *Very Large Ships*, p. 10, and *Steam and Steam Navigation*, p. 294; Napier, *Life of Robert Napier*, p. 80; Halsted, *Iron-Cased Ships*, pp. 73-74.

5. Russell, *The Fleet of the Future, Iron or Wood?*, pp. 55-56.

6. Smith, *History of Naval and Marine Engineering*, pp. 62, 122; Zimmermann, *Ocean Shipping*, pp. 137-138; Lubbock, *China Clippers*, p. 37; Scott's, *Two Centuries of Shipbuilding*, p. 12; *TINA*, 1941, p. 83; Rennie, *Autobiography*, p. 62. See Barry's typical contempt for science. *Dockyard Economy*, pp. 119-122.

7. Pollock, *Shipbuilding Industry*, p. 85; Abell, *Shipwright's Trade*, p. 157; Bruce, *William Denny*, pp. 104-105; *TINA*, 1941, pp. 85-86, 105.

8. Pollock, *Modern Shipbuilding*, p. 119; Schwarz and Halle, *Die Schiffbauindustrie*, p. 251; *Shipbuilder*, no. 39; Bruce, *William Denny*, pp. 169, 195; Gilfillan, *Inventing the Ship*, p. 257.

9. Pollock, *Shipbuilding Industry*, p. 86; *TINA*, 1901, pp. 264ff.; 1904, pp. 39-52; 1909, pp. 144-146; 1910, p. 22; *TNECI*, 1891-92, p. 288; Lady Yarrow, *Alfred Yarrow*, p. 208; *Shipbuilder*, no. 21.

10. Blake, *Down to the Sea*, p. 236.

11. Bruce, *William Denny*, pp. 144, 147; Denny, "The Trial of Screw Steamships," *British Association Reports*, 1875, p. 246; *TINA*, 1941, p. 108.

12. The common style of hull was a straight bow, but slow vessels for which capacity was more important than speed had round bows, and very fast vessels had slightly hollow bows. Pollock, *Shipbuilding Industry*, p. 83; Smiles, *Men of Invention and Industry*, pp. 306-310; *Economist*, December 3, 1910; Peirson, *Great Shipbuilders*, pp. 12-13; Reed, *Our Ironclad Ships*, p. 196. Compare *TIESS*, 1907-08, p. 237; and *TINA*, 1941, p. 90.

13. Holmes, *Ancient and Modern Ships*, II, 39, 65; *TINA*, 1911, II, 153; Pollock, *Modern Shipbuilding*, pp. 13-17, and *Shipbuilding Industry*, pp. 53-54; *TNECI*, 1934-35, p. 407; British Association, *Trade and Manufactures*, p. 124.

14. Bruce, *William Denny*, pp. 78-81; Barry, *Dockyards and Private Shipyards*, p. 11.

15. Haldane, *Steamships and their Machinery*, pp. 233-234; *Transactions of the Institute of Marine Engineers*, 1912-13, p. 212; *TINA*, 1909, pp. 10-17; Jeans, *Iron Trade*, p. 107. In 1905 even patterns, hitherto separate and different for each engine, could be lent to one firm on the Clyde by others at short notice in order to break a local patternmakers' strike. *Statist*, May 6, 1905.

16. Bruce, *William Denny*, pp. 88-89; Scott's, *Two Centuries of Shipbuilding*, p. 88; Gilfillan, *Sociology of Invention*, pp. 83, 95, and *Inventing the Ship*, p. 182.

17. Appleyard, *Charles Parsons*, p. 101; Smellie, *Shipbuilding in Dublin*, p. 68; *Jahrbuch der Schiffbautechnische Gesellschaft*, 1901, p. 434.

18. *TINA*, 1903, p. 56; *TNECI*, 1903-04, p. 184; 1907-08, p. 109.

19. *TNECI*, 1907-08, p. 112; 1908-09, p. 250.

20. *TNECI*, 1907-08, pp. 67-68, 70-71; 1908-09, p. 103.

21. *TINA*, 1881, pp. 155, 157; *TNECI*, 1903-04, p. 82; 1915-16, pp. 218, 250; R. C. on the Poor Law, 1910, ev. Browne, qs. 86,295, 86,365-86,381.

22. Bray, "The Apprenticeship Question."

23. *TINA*, 1903, p. 53; *TNECI*, 1915-16, pp. 194-197 and the discussion following.

24. *TIESS*, 1880-81, p. 153; *TNECI*, 1915-16, pp. 210, 275.

25. *TNECI*, 1908-09, pp. 182, 192; 1915-16, pp. 229, 246.

26. *TNECI*, 1899-1900, pp. 131, 135; 1903-04, p. 139; 1908-09, pp. 218-219, 251; *TINA*, 1903, pp. 54-55, 62, 64.

27. Among them were two future Chief Constructors, E. J. Reed and N. Barnaby; also Barnes, Crosland, and Pearce.

28. Abell, *Shipwright's Trade*, p. 152. "Only a score of students [was] trained, but they were able to guide the Royal Navy in the drastic changes from the wooden fighting ship to the iron-clad."

29. *TINA*, 1867, pp. 223-239; 1877, pp. 361-362; 1878, pp. 120-121.

30. *TINA*, 1889, p. 65; 1905, II, 251-252; *TNECI*, 1908-09, pp. 185-187; *Shipbuilder*, no. 48; *Liverpool Courier*, January 8, 1909; Sanderson, *The Universities and British Industry 1850-1970*, p. 45.

31. *TINA*, 1877, pp. 361-362; 1878, pp. 120-121.

32. Compiled from *TINA*, 1905, II, 242-246.

33. *TINA*, 1878, pp. 212-213; 1905, II, 251.

34. *TINA*, 1905, II, 253-254.

35. *TINA*, 1878, pp. 130-133; 1881, p. 161; *TNECI*, 1908-09, pp. 220, 245-249.

36. *TNECI*, 1903-04, pp. 73, 88-90; 1908-09, pp. 187-188.

37. *TINA*, 1903, p. 37; 1911, II, 210; *TNECI*, 1903-04, pp. 19-21.

38. *TNECI*, 1907-08, pp. 65, 73, 81, 100; 1915-16, pp. 208, 228, 234-235, 240.

39. *TNECI*, 1907-08, p. 111; 1915-16, pp. 220-221.

40. Russell, "On the Education of Naval Architects in England and France."

41. See the following articles and the accompanying discussions: Dalby, "The Training of Engineers in the United States"; Wurl, "Technical Education in Germany"; Flamm, "The Scientific Study of Naval Architecture in Germany"; Smith, "The Future of British Engineering and Shipbuilding."

42. Direct estimates suggest that productivity may, in fact, have continued to rise rapidly until as late as 1908 (see chapter 9).

43. *TNECI*, 1914-15, p. 184; Pollard, "British and World Shipbuilding," table 3, p. 438. Although the productivity figures are crude, the margin of error would have to be very large to reverse the basic finding that the productivity of the average British shipyard worker was greater than that of his nearest competitors.

44. Talbot, *Steamship Conquest*, p. 336.

45. Quoted in *TNECI*, 1915-16, p. 254.

46. *TNECI*, 1903-04, pp. 42, 154.

47. For data on American educational expenditures, see *TINA*, 1903, p. 39.

48. *TINA*, 1911, II, 5-15.

49. For the interwar years see in particular Jones, *Shipbuilding in Britain, Mainly between the Two World Wars*; and Buxton, "The Scottish Shipbuilding Industry between the Wars: A Comparative Study."

8. Labor and Labor Relations

1. Demarcation Committee, 1894, *Shorthand Notes of Reference Court*, p. 6; R. C. on Trade Unions, 1867-68, q. 2968. See the table of 90 different shipyard trades in Smellie, *Shipbuilding in Dublin*, pp. 63-65.

2. Bowley and Wood, "Statistics of Wages in the U.K. for the Last Hundred Years, Engineering and Shipbuilding," pp. 171, 178; Yarrow, *Alfred Yarrow*, p.

186; Jefferys, *Engineers*, p. 143; Blake, *Down to the Sea*, p. 294; MacArthur, *History of Port Glasgow*, p. 113.

3. Abell, *Shipwright's Trade*, p. 185; R. C. on Labour, 1892, Group A, qs. 21,400-21,401.

4. R. C. on Trade Unions, 1867-68, q. 17,165.

5. S. C. on the Navigation Acts, 1848, qs. 6056, 6265-6266; also Young before the same committee, 1847, q. 5044; *Clyde Arbitration, 1877*, pp. 99, 124; S. C. on Combination Laws, 1825, pp. 782-783, 799; Associated Shipwrights' Society, *Quarterly Report*, 1905, first quarter; Webb MSS., II, 122.

6. S. C. on Government Contracts, 1897, qs. 2662, 2678; Associated Shipwrights' Society, *Quarterly Report*, 1894, second quarter (Greenock), third quarter (Bristol Channel); Webb MSS., I, 187, and II, letter by Cummings, 25 May 1894; S. C. on Artizans, 1824, pp. 204-206; Rathbone, "Shipwrights' Trade Societies," p. 480.

7. *Clyde Arbitration, 1877*, p. 125; R. C. on Labour, 1892, Group A, qs. 25,704-25,705; Bowley and Wood, "Statistics of Wages," 1905, pp. 594-595.

8. R. C. on the Depression of Trade, 1886, Trade Union Questionnaires, Report of Ayr Branch, p. 18; Robert Knight, "Agreement between Employers' Associations and the Boilermakers and Iron and Steel Shipbuilders Society," 1894, p. 6; Cummings, *History of the Boilermakers' Society*, pp. 36, 127-128; "Agreement between Laird Brothers and the Boilermakers," September, 1891 (Webb MSS.); R. C. on Labour, 1892, Group A, Reply to Trade Union Questionnaires, pp. 121-125; Cormack, "An Economic History of Shipbuilding," p. 224; Webb MSS., II, 190; Inglis, *The Apprentice Question*, pp. 11ff.

9. Resolution of the Liverpool Shipwrights, 4 July 1851, Webb MSS., I, 372; R. C. on Labour, 1892, Group A, Reply to Trade Union Questionnaires, pp. 30-31; Associated Shipwrights' Society, *Quarterly Report*, 1894, second quarter (Hull); Proceedings at Court of Sessions, Edinburgh, May 17-18, 1904, Peter Houston vs. A.S.S., p. 20. In the Dockyards the apprenticeship period was lowered from seven years to six in 1891. R. C. on Labour, 1892, Group A, q. 23,983.

10. R. C. on Labour, 1892, Group A, qs.20,683, 21,219, 22,794-22,795, 23,288-23,289, 23,588.

11. *TIESS*, 1880-81, pp. 150-155, 160-161. In 1881 the Glasgow engineer David Rowan noted that poor apprenticeship conditions "remained much the same as they were forty years ago." Sir Bernhard Samuelson, for many years an ironmaster, stated in 1909, "There has been very little teaching by skilled workmen to apprentices within my time." R. C. on the Poor Law, 1909, *Report*, pp. 348-349.

12. R. C. on Labour, 1892, Group A, qs. 20,988-20,990, 21,403-21,409.

13. S. C. on Combination Laws, 1825, pp. 250-275; Webb, *Industrial Democracy*, pp. 73-74; Clegg, Fox, and Thompson, *A History of British Trade Unions since 1889*, I, 132, 443.

14. Webb, *The History of Trade Unionism*, pp. 204-215; R. C. on Labour, 1892, Group A, q. 22,655.

15. Mortimer, *History of the Boilermakers Society*, I, passim; Cummings, *History of the Boilermakers' Society*, pp. 13, 111-112, 147; Clegg, Fox, and Thompson, *British Trade Unions*, I, 132.

16. Lynch, "Skilled and Unskilled Labour in the Shipbuilding Trade," *Report of the Industrial Remuneration Conference*, pp. 114, 116-117; R. C. on Labour, 1892, Group A, qs. 20,449, 20,566, 20,632-20,647, 20,815-20,817, 20,998, 21,-244-21,245, 21,256; T. U. C., *Annual Reports*, 1877, pp. 26, 28; 1878, pp. 10, 15, 16; Webb MSS., II, 41-43, 86, 131, 212-217.

17. Shipbuilding Trades Committee, *Proposed Machinery*, 1908; "Verbatim Report of the Edinburgh Conference," p. 1.

18. "Agreement" between the Shipbuilding Employers' Federation and the Shipyard Trade Unions, 1909, sects. II-V; *Statist*, March 11, 1911; November 30, 1912.

19. R. C. on Trade Unions, 1867-68, qs. 17,281-17,282; Smith and Holden, *Where Ships Are Born*, p. 78; Shipbuilding Trades Agreement, *Report*, p. 18; A. S. S., *Quarterly Report*, 1889, third quarter; *Labour Gazette*, December 1895; A. S. E., *Jubilee Souvenir*, 1901, pp. 104-105; Jefferys, *Engineers*, p. 143; *Economist*, April 29, 1905; R. C. on Labour, 1892, Group A, qs. 25,504, 26,512, 26,548-26,549, 26,591.

20. *Ninth Report by the Chief Correspondent of the Board of Trade on Trade Unions*, 1896, C.- 8644 (1897), pp. 38-45; *Report on Trade Unions in 1908-1910*, Cd. 6109 (1912), pp. 16-19.

21. The sources for this and the following paragraphs are widely scattered. Reliable information can be found in the following references: R. C. on Labour, 1892-94, Replies from Trade Societies, appendix XLIV; Replies from Employers' Associations, pp. 311-315; Replies from Employers, pp. 274-287; also evidence before the Commission, 1892, Group A, qs. 20,550-20,557, 20,662-20,665, 24,139; R. C. on the Depression in Trade, 1886, appendix A.IV, p. 299; appendix A.V, p. 302; appendix D, Replies from Trade Societies; Board of Trade, *Labour Gazette;* and *Return Showing the Number of Hours Worked*, 1890, pp. 54-55; and *Returns of Wages, 1830-1886*. C. 5172 (1887), pp. 39-42, 207-230; Bowley and Wood, "Statistics of Wages," 1905, p. 602; *Clyde Arbitration, 1877*, pp. 12, 38, 54, 144; Webb MSS., II, 160-167; Smiles, *Men of Invention*, p. 292; Rutherford, *The Man Who Built the "Mauretania"*, p. 34.

22. *Beehive*, September 23, October 7, 1865; June 3, August 26, September 16, 23, October 14, 1871; Cormack, "An Economic History of Shipbuilding," p. 261; Jefferys, *Engineers*, pp. 86, 96-99; Wilkinson, *The Town that Was Murdered*, pp. 89-91; Burnett, *History of the Engineers' Strike*, 1872; Bowley and Wood, "Statistics of Wages," 1905, p. 614; R. C. on Labour, 1892, Group A, q. 25,599; *Sunderland Chronicle*, December 5, 1891; Jefferson, *Pirrie*, p. 95.

23. Young to the Commons S. C. on the Navigation Acts, 1847, qs. 5646-5647, 6009, and also to the Lords S. C. on the Navigation Acts, 1848, q. 8094; R. C. on Trade Unions, 1867-68, qs. 18,003-18,008; R. C. on Labour, 1892, Group A, qs. 21,085, 25,589; Bowley and Wood, "Statistics of Wages," 1906, pp. 179-180. The *Statist*, on April 25, 1914, still complained that the Clyde worker would not work five and a half days if he could earn a satisfactory wage in four.

24. The employers' view (for example, Seaton to the R. C. on Labour, 1892, Group A, q. 25,511) that overtime was worked only under pressure and that therefore no unemployment could result, was misleading. The London Bargebuilders found that after their successful strike in 1890, which reduced "forced

overtime," 30 additional members found employment and could be taken off their funds. Ibid., qs. 20,293, 20,324. Fear of unemployment was also the motivation behind union rules limiting output. Webb MSS., I, 273, 367.

25. R. C. on Labour, 1892, Group A, qs. 20,111, 20,120-20,122, 20,870-20,872, 25,194, 25,273-25,278, 25,332; R. C. on the Depression in Trade, 1886, qs. 11,030, 11,051-11,052.

26. The confidential reports of a number of the meetings between July 31, 1909 and October 3, 1911 are available in the Webb collection. Also, R. C. on Labour, 1892, Group A, q. 22,387; Boilermakers' Society, *Annual Report*, 1912, p. v.

27. R. C. on Labour, 1892, Group A, q. 24,906; Smiles, *Men of Invention*, p. 303.

28. *Capital and Labour*, November 8, 15, December 9, 16, 23, 1874; January 6, February 3, March 17, April 7, 1875; March 14, May 16, 30, August 29, September 26, October 17, November 14, December 5, 1877; *Clyde Arbitration, 1877*, pp. xi-xii, 5, 14, 22, 29-30, 64, 141-143; R. C. on Labour, 1892, Group A, qs. 21,414-21,415, 21,660, 21,699-21,675; Cormack, "An Economic History of Shipbuilding," p. 267; Bowley and Wood, "Statistics of Wages," 1905, pp. 582-583; Webb and Webb, *History of Trade Unionism*, p. 343; Cummings, *Boilermakers'*, p. 106; Webb MSS., II, 31.

29. Jefferys, *Engineers*, pp. 140ff.; *Statist*, January 25, 1896; *Labour Gazette*, November, December, 1895; January, February, 1896; A. S. E., *The Engineering Trade Lock-out*, 1897-98, and documents appended; A. S. E., *Jubilee Souvenir*, pp. 105-106; Cummings, *Boilermakers'*, p. 149; Yarrow, *Alfred Yarrow*, pp. 182-185.

30. *Statist*, July 9, August 17, September 10, 17, 24, November 5, 12, December 17, 1910; *Economist*, September 10, 24, October 15, November 12, 1910; *Labour Gazette*, December, 1910; Shipyard Standing Committee, Minutes, August 12, 16, September 10, 20, 21, 22, October 6, November 21, 1910; *Syren and Shipping*, January 4, 1911; Boilermakers' Society, *Annual Report*, 1911, p. v.

31. R. C. on Trade Unions, 1867-68, q. 18,077; *TINA*, 1899, p. 121.

32. R. C. on the Depression in Trade, 1886, q. 11,066; Wilkinson, *The Town that Was Murdered*, pp. 73, 101-102; R. C. on Trade Unions, 1867-68, q. 17,395; *Clyde Arbitration, 1877*, p. 136.

33. *Clyde Arbitration, 1877*, pp. 87-88, 93, 124; Burnett, *Engineers' Strike*, pp. 32ff.; *Beehive*, July 29, 1871; also October 25, 1862; April 18, 1868; April 29, 1874. On the other hand, Harland broke the strike in Belfast, called after his arrival, by importing his old foremen and craftsmen from the Tyne.

34. Phelps Brown, *The Growth of British Industrial Relations*, p. 236. As early as 1897 the Webbs chose shipbuilding as their primary example of interunion warfare in their discussion of "the right to a trade." *Industrial Democracy*, pp. 508-527. More recent treatments of demarcation disputes in shipbuilding include Roberts, *Demarcation Rules in Shipbuilding and Shiprepairing*, ch. 2; and Brown, Brannen, Cousins, and Samphier, "The Contours of Solidarity: Social Stratification and Industrial Relations in Shipbuilding."

35. S. C. on Combination Laws, 1825, pp. 192-194, 217, 222; R. C. on Trade Unions, 1867-68, qs. 16,591-16,608.

36. S. C. on Artizans and Machinery, 1824, pp. 190, 192-193.

37. R. C. on Trade Unions, 1867-68, qs. 16,616-16,630, 16,711-16,712, 17,050-17,066, 17,076-17,095, 17,131, 17,193, 17,226, 17,335, 17,642; R. C. on Labour, 1892, Group A, qs. 20,798-20,800. Since 1945 a revised version of the contract system, under which the members of a trade, although employees of the builder, are paid a fixed price for all of the work on a ship, has been introduced. Brown et al., "The Contours of Solidarity," pp. 19-20.

38. R. C. on Labour, 1892, Group A, qs. 20,448-20,456, 20,511-20,529, 20,561, 20,608-20,620, 20,632, 20,668-20,671, 20,775-20,800, 20,853, 20,995-20,999; Lynch, "Skilled and Unskilled Labour," pp. 114-119; Hobsbawm, "The Labour Aristocracy in Nineteenth Century Britain," table V, p. 228.

39. *TNECI*, 1888-89, pp. 46-48.

40. Roberts, *Demarcation Rules*, ch. 2; R. C. on Trade Unions, 1867-68, q. 16,852.

41. Brown, *British Industrial Relations*, p. 102. The attention given to job security is illustrated both by the long struggles of the unions to regulate hours and overtime and by the very high proportion of their statements that witnesses before the R. C. on Labour in 1892 chose to devote to matters of apprenticeship and hours rather than to wages.

42. A dispute between caulkers and shipwrights was reported in 1825, for instance. S. C. on Combination Laws, 1825, p. 185. Also, R. C. on Labour, 1892, Group A, qs. 20,448-20,457, 26,051-26,052, 26,284, 26,419-26,421; Clegg, Fox, and Thompson, *British Trade Unions*, p. 131; Cummings, *Boilermakers'*, p. 111; Webb and Webb, *Industrial Democracy*, pp. 516-518.

43. Webb and Webb, *Industrial Democracy*, pp. 509, 513.

44. R. C. on Labour, 1892, Group A, qs. 21,434-21,437.

45. Ibid., qs. 26,037, 26,056, 21,438-21,445; Webb and Webb, *Industrial Democracy*, pp. 510-512.

46. R. C. on Labour, 1892, Group A, q. 21,453.

47. *Report on the Strikes and Lockouts of 1900*, Cd. 689 (1901), p. liii; R. C. on Labour, 1892, Group A, qs. 26,132, 26,287-26,289; *Industrial Council Inquiry into Industrial Agreements*, Cd. 6953 (1913), qs. 2819-2821, 3556, and appendix V, p. 589; *Fairplay*, November 16, 1911, p. 696; *Second Report on Rules on Voluntary Conciliation and Arbitration Boards and Joint Committees*, Cd. 5346 (1910), pp. 196, 199, 202; Aves, "Labour Notes: The Dispute in the Engineering Trade, &c.," p. 630.

48. R. C. on Labour, 1892, Group A, q. 26,284.

49. Ibid., qs. 24,139, 24,167, 21,819. Barnaby held that, while more skilled than the men in private yards, they did not work as hard, since they did not have to fear instant dismissal. Committee on Dockyard Management, 1886-87, q. 1167.

50. *Beehive*, September 12, 1863; R. C. on Control and Management of Naval Yards, appendix 53 and q. 1394; Barry, *Dockyard Economy*, p. 58n. In Sheerness, weekly earnings within 13 weeks ranged from 3/10 per day to 7/2, in Deptford from 4/2 to 7/6, etc. R. C. on Control and Management of Naval Yards, 1861, appendix 23, gives detailed statistics.

51. S. C. on Admiralty Money and Accounts, 1868, qs. 3404, 3412-3416, 3434.

52. R. C. on Labour, 1892, Group A, qs. 21,697-21,701, 21,789-21,792, 24,308, 24,037, 25,821, and Replies from Trade Union Questionnaires, pp. 121-122; A.S.S., *Quarterly Report*, 1891, first quarter, report on the conference held in London, February 18-20, 1891; Webb MSS., I, 57, 62, 106; II, 243.

53. Admiralty letter, March 25, 1891, in B.P.P., 1890-91, LII (173.), p. 2; also R. C. on Labour, 1892, Group A, qs. 25,811-25,812, 25,821.

54. R. C. on Labour, 1892, Group A, qs. 21,483-21,490, 21,729, 21,738-21,740, 24,236, 24,405, 24,456, 24,201, 24,138-24,139; Webb MSS., I, 62; II, 208-209; S.C.A., Devonport Branch Meetings, *Reports*, 1891, first, second, and third quarters.

55. B.P.P., 1893, LV (386.) and (429.); 1906, LXX (Cd. 2861), p. 75; *Labour Gazette*, July 1894; A.S.S., *Quarterly Report*, 1897, third quarter; 1898, first quarter; 1906, fourth quarter; 1909, first quarter; 1912, second quarter.

56. S. C. on Government Contracts, 1897, *Report*, pp. iii-vii and qs. 4246-4250; Hills, *The Thames Ironworks vs. the Admiralty*, 1911, p. 18.

9. Wages, Employment, and Productivity

1. Bowley and Wood, "Wages in Shipbuilding and Engineering."

2. Webb MSS., I, 200; *Clyde Arbitration, 1877*, pp. 5, 49, 132.

3. The series were calculated from raw data by Denny employees and appear in their final form in the firm's "wages notes."

4. The formula is: (Total annual wage bill/number of fortnightly pays)/(average number of workers per pay period) = "average earnings per man per pay."

5. Derived from ledger entries for Charles Connell and Company, July 1902-June 1909. Over the period, each month's percentage of the total annual wages bill was as follows: January, 5.6; February, 8.7; March, 8.7; April, 8.3; May, 8.5; June, 9.8; July, 7.0; August, 8.8; September, 8.2; October, 7.4; November, 8.1; December, 11.0.

6. See for example, "Number of Employees in Fairfield Yard as per Weekly Returns Commencing January, 1895" which gives figures for both the shipyard and the engine works down to the end of 1913. Fairfield Collection, UCS2/54/1.

7. Wilkinson, *The Town that Was Murdered*, p. 103; Jewkes and Winterbottom, *Industrial Survey of Cumberland and Furness*, 1933, pp. 3, 67.

8. The following trends were calculated: (1) for the Bowley and Wood index of time wages, $y = 128 + 1.31x$; (2) for the money earnings of all workers in the Leven Shipyard, $y = 1.20 + 0.017x$; (3) for the real earnings of all workers in the Leven Shipyard, $y = 1.25 + 0.024x$; (4) for tonnage launched on the Clyde, $y = 335 + 9.10x$; (5) for Denny's and Connell's productivity, calculated with the Feinstein price index, $y = 5.57 + 0.051x$; (6) for Denny's and Connell's productivity, calculated with the Pollard-Robertson price index, $y = 5.98 + 0.093x$. In equations 1-3, y denotes the amount or index number of wages or earnings; in 4, y denotes tonnage launched; and in 5 and 6, y denotes productivity (output by value per man-week). Throughout, x denotes the year.

9. R. C. on Labour, 1892, Group A, qs. 21,307, 22,288; Replies from Employers, pp. 274-287; Replies from Trade Unions.

10. R. C. on Labour, 1892, Group A, qs. 25,104-25,153, 22,077-22,084.

11. R. C. on Labour, 1892, Group A, qs. 20,821-20,829; 26,080, 26,304;

Bruce, *William Denny*, pp. 129-130; *Glasgow Herald*, letters from "Shipbuilders," July 9, 12, 17, 18, 19, 1900, alleging that drunkenness was peculiar to Scotland because of the whisky drunk there.

12. Board of Trade, *Report on Wages*, C.6889 (1893), p. 99, and *Enquiry into Earnings*, 1911, p. xxxiv; *Labour Gazette*, October 1912; *Statist*, October 13, 1906.

13. R. C. on Labour, 1892, Group A, q. 20,561; Replies from Employers' Associations (311-315) and from Employers (274-287); *Statist*, October 13, 1906.

14. It is likely that some of the labor would have been performed in the preceding year ($x - 1$), especially for vessels launched in the early months of the year x, or in the following year ($x + 1$) for vessels launched toward the end of the year x. As most of the vessels constructed by Denny's and Connell's were relatively small and required less than a year to build, it is reasonable to assume that one-half of the total labor input for the aggregate tonnage launched during a particular year was performed in that year, and one-quarter in each of the preceding and following years.

Because of the inaccuracy of the date of launching as an indicator of when the work was performed, three-year moving averages are used in the charts.

15. Maywald, "Construction Costs," p. 50; Feinstein, *National Income*, pp. 187-188 and table 63, pp. T136-137.

16. The GNP deflator is given in Brown and Browne, *A Century of Pay*, pp. 442-443, column 5. To allow for the extreme cyclical fluctuations in shipbuilding, the deflator has been altered to double the amplitude of the year-to-year changes.

17. Maywald, "Construction Costs," pp. 53-55.

18. Brown and Browne, *A Century of Pay*, pp. 442-443, column 9.

19. Ibid., column 3.

20. Pollard, "The Decline of Shipbuilding on the Thames," pp. 72-89.

21. Industrial Council, *Enquiry into Industrial Agreements*, Cd. 6953 (1913), qs. 2753-2755.

22. Weekly time wages for joiners and shipwrights are in Robertson, "Management of Manpower in British Shipbuilding," tables C.6 and C.7, pp. 277-279.

23. R. C. on Trade Unions, 1867-68, qs. 17,179-17,180, 17,395, 17,555-17,564, 17,842-17,846; Webb and Webb, *Industrial Democracy*, p. 74; Cummings, *Boilermakers'*, p. 31; R. C. on Labour, 1892, Group A, q. 20,724.

24. R. C. on Labour, 1892, Group A, q. 20,698; Bowley and Wood, "Wages in Shipbuilding," 1905, p. 565.

25. Salaries are given in Robertson, "Management of Manpower in British Shipbuilding," tables XXI and XXII, pp. 208-209.

10. The Influence of the State

1. See the discussion and extensive bibliography in Taylor, *Laissez-Faire and State Intervention in Nineteenth-Century Britain*.

2. Parris, "The Nineteenth-Century Revolution in Government: A Reappraisal Reappraised," in Stansky, ed., *A Victorian Revolution*, p. 54.

3. Baxter, *The Introduction of the Ironclad Warship*, pp. 196, 201-204.

4. Reed, *Our Ironclad Ships*, pp. 45-46; Henwood, "Ironclads, Present and Future," address to the R.U. Services Institution, p. 8.

5. Hovgaard, *Modern History of Warships*, pp. 387-390; Wilmot, *Our Fleet Today*, pp. 138-139, 171; Admiralty Committee on Designs, 1872, p. XIV, C.477, 489, and letter from Whitworth, p. xxvi.

6. Hovgaard, *Modern History*, pp. 168-181, 203-215; *Proceedings of the Institution of Civil Engineers*, 1903-04, p. 152; Reed, *Fast Cruisers*, pp. 2-12.

7. Manning, *Life of Sir William White*, p. 345; Wood, *The Battleship*, p. 235; Hovgaard, *Modern History*, pp. 74-83, 214, 264-265; *TINA*, 1908, pp. 59-66; Jackson, *Ship under Steam*, p. 214.

8. Levert, *Fundamentals of Naval Warfare*, pp. 28, 43; Lewis, *The Navy of Britain*, p. 598; Tramond and Reussner, *Éléments d'histoire maritime et coloniale*, pp. 517-520, 700-702; *Statist*, October 1, 1904; March 4, 1905; Bacon, *The Life of Lord Fisher*, pp. 260-261.

9. Hovgaard, *Modern History*, pp. 140-142, 392-393; *TINA*, 1911, II, 310; Richmond, *Sea Power in the Modern World*, pp. 83-88.

10. Levert, *Fundamentals*, p. 45; *TINA*, 1911, II, 323; Bacon, *Lord Fisher*, p. 256; Hovgaard, *Modern History*, p. 226.

11. Such were the *Boadicea* class, 1908-1912 (3,500 t., 26 kn.), the *Bristol* class, 1909-1910 (4,800 t.), the *Dartmouth* class, 1910-1913 (5,250-5,500 t., 27 kn.), and the *Arethusa* class, 1913-1914 (3,500 t., 30 kn.). Hovgaard, *Modern History*, p. 193; *Economist*, January 9, 1909.

12. *Proceedings of the Institution of Civil Engineers*, 1861-62, pp. 192, 195-196; Manning, *White*, p. 246; Admiralty Committee on Contracts, 1884, qs. 287-288; *Return of the Admiralty on the Design of Ships*, 1868; R. C. on the Control and Management of Naval Yards, q. 343.

13. Russell, *Reply to Statement on Iron Shipbuilding*, pp. 2-5; Rennie, *Autobiography*, p. 389; *Steamship Chronicle*, October 25, 1861; Grantham, *Iron Shipbuilding*, p. 74; and the writings of Barry in those years. For the material covered in this and later sections, see also Pollard, "*Laissez-Faire* and Shipbuilding."

14. Pollock, *Modern Shipbuilding*, p. 86.

15. Ashworth, "Economic Aspects of Late Victorian Naval Administration," pp. 496-499, 501; Committee on Contracts, 1884, *Report*, p. 1, and appendix 17, p. 218.

16. Ashworth, "Naval Administration," p. 492; Wilmot, *Our Fleet*, p. 258.

17. S. C. on Admiralty Money and Accounts, 1867-68, qs. 337, 1665-1671, 3402-3403, 4361; Robinson, *Report on Dockyards*, 1874; Committee on Contracts, 1884, *Report*, p. 2, and qs. 1837, 2910, 2941-2952, 2392, 2458-2463, 2192-2199, 2213-2219, 2160-2169, 2184, 2717-2719.

18. Baxter, *Ironclad Warship*, p. 116.

19. Committee on Contracts, 1884, q. 2511; Admiralty, *Statement of Policy*, 1905, p. 7.

20. Barry, *Dockyard Economy*, p. 158. See table 10.3.

21. R. C. on Naval Yards, 1861, *Report*, pp. V-VI, and qs. 278-283, 292, 9885-9889, 11,020-11,021, 11,030-11,031.

22. S. C. on the Admiralty Board, qs. 439, 446, 546; Hamilton, *Naval Administration*, pp. 23-25, 51, 133.

23. Manning, *White*, pp. 29, 189; Hamilton, *Administration*, pp. 25-26, 79-82; Lewis, *The Navy*, p. 380; Bacon, *Lord Fisher*, pp. 97-100.

24. Manning, *White*, pp. 89-96, 104, 61, 100, 190; Hamilton, *Administration*, p. 70.

25. Manning, *White*, pp. 179, 194-204, 213; Hamilton, *Administration*, pp. 68-75, 85-90.

26. Orders in Council of 1 January, 10 August, 20 October, 1904, in B.P.P. 1905, XLVIII, Cd. 2416 and 2417; *Statist*, February 22, 1902; Admiralty, *Statement on Policy*, 1905, pp. 32-33.

27. Committee on Contracts, 1884, qs. 428, 588; Committee on Marine Engines, 1859, p. 20.

28. Forwood's Committee on Contracts, 1887, suggested all purchases go through one department, but this was never carried out. *Report*, p. XIV; Hamilton, *Administration*, p. 133.

29. Committee on Contracts, 1884, appendix 1, p. 189, and appendix 2; also qs. 15-16; R. C. into Control and Management of the Naval Yards, 1861, qs. 10,914, 10,901-10,904, 10,916-10,922; Committee on Marine Engines, 1859, pp. 27-28; *Statist*, October 6, 1900; Committee on Contracts, 1887, qs. 3475-3479, 3495-3498, 3625-3633.

30. Committee on Dockyard Administration, 1887, qs. 5453-5456; Committee on Contracts, 1887, *Report*, pp. IX-XI; qs. 154-157, 3255-3260, 3329-3369, 3486, 3792, 3985, 4025-4030; *Return on the Design of Ships*, 1868, p. 21; *Economist*, March 5, 1898.

31. The manager of Palmer's assessed the cost of erecting the ironwork of a vessel at £3 10s. per ton for merchant ships and £12 per ton for warships. Committee on Contracts, 1884, q. 1731; also, qs. 39-45, 305, 325, 488-500, 710-712, 757, 771, 1054-1070, 1115, 1159, 1352, 1390-1415, 1484ff., 1601-1603, 1646-1665, 1766-1768, 1801, 2624-2628. S. C. on Admiralty Money and Accounts, 1868, qs. 1128-1172, 4051, 4893, 5040; Committee on Contracts, 1887, *Report*, p. IX; qs. 2610-2650, 3250-3252.

32. Committee on Contracts, 1884, qs. 144-148, 328; Committee on Marine Engines, 1859, *Report*, pp. 28-29, 33; Committee on Contracts, 1887, qs. 3421-3424.

33. Committee on Contracts, 1884, qs. 64-65, 178, 192, 785, 1593; Committee on Dockyard Administration, 1887, qs. 3264-3267; Hamilton, *Administration*, pp. 179-180.

34. Committee on Contracts, 1884, qs. 262, 801, 1758, 1780, 1820-1822, 2475; Committee on Contracts, 1887, qs. 2395, 2649; *Statist*, April 16, 1896; July 1, December 9, 1899. A sharply contrary interpretation is given in Trebilcock, "Legends of the British Armament Industry, 1890-1914: A Revision."

35. Committee on Contracts, 1884, qs. 69-73, 85, 275-277, 615, 665, 1531, 1643, 1824, 2467, 2599.

36. Committee on Contracts, 1884, qs. 1892, 2147-2149; Committee on Contracts, 1887, qs. 3522-3529, 4237-4256, 4261-4270.

37. Committee on Contracts, 1884, qs. 2186, 2923-2926, 3523-3524; Committee on Dockyards, 1887, Grant's Sub-committee, qs. 3179-3185, 3338; Manning, *White*, pp. 411-412; *Economist*, January 6, 1912; *Statist*, March 3, 1900; March 16, 1901.

38. *Statist*, July 28, 1900; March 26, 1904; *Economist*, November 28, 1908; April 12, 1902; Hills, *Cause of the Thames Ironworks Collapse*.

39. *TNECI*, 1909-10, p. 125; *Economist*, April 12, 1902; *Industrial Survey of the North-East Coast*, p. 242; League of Nations, *Memorandum on Shipbuilding*, pp. 7-8, 15; Jones, *Is Unemployment Inevitable?*, p. 286. The Dockyards did not expand as rapidly as the warship building program, so the share of naval building in private yards rose from 19 percent for the years 1870-1874 to 50 percent in 1910-1914.

40. Memoranda in Manning, *White*, pp. 77-81.

41. Wood, *Battleships*, p. 272.

42. Manning, *White*, p. 161.

43. Falkner, reported in the *Statist*, April 25, 1914.

44. The *Economist*, under Radical editorship, was almost alone in opposing the gigantic armament burden. It also pressed for the expansion of private, and the reduction of Dockyard, work. March 3, 1900; February 23, March 9, 1901; April 22, 1905; March 14, 1914; *Statist*, March 18, 1905; March 27, April 3, 1909; March 7, July 5, 1913.

45. Committee on Contracts, 1884, q. 1917. A number of other services rendered to the firm at the expense of the country were denied by White. Manning, *White*, p. 212.

46. Manning, *White*, pp. 97, 104, 180, 187, 207; D'Eyncourt, *A Shipwright's Yarn*, pp. 29, 41, 43, 48, 59; Bacon, *Lord Fisher*, p. 156.

47. *Economist*, January 4, 1913; *TINA*, 1913, memoir of Dunn.

48. *Economist*, January 4, 1913; *Shipbuilder*, nos. 19, 37; *Directory of Shipowners*, 1914; Wilmot, *Our Fleet*, pp. 125-126n.; *TIESS*, LII, pp. 421-422.

49. Board of Ordnance Committee on the Capabilities of the Merchant Navy for Purposes of War, 1852-53, *Report*. The committee found that privately owned vessels could be used, at best, as auxiliaries only. Of 91 vessels examined, only 16 were suitable for naval uses. See also, Bacon, *Ship Subsidies*, pp. 20-21; Michon, *Les grandes compagnies anglaises*, p. 20; *TINA*, 1877, p. 1; Brassey, *Steamship Subsidies*, p. 223.

50. B.P.P., 1886, C.5006; Bowen, *A Century of Atlantic Travel*, pp. 169-170; Barnaby, *Naval Development*, pp. 352-353, and *TINA*, 1878, pp. 11-22; *Proceedings of the Institution of Civil Engineers*, 1898-99, p. 447; Holmes, *Ancient and Modern Ships*, II, 56, 71.

51. Colin, *Navigation commerciale*, p. 314; Bates, *American Marine*, pp. 89-92; S. C. on Steamship Subsidies, 1901-02, qs. 1318-1321; Ginsburg in *Journal of the Royal Statistical Society*, 1901, p. 462; *TINA*, 1887, Barnaby; Brassey, *TINA*, 1901, pp. 226-230, and 1904, pp. 32-33.

52. *Final Report*, 1902; also *Economist*, December 20, 1902; *Statist*, December 20, 1902; March 7, 1903.

53. Mercantile Cruiser Committee, 1902, *Report*, pp. ii-iv.

54. *Agreement of Subsidy to the Cunard Company*, 1902.

55. The *Statist* calculated the future profits on those two vessels at 9 percent, while the rest of the Cunard fleet earned only 4 percent. August 8, September 5, 1903. Colin, *Navigation commerciale*, pp. 133ff., 251; B.P.P., 1903, Cd. 1704.

56. Bacon, *Manual*, p. 9; Jones, *Government Aid to Merchant Shipping*, 2nd ed., p. 22; Fayle, *War and the Shipping Industry*, p. 22; Siegert, *Die Subventionen der Weltschiffahrt*, p. 3.

57. For the profits and subsidies of the world's chief lines, see *TIESS*, 1892-93, p. 6.

58. Trebilcock, "A 'Special Relationship'—Government, Rearmament, and the Cordite Firms."

59. Jack, "The Purchase of British Government Shares in the British Petroleum Company 1912-1914."

60. Trebilcock, "Special Relationship," p. 374.

61. See Trebilcock, " 'Spin-off' in British Economic History: Armaments and Industry, 1760-1914"; and "British Armaments and European Industrialisation, 1890-1914." It must be stressed, of course, that these beneficial side-effects and spinoffs arose from having the technical capacity to produce armor, and not from the organization of the industry into a monopolistic ring.

62. Ashworth, "Naval Administration," p. 494.

63. Trebilcock, "Special Relationship," p. 366.

64. Hutchins, *American Maritime Industries*, pp. 476, 510-513. Leckebusch discusses the German government's attempts to encourage shipbuilding and the way the Germans employed legislation to build up the demand for steel plates and armor to a level at which domestic steelmakers found it profitable to produce them at prices competitive with those of British imports. *Die Beziehungen.*

65. Hutchins, *American Maritime Industries*, pp. 456-478, 540-541.

11. Conclusion

1. See, for example, most of the studies in McCloskey, ed., *Essays on a Mature Economy*, as well as his own *Economic Maturity*. A vigorous defense of British entrepreneurial behavior in general is McCloskey and Sandberg, "From Damnation to Redemption: Judgments on the Late Victorian Entrepreneur."

Appendix A: A Note on Tonnage Measurements

1. Gibson, *The Story of the Ship*, p. xiii.

2. A history of the development of British tonnage statistics is in van Driel, *Ship Tonnage Measurement*, pp. 49-76.

3. Ibid., p. 54.

4. *Engineering*, February 12, 1904, pp. 230-231.

ABBREVIATIONS

d.t.	Displacement tons*
dwt.	Deadweight tons*
g.t.	Gross tons*
n.t.	Net tons*
SNAME	*Transactions of the Society of Naval Architects and Marine Engineers*
TIESS	*Transactions of the Institution of Engineers and Shipbuilders in Scotland*
TINA	*Transactions of the Institution of Naval Architects*
TNECI	*Transactions of the North-East Coast Institution of Engineers and Shipbuilders*

* For tonnage definitions, see Appendix A.

Selected Bibliography

Unpublished Sources

Five extensive sets of company records were used in the preparation of this book. Those of William Denny and Brothers and of Alexander Stephen and Sons (Stephen of Linthouse) are owned by the Department of Economic History of the University of Glasgow. The records of John Brown and Company, which are also stored at the University of Glasgow, are under the control of the Scottish Record Office, as are those of the Fairfield Shipbuilding and Engineering Company, which are stored in the Glasgow City Archives. Finally, the records of Charles Connell and Company are owned by Mr. Charles Connell but stored in the Glasgow City Archives.

In addition, we used the following unpublished manuscripts:

Cormack, William S. "An Economic History of Shipbuilding and Marine Engineering, with Special Reference to the West of Scotland." Ph.D. dissertation, University of Glasgow, 1930.

Milne, Thomas E. "A British Shipowning Company in the Late Nineteenth and Early Twentieth Centuries." B.Litt. thesis, University of Glasgow, 1965.

Pollard, Sidney. "The Economic History of British Shipbuilding, 1870-1914." Ph.D. dissertation, University of London, 1950.

Robertson, Paul L. "The Management of Manpower in British Shipbuilding, 1870-1914: A Study in the Organization of Human Resources under Conditions of Changing Technology." Ph.D. dissertation, University of Wisconsin-Madison, 1972.

Webb, Sidney, and Webb, Beatrice. Manuscript Trade Union Collection, Shipbuilding Trades: Vol. I, Shipwrights; Vol. II, Other Trades. British Library of Economics and Political Science, London.

Official Publications

A. Periodic Publications

Annual Statement of the Shipping and Navigation of the United Kingdom.

Board of Trade: Changes in Rates of Wages, Hours of Labour

Board of Trade: Annual Gazette of Trade Unions
Board of Trade: Labour Gazette (Monthly), 1893 to 1950
Census of the Population, 1861, 1871, 1881, 1891, 1901, 1911
Statistical Abstract of the United Kingdom (Annual)
Statutes of the Realm

B. British Official Publications

Select Committee on Artizans and Machinery	1824.V.(51)
Select Committee on Combination Laws	1825.IV.(437)
Select Committee on Combinations of Workmen, Second Report	1837-8.VIII.(646)
Select Committee on British Shipping, 1844	1844.VIII.(545)
Select Committee on Navigation Laws (Commons)	1847-8.XX.I.(7)
Select Committee on Navigation Laws (Lords)	1847-8.XX.II.(341,431,754)
Board of Ordnance Committee on the Mercantile Steam Navy for Purposes of War	1852-3.LXI.(687)
Select Committee on Emigrant Ships	1854.XIII.(163,349)
Royal Commission on the Manning of the Navy	1859.VI.(2469)
Lindsay's Reasons for Dissenting from Above	1859.XV.(99)
Admiralty Committee on Marine Engines	1859.XV.(183)
Select Committee on Merchant Shipping	1860.XIII.
Royal Commission into the Control and Management of the Naval Yards	1861.XXVI.(2790)
Correspondence on the *Warrior*	1861.XXXVIII.(207)
Correspondence on the School of Naval Architecture	1864.XXXVII.(407)
Royal Commission on Trade Unions	1867.XXXII. 1867-8.XXXIX.
Admiralty Committee on Metals	1867.XLIV.(394)
Select Committee on Admiralty Money and Accounts	1867-8. VI. (469)
Return of Correspondence of the Admiralty	1867-8.XLV.(26)
Select Committee on the Board of the Admiralty	1871.VII.(80)
Royal Commission on the Truck System	1871.XXXVI.(C.326,C.327)
Return on Factories and Workshops	1871.LXII.(440)
Admiralty Committee on Designs	1872.XIV.(C.477,C.489)
Robinson, Admiral Sir Spencer: Letter to the First Lord	1872.XXXIX.C.487
Royal Commission on Unseaworthy Ships	1873.XXXVI.C.853 1874.XXXIV.C.1027
Select Committee on the Merchant Shipping Bill	1874.X.(309)
Robinson, Admiral Sir Spencer: Report on Naval Expenditure	1874.XXXVIII.(109)
Royal Commission on the Spontaneous Combustion of Coal	1876.XLI.(1586-1)
Admiralty Memorandum	1877.LII.(130)
Royal Commission on the Measurement of	1881.XLIX.(C.3074)

Tonnage	1882.XXI.(C.3380)
Board of Trade: Proposals on Merchant Shipping Legislation	1884.LXXI.(74)
Royal Commission on Technical Instruction, Second Report	1884.XXXIX.
Committee on Contracts	1884-5.XIV.C.4219
Return Showing Action on Above	1884-5.XLVIII.C.4498
Board of Trade: Load Line Committee	1884-5.LXIX.C.4566
Admiralty Committees on Dockyard Administration	1886.XIII.C.4615
Admiralty Committee on Dockyard Management	1887.XVI.C.4979
Royal Commission on the Depression in Trade	1886.XXI-XXIII.
Admiralty Committee on Contracts	1887.XVI.C.4987
Statement Showing Action on Above	1887.XVI.C.5231
Correspondence on Merchant Steamers	1887.LII.C.5006
Board of Trade: Returns of Wages, 1830-66	1887.LXXXIX.C.5172
Admiralty Board: Action on Designs	1889.L.C.5635
Commercial Reports on Subsidies, Commercial No.14 (1889), No.2 (1898), No.8 (1895)	
Board of Trade: Return of Hours Worked	1890.LXVIII.(375)
Board of Trade: Committee on Bulkheads	1890-1.LXXVII.C.6405
Departmental Committee on the Trans-Atlantic Cattle Trade	1890-1.LXXVIII.C.6350
Board of Trade: Wages and Costs	1891.C.6535
Royal Commission on Labour	1892.XXXIV-XXXVI. 1893.XXXII-XXXIX. 1895.XXXV.
Board of Trade: Report on Wages	1893.LXXXIII.II.C.6889
Admiralty Committee on Boilers	1893-4.LIV.C.6991
Select Committee on Government Contracts	1897.X.(334)
Ninth Report by the Chief Correspondent of the Board of Trade on Trade Unions	1897.C.8644
Load Line Committee	1899.LXXVIII.(126,283)
Select Committee on Steam Ship Subsidies	1901.VIII.(300) 1902.IX.(385)
Report on the Strikes and Lockouts of 1900	1901.Cd.689
Admiralty Committee on Boilers	1901.XLII.Cd.503 1902.LX.Cd.1158 1904.LIII.Cd.2207
Admiralty Committee on Mercantile Cruisers	1902.XCII.Cd.1379
Factory Inspector's Annual Report, 1902	1903.X.Cd.1816 1903.XII.Cd.1610
Copy of Agreement with the Cunard Steamship Company	1903.XXVI.Cd.1703
Copy of Agreement with the International Mercantile Marine Company	1903.LXIII.Cd.1704
Select Committee on Foreign Ships, 1st Report	1904.VI.(229)
2nd Report	1905.VII.(269)

Board of Trade: Memoranda, Tables, Charts	1905.LXXXIV.Cd.2337
Admiralty: Statement of Policy	1906.LXX.Cd.2791
Royal Commission on Trade Disputes and Trade Combinations, 1903-06	1906.Cd.2826
Board of Trade Committee on Tonnage Measurement	1906.CIX.Cd.3045-6
Select Committee on the Merchant Shipping Bill	1907.VI.(256)
Royal Commission on Shipping Rings	1909.XLVII.Cd.4668-70, 4685
Royal Commission on the Poor Law, Report	1909.Cd.4499
appendix vol. VIII	1910.Cd.5066
"British Foreign Trade and Industry"	1909.Cd.4954
Census of Production, Preliminary Tables, part 2	1910.CIX.Cd.5005
Second Report on Rules on Voluntary Conciliation and Arbitration Boards and Joint Committees	1910.Cd.5346
Board of Trade Enquiry into Earnings, VI	1911.LXXXVIII
Report on Trade Unions in 1908-1910	1912.Cd.6109
Merchant Shipping Advisory Committee	1912-3.XXXVIII.Cd.6353
Select Committee on Estimates	1913.VI.(231)
The Industrial Council Enquiry into Industrial Agreements	1913.Cd.6953
Report on Bounties in Foreign Countries	1913.XL.Cd.6899
Agreement with the Anglo-Persian Oil Company	1914.LIV.Cd.7419
Board of Trade Committee on the Subdivision of Merchant Ships, 1st Report	1914-6.XXVII.Cd.7743
2nd Report	1914-6.XXVII.Cd.8080
Board of Trade Committee on the Load Line	1916.XIII.Cd.8204
Departmental Committee on Shipping and Shipbuilding after the War	1918.XIII.Cd.9092
Committee on Trade and Industry ("Balfour Committee"), part IV	1928

C. Other Official Publications

Hall, Henry: Shipbuilding in the U.S.A., Washington, 1884

Hichborn, Philip: Report on European Dockyards, Washington, 1886

Jones, Grosvenor M.: Government Aid to Merchant Shipping, 2nd Ed. Washington, 1923

League of Nations: Memorandum on Shipbuilding, Geneva, 1927

Melville, George W.: Report of the United States Naval Liquid Fuel Board, Washington, 1904

Periodical Publications

Associated Shipwrights' Society, Annual Reports and Quarterly Reports

Barrow Herald

Barrow Naturalists' Field Club, Proceedings

Barrow News and Dalton Chronicle, biweekly

Barrow Pilot
Beehive, weekly
Brassey, Naval Annual
British Association, Reports of Meetings, annual
Capital and Labour, weekly
Cassier's Magazine, monthly
Chamber of Shipping, Report of Proceedings of Annual Meetings
Comité Maritime Internationale, Bulletin
Directory of Shipowners
Economist, weekly
Engineering, weekly
Engineering Magazine, monthly
Fairplay, weekly
Federation of Engineering and Shipbuilding Trades, Reports of Annual Meetings
Franklin Institute, Journal
Institute of Marine Engineers, Transactions
Institution of Civil Engineers, Minutes of Proceedings
Institution of Engineers and Shipbuilders in Scotland, Transactions
Institution of Mechanical Engineers, Proceedings
Institution of Naval Architects, Transactions
Iron and Steel Institute, Journal
Lloyd's Register of Shipping, Annual Statistical Returns
Lloyd's Register of Shipping, Quarterly Statistical Returns
Marine Engineering, monthly
Mariner's Mirror, quarterly
Mechanic's Magazine
Mid-Tyne Link, quarterly
Nautical Magazine, monthly
Naval Science, quarterly
Newcomen Society, Transactions
North-East Coast Institution of Engineers and Shipbuilders, Transactions
Palmer Record
Penny Magazine for the Propagation of Useful Knowledge: Days at the Factory
Quarterly Review
Royal Statistical Society, Journal
Schiffbautechnische Gesellschaft, Jahrbücher
Shipbuilder, quarterly
Shipping, monthly
Shipping World, monthly (weekly from 1896)
Society of Naval Architects and Marine Engineers, Transactions
Stahl und Eisen
Statist, weekly
Steamship, monthly (1889-1914)
Steamship, semimonthly (1883)
Steam Shipping Chronicle, weekly
Syren and Shipping
Thames Ironworks, Quarterly Gazette

Thomson's Annals of Philosophy
Times, daily
Trades Union Congress, Annual Reports
United Society of Boilermakers, Annual Reports
Vulcan, weekly
Westminster Review, monthly
White, John, Annual Steamship Circular

IV. Books and Articles

Abell, Sir Westcott. *The Safe Sea.* Liverpool: C. Birchall and Sons, 1932.

______. *The Shipwright's Trade.* Cambridge: Cambridge University Press, 1948.

Aberconway, 1st Baron. *The Basic Industries of Great Britain.* London: Ernest Benn, 1927.

Adamson, John William. *English Education 1789-1902.* Cambridge: Cambridge University Press, 1930.

Albion, Robert G. *Forests and Seapower.* Cambridge, Mass.: Harvard University Press, 1926.

Aldcroft, Derek H. "Economic Progress in Britain in the 1920s." *Scottish Journal of Political Economy* 13 (1966): 297-316. Reprinted in D. H. Aldcroft and H. W. Richardson, *The British Economy, 1870-1939.* London: Macmillan, 1969.

———. "The Entrepreneur and the British Economy, 1870-1914." *Economic History Review*, n.s. 17 (1964): 113-134.

———. "McCloskey on Victorian Growth." *Economic History Review*, n.s. 27 (1974): 271-274.

———. "Technical Progress and British Enterprise, 1875-1914." *Business History* 8 (1963): 122-139. Reprinted in D. H. Aldcroft and H. W. Richardson, *The British Economy, 1870-1939.* London: Macmillan, 1969.

Ames, Edward, and Rosenberg, Nathan. "Changing Technological Leadership and Industrial Growth." *Economic Journal* 73 (1963): 13-31.

———. "The Progressive Division and Specialization of Industries." *Journal of Development Studies* 1 (1965): 363-383.

Appleyard, Rollo. *Charles Parsons, His Life and Work.* London: Constable, 1933.

Archer, R. L. *Secondary Education in the Nineteenth Century.* Cambridge: Cambridge University Press, 1921.

Ashworth, William. "Economic Aspects of Late Victorian Naval Administration." *Economic History Review*, n.s. 22 (1969): 491-505.

———. "The Late Victorian Economy." *Economica*, n.s. 33 (1966): 17-33.

Aves, Ernest. "Labour Notes: The Dispute in the Engineering Trade, &c." *Economic Journal* 7 (1897): 623-630.

Bacon, Edwin M. *A Manual of Ship Subsidies.* Chicago: A. C. McClurg, 1911.

Bacon, Sir Reginald C. *Life of Lord Fisher of Kilverstone.* London: Hodder and Stoughton, 1929.

Ballard, Maxwell. "The Relation between Shipbuilding Production, Prices and the Freight Market." *TNECI* 37 (1920-21): 152-184.

Barnaby, K. C. *100 Years of Specialized Shipbuilding and Engineering*. London: Hutchinson, 1964.

Barnaby, Nathaniel. *Naval Development in the Century*. London and Edinburgh: W. and R. Chambers, 1902.

Barry, Patrick. *Dockyard Economy and Naval Power*. London: S. Low, Son and Co., 1863.

———. *Dockyards and Private Shipyards*. London: T. Danks, 1863.

———. *Dockyards of France*. London: Simpkin, Marshall and Co., 1864.

Barry, William J. *History of the Port of Cork Steam Navigation*. Cork, 1915.

Basset-Lowke, W. J. and Holland, George. *Ships and Men*. London: G. G. Harrap, 1946.

Bates, W. W. *American Marine*. Boston and New York: Houghton Mifflin, 1893.

———. *American Navigation*. Boston and New York: Houghton Mifflin, 1902.

Baxter, James P., III. *The Introduction of the Ironclad Warship*. Cambridge, Mass.: Harvard University Press, 1933.

Bell, I. Lowthian. *Principles of the Manufacture of Iron and Steel*. London: G. Routledge and Sons, 1884.

Bevan, G. P. "The Industrial Resources of Ireland." *Journal of the Royal Statistical Society* 44 (1881).

Biles, John H. *Design and Construction of Ships*. London: C. Griffin and Co., 1908 and 1911.

———. "Fifty Years of Warship-Building on the Clyde." *TIESS* 52 (1908-09): 347-370.

Blake, George. *Down to the Sea*. London: Collins, 1937.

Board of Trade (Armstrong College, Newcastle-upon-Tyne). *Industrial Survey of the North-East Coast Area*. London: H.M.S.O., 1932.

Bonsor, N. R. P. *North Atlantic Seaway*. Prescot: T. Stephenson and Sons, 1955.

Borrowman, W. C. "Some Considerations Affecting the Training of Young Engineers." *TNECI* 16 (1899-1900): 113-134.

Bowen, Frank C. *A Century of Atlantic Travel, 1830-1930*. London: Sampson Low, Marston, 1932.

Bowley, A. L., and Wood, George H. "Wages in Shipbuilding and Engineering." *Journal of the Royal Statistical Society* 68 (1905): 104-137, 373-394, 563-614; 69 (1906): 148-192.

Bray, R. A. "The Apprenticeship Question." *Economic Journal* 19 (1909): 404-415.

British Association for the Advancement of Science. *Trade and Manufactures on the Tyne, Wear and Tees*. Newcastle and London, 1863.

Brown, E. H. Phelps. *The Growth of British Industrial Relations*. London: Macmillan, 1959.

———. and Browne, Margaret H. *A Century of Pay*. London: Macmillan, 1968.

Brown, R. K., Brannen, P., Cousins, J. M., and Samphier, M. L. "The Conditions of Solidarity: Social Stratification and Industrial Relations in Shipbuilding." *British Journal of Industrial Relations* 10 (1972): 12-41.

Bruce, A. B. *Life of William Denny, Shipbuilder, Dumbarton*. London: Hodder and Stoughton, 1889.

Brunel, Isambard. *Life of Isambard Kingdom Brunel*. London: Longmans, Green, 1870.

Burn, Duncan. *The Economic History of Steelmaking 1867-1939*. Cambridge: Cambridge University Press, 1940.

Burnett, John. *History of the Engineers Strike*. Newcastle: J. W. Swanston, 1872.

Burstall, A. R. "Shipbuilding in Liverpool." *Sea Breezes*, April, 1936.

Buxton, Neil K. "The Scottish Shipbuilding Industry between the Wars: A Comparative Study." *Business History* 13 (1968): 101-120.

Byres, T. J. "Entrepreneurship in the Scottish Heavy Industries, 1870-1900." In Peter L. Payne, ed., *Studies in Scottish Business History*. London: Frank Cass, 1967.

Cardwell, D. S. L. *The Organization of Science in England*. 2nd ed. London: Heinemann, 1972.

Carvel, John L. *Stephen of Linthouse*. Glasgow: Robert Maclehose, 1950.

Casson, Robert. *A Few Furness Worthies*. Ulverston: J. Atkinson, 1889.

Chandler, Alfred D., Jr., ed. *Giant Enterprise*. New York: Harcourt, Brace and World, 1964.

———. *Strategy and Structure*. Cambridge, Mass.: M.I.T. Press, 1962.

Chapelle, H. I. "The First Clipper." *Mariner's Mirror*, 1948.

Chapman, Dennis. "The New Shipwright Building Co. of Dundee." *Economic History Review*, o.s. 10 (1940): 148-151.

Clapham, J. H. *Free Trade and Steel*. Vol. II of *An Economic History of Modern Britain*. Cambridge: Cambridge University Press, 1932.

Clark, Arthur H. *The Clipper Ship Era, 1843-1869*. London: G. P. Putnam's Sons, 1911.

Clegg, H.A.; Fox, Alan; and Thompson, A. F. *A History of British Trade Unions since 1889*. Vol. I. Oxford: Oxford University Press, 1964.

Clough, Shepard B. *European Economic History*. 2nd ed. New York: McGraw-Hill, 1968.

Clyde Shipbuilders' and Engineers' Association and Clyde Associated Shipwrights. *Arbitration Proceedings*, 1877. Glasgow: W. Macrone, 1878.

Coleman, D. C. "Gentlemen and Players." *Economic History Review*, n.s. 26 (1973): 92-116.

Colin, Ambroise. *La navigation commerciale au XIXe Siècle*. Paris: A. Rousseau, 1901.

Corson, F. Reid. *The Atlantic Ferry in the Nineteenth Century*. London: Sampson Low, Marston and Co., 1930.

Cotgrove, Stephen F. *Technical Education and Social Change*. London: George Allen and Unwin, 1958.

Course, A. G. *The Deep Sea Tramp*. London: Hollis and Carter, 1960.

Crammond, Edgar. *The British Shipping Industry*. London: Constable, 1917.

Critchell, J. T., and Raymond, J. *History of the Frozen Meat Trade*. London: Constable, 1912.

Crouzet, F. "Essai de construction d'un indice annuel de la production industrielle française au XIXe siècle." *Annales (Economies-Sociétés-Civilisations)* 25 (1970): 56-99.

Cummings, D. C. *History of the United Society of Boilermakers and Iron and Steel Ship Builders*. Newcastle-upon-Tyne: R. Robinson and Co., 1904.

Curtis, S. J. *History of Education in Great Britain*. London: University Tutorial Press, 1965.

Dalby, W. E. "The Training of Engineers in the United States." *TINA* 45 (1903).

Davies, P. N. "The African Steam Ship Company." In J. R. Harris, ed., *Liverpool and Merseyside*. London: Frank Cass, 1969.

Denny, Dumbarton. London and Cheltenham: Ed. J. Burrow, 1932.

Denny, William. "The Trial of Screw Steamships." *British Association Reports*, 1875.

Dillon, Malcolm. *Palmer's Shipbuilding and Iron Company Limited*. 2nd ed. Newcastle-upon-Tyne, 1904.

Dougan, David. *The History of North East Shipbuilding*. London: George Allen and Unwin, 1968.

Driel, A. van. *Ship Tonnage Measurement*. The Hague: Government Printing Office, 1925.

Duff, Peter. *British Ships and Shipping*. London: G. G. Harrap, 1949.

Dunmore, W. T. *Ship Subsidies*. Boston and New York: Houghton Mifflin, 1907.

Dunn, James. "The Arrangement and Equipment of Shipbuilding Works." *Engineering*, August 2, 1901, pp. 169-170; August 9, 1901, pp. 183-186; August 16, 1901, pp. 215-218.

D'Eyncourt, Sir E. W. H. T. *A Shipbuilder's Yarn*. London: Hutchinson, 1948.

Einarsen, Johan. *Reinvestment Cycles*. Oslo: J. Chr. Gerndersens boktrykkeri, 1938.

Eucken, Walter. *Die Verbandsbildung in der Seeschiffahrt*. Munich and Leipzig: Duncker and Humblot, 1914.

The Fairfield Shipbuilding and Engineering Works. London: Offices of *Engineering*, 1909.

Fayle, C. Ernest. *A Short History of the World's Shipping Industry*. London: George Allen and Unwin, 1933.

———. *The War and the Shipping Industry*. New Haven: Yale University Press, 1927.

Feinstein, Charles H. *National Income, Expenditure and Output of the United Kingdom, 1855-1965*. Cambridge: Cambridge University Press, 1972.

Fincham, John. *An Outline of Shipbuilding*. 3rd ed. London: Whittaker and Co., 1852.

Fisher, Joseph. *A Popular History of Barrow-in-Furness*. Bournemouth: E. J. Frampton, 1891.

Fitger, Emil. "Schiffbau und Seeschiffahrt in den Letzen Jahren." *Volkswirtschaftliche Gesellschaft* 14, no. 1 (1892).

Flamm, O. "The Scientific Study of Naval Architecture in Germany." *TINA* 53, part 2 (1911).

Furness, Christopher. *The American Invasion*. London: Simpkin, Marshall, 1902.

Fry, Henry. *History of the North Atlantic Steam Navigation*. London: Sampson Low, Marston and Co., 1896.

Gibson, Charles E. "The Ship and Society." *Modern Quarterly*, Spring 1947.

———. *The Story of the Ship*. New York: H. Schuman, 1948.

Gilfillan, S. Colum. *Inventing the Ship*. Chicago: Follett Publishing Co., 1935.

———. *The Sociology of Invention*. Chicago: Follett Publishing Co., 1935.

Glover, John. "Statistics of Tonnage during the First Decade under the Navigation Law of 1849." *Journal of the Royal Statistical Society* 22 (1863).

———. "Tonnage Statistics for the Decade 1860-70." *Journal of the Royal Statistical Society* 31 (1872).

Gradon, W. M. *The Furness Railway, Its Rise and Development, 1846-1923*. Manchester: James Collins and Co., 1946.

Graham, Gerald S. "The Ascendancy of the Sailing Ship 1850-85." *Economic History Review*, n.s. 9 (1956-57): 74-88.

Grant, Allan. *Steel and Ships*. London: Michael Joseph, 1950.

Grantham, John. *Iron as a Material for Shipbuilding*. London: Virtue and Co., 1842.

———. *Iron Shipbuilding*. 5th ed. London: Lockwood and Co., 1868.

Greeley, Horace. *Essays Designed to Elucidate the Science of Political Economy*. Boston: Fields, Osgood and Co., 1870.

Gregg, E. S. "Decline in Tramp Shipping." *Quarterly Journal of Economics* 40 (1926): 338-346.

Guthrie, James. *The River Tyne: Its History and Resources*. Newcastle-upon-Tyne: A. Reid, 1880.

Haack, Rudolph. "The Development of German Shipbuilding." *Engineering Magazine*, August 1899.

Haarmann, Hermann J. *Die Ökonomische Bedeutung der Technik in der Seeschiffahrt*. Leipzig: W. Klinkhardt, 1908.

Haji, Sarabhai N. *The Economics of Shipping*. Bombay: Commercial Printing Press, 1924.

Haldane, John Wilton Cunningham. *Steamships and Their Machinery*. London: E. and F. N. Spon, 1893.

Hall, A. Rupert. "Scientific Method and the Progress of Techniques." In E. E. Rich and Charles H. Wilson, eds., *Cambridge Economic History of Europe*. Vol. IV. Cambridge: Cambridge University Press, 1967.

Halle, Ernst L. von. "Die Volkswirtschaftliche Entwicklung des Schiffbaues in Deutschland und in der Hauptländern." *Jahrbuch der Schiffbautechnische Gesellschaft*, 1902.

Halsted, E. P. "Iron-Cased Ships." A Lecture Delivered at the Royal United Service Institution, March 4th, 1861.

Hamilton, Sir Richard V. *Naval Administration*. London: G. Bell and Sons, 1896.

Harcourt, L. F. V. *Achievements in Engineering in the Last Half Century*. London: Seeley and Co., 1891.

Hardy, Alfred Cecil. *From Ship to Sea*. Glasgow: Brown and Son, 1926.

Harley, Charles Knick. "On the Persistence of Old Techniques: The Case of North American Wooden Shipbuilding." *Journal of Economic History* 33 (1973): 372-398.

———. "The Shift from Sailing Ships to Steamships, 1850-1890: A Study in Technological Change and Its Diffusion." In Donald N. McCloskey, ed., *Essays on a Mature Economy: Britain after 1840*. London: Methuen, 1971.

———. "Skilled Labour and the Choice of Technique in Edwardian Industry." *Explorations in Economic History* 11 (1974): 391-414.

Hesketh, E. *J. & E. Hall Ltd. 1785-1935*. Glasgow: Glasgow University Press, 1935.

Hills, A. F. *The Thames Ironworks vs. the Admiralty*. London, 1911.

Hobsbawm, Eric J. "The Labour Aristocracy in Nineteenth-Century Britain." In *Labouring Men*. New York: Basic Books, 1964.

Holmes, George C. V. *Ancient and Modern Ships*. London: South Kensington Museum, 1906.

Hovgaard, William. *Modern History of Warships*. London: E. and F. N. Spon, 1920.

Hughes, J. R. T., and Reiter, Stanley. "The First 1,945 British Steamships." *Journal of the American Statistical Association* 53 (1958): 360-381.

Huldermann, Bernhard. *Albert Ballin*. Berlin: Oldenburg, 1922.

Hume, John R., and Moss, Michael S. *Clyde Shipbuilding from Old Photographs*. London: Batsford, 1975.

Hunt, Bishop Carlton. *The Development of the Business Corporation in England, 1800-1867*. Cambridge, Mass.: Harvard University Press, 1936.

Hyde, F. E. *Blue Funnel*. Liverpool: Liverpool University Press, 1956.

Inglis, J. *The Apprenticeship Question*. Glasgow, 1894.

Irving, R. J. "New Industries for Old? Some Investment Decisions of Sir W. G. Armstrong, Whitworth and Co. Ltd., 1900-14." *Business History* 17, no.2, July 1975.

Jack, Marian. "The Purchase of British Government Shares in the British Petroleum Company 1912-1914." *Past and Present*, no. 39 (April 1968): 139-168.

Jackson, George Gibbard. *The Ship under Steam*. London: T. F. Unwin, 1927.

James, F. Cyril. *Cyclical Fluctuations in the Shipping and Shipbuilding Industries*. Philadelphia: University of Pennsylvania, 1927.

Jamieson, Andrew. "On the Technical Education of Our Young Engineers, Shipbuilders, and Artizans." *TIESS* 24 (1880-81): 149-160.

Jeans, J. S. *The Iron Trade of Great Britain*. London: Methuen, 1885.

———. *Pioneers of the Cleveland Iron Trade*. Middlesbrough-on-Tees: *Gazette* Publishing Offices, 1875.

Jefferson, H. *Viscount Pirrie of Belfast*. Belfast and Newry: Wm. Mullan and Son, 1948.

Jefferys, James B. *The Story of the Engineers*. London, 1945.

Jewkes, John, and Winterbottom, A. *Industrial Survey of Cumberland and Furness*. Manchester: Manchester University Press, 1933.

John, T. G. "Shipbuilding Practice of the Present and Future." *Engineering*, July 10, 1914, pp. 68-71.

Jones, Clement. *British Merchant Shipping*. London: Edward Arnold, 1922.

Jones, David Caradog. *Industrial Survey of Merseyside*. Liverpool, 1934.

Jones, John H. *Is Unemployment Inevitable*? London, 1924.

Jones, Leslie. *Shipbuilding in Britain, Mainly between the Two World Wars*. Cardiff: University of Wales Press, 1957.

Jordan, A. E. "Shipping on the Great Lakes." *Transactions of the Institution of Marine Engineers*, 1911-12.

Kenny, E. F. "Steel for Shipbuilding." *Transactions of the Society of Naval Architects and Marine Engineers*, 1945 (Historical).

Kindleberger, Charles P. *Economic Growth in France and Britain: 1851-1950*. Paperback ed. New York: Simon and Schuster, 1969.

———. "Germany's Overtaking of England, 1806-1914, Part II." *Weltwirtschaftliches Archiv* 4 (1975): 477-504.

Kirkaldy, Adam W. *British Shipping*. London: K. Paul, Trench, Trübner, 1914.

Knight, Robert. "Agreement between Employers' Associations and the Boilermakers and Iron and Steel Shipbuilding Society." Newcastle, 1894.

Koopmans, T. *Tanker Freight Rates and Tankship Building*. Haarlem: Deerven F. Bohn, 1938.

Kreuter, Alexander. *Preisbildung in der Linienreederei*. Munich: Ludwig-Maximilians Universität Dr. Phil. Thesis, 1909.

Laeisz, C. F. "Shipbuilding in Germany." *TINA* 39 (1895).

Laeisz, Walter. *Technik und Wirtschaftlichkeit im Schiffahrtsbetriebe*. Jena: Academia Jenensis, 1910.

Landes, David S. "Entrepreneurship in Advanced Industrial Countries: The Anglo-German Rivalry." Paper presented to the Conference on Entrepreneurship and Economic Growth, Cambridge, Mass., November, 1954.

———. "The Structure of Enterprise in the Nineteenth Century: The Cases of Britain and Germany." International Congress of Historical Science, Stockholm, 1960. *Rapports*. Vol. V.

———. *The Unbound Prometheus*. Cambridge: Cambridge University Press, 1969.

Lange, J. "Stettiner Vulkan." *Jahrbuch der Schiffbautechnische Gesellschaft*, 1907.

Leach, F. *Barrow-in-Furness: Its Rise and Progress*. Barrow-in-Furness: *Daily Times* Office, 1872.

Leach, H. *The Duke of Devonshire: A Personal and Political Biography*. London: Methuen, 1904.

Leckebusch, G. "Der Beginn des deutschen Eisenschiffbaues 1850-1890." In Karl Erich Born, ed., *Moderne deutsche Wirtschaftsgeschichte*. Cologne, 1966.

———. *Die Beziehungen der deutschen Seeschiffswerften zur Eisenindustrie an der Ruhr in der Zeit von 1850 bis 1930*. Cologne: Rheinisch-Westfälischen Wirtschaftsarchiv zu Köln, 1963.

Lehmann-Felkowski, G., ed. *The Shipbuilding Industry of Germany*. London: Crosley Lockwood and Son, 1904.

Levert, Lee J. *Fundamentals of Naval Warfare*. New York: Macmillan, 1947.

Levy, Hermann. *Monopolies, Cartels and Trusts in British Industry*. 2nd ed. London: Macmillan, 1927.

———. *Monopoly and Competition*. London: Macmillan, 1911.

———. "Trust und Kartellenwicklung in Grossbritannien." *Schmoller's Jahrbuch* 32 (1908).

Lewis, Michael A. *The Navy of Britain*. London: George Allen and Unwin, 1948.

Lewis, W. Arthur. "International Competition in Manufactures." *American Economic Review* 47, Supplement (1957): 578-587.

Lindsay, W. S. *History of Merchant Shipping*. London: Sampson Low, Marston, 1874-1883.

Lisle, B. Orchard. *Tanker Technique 1700-1936*. London: World Tankship Publications, 1936.

Lloyd's Register. *Annals of Lloyd's Register*. London, 1934.

Lubbock, Basil. *China Clippers*. Glasgow: James Brown and Son, 1916.

———. *Colonial Clippers*. Glasgow, James Brown and Son, 1924.

———. *Last of the Windjammers*. Glasgow: Brown, Son, and Ferguson, 1927.

———. *Opium Clippers*. Glasgow: Brown, Son and Ferguson, 1933.

Lynch, J. "Skilled and Unskilled Labour in the Shipbuilding Trade." In *Report of the Industrial Remuneration Conference*, pp. 114-119. London: Cassell, 1885.

MacArthur, William Forrest. *History of Port Glasgow*. Glasgow: Jackson, Wylie and Co., 1932.

McCloskey, Donald N. "Did Victorian Britain Fail?" *Economic History Review*, n.s. 23 (1970): 446-459.

———. *Economic Maturity and Entrepreneurial Decline: British Iron and Steel 1870-1913*. Cambridge, Mass: Harvard University Press, 1973.

———. *Essays on a Mature Economy: Britain after 1840*. London: Methuen, 1971.

———. "Victorian Growth: A Rejoinder." *Economic History Review* 27 (1974): 275-277.

———. and Sandberg, Lars. "From Damnation to Redemption: Judgments on the Late Victorian Entrepreneur." *Explorations in Economic History* 9 (1971): 89-109.

MacDonagh, Oliver. *A Pattern of Government Growth: The Passenger Acts and Their Enforcement*. London: Macgibbon and Kee, 1961.

Mackinnon, James. *Social and Industrial History of Scotland*. London: Longmans, Green and Co., 1921.

Macrosty, H. W. *The Trust Movement in British Industry*. London: Longmans, Green and Co., 1907.

Manning, Frederic. *Life of Sir William White*. London: J. Murray, 1923.

Marder, Arthur J. *From the Dreadnought to Scapa Flow*. Vol. I. New York: Oxford University Press, 1961.

Marshall, J. D. *Furness and the Industrial Revolution*. Barrow-in-Furness: J. Milner, 1958.

Marwick, W. H. "The Limited Company in Scottish Economic Development." *Economic History* 2 (1937): 415-429.

Maywald, Karel. "The Construction Costs and Value of the British Merchant Fleet, 1850-1938." *Scottish Journal of Political Economy* 3 (1956): 44-66.

Meeker, Royal. *History of Shipping Subsidies*. New York: Macmillan for the American Economic Association, 1905.

Melville, James, and Hobbs, James Leslie. *Some Early Railway History in Furness*. Kendal: Titus Wilson and Son, 1951.

Michon, Georges. *Les grandes compagnies anglaises de navigation*. Paris: A. Rousseau, 1913.

Milne, Thomas E. "British Shipping in the Nineteenth Century: A Study of the Ben Line Papers." In Peter L. Payne, ed., *Studies in Scottish Business History*. London: Frank Cass, 1967.

Mitchell, B. R., and Deane, Phyllis. *Abstract of British Historical Statistics*. Cambridge: Cambridge University Press, 1962.

Moody, J. *The Truth about Trusts*. New York: Moody Publishing Co., 1904.

Mortimer, J. E. *History of the Boilermakers' Society*. Vol. I. London: George Allen and Unwin, 1973.

Moss, Michael S., and Hume, John R. *Workshop of the British Empire: Engineering and Shipbuilding in the West of Scotland*. London: Heinemann, 1977.

Mott, C. G. *Reminiscences of Birkenhead*. Liverpool: H. Young and Sons, 1900.

Napier, David D. *Autobiographical Sketch of David Napier, Engineer, 1790-1869*. Glasgow: James Maclehose and Sons, 1912.

Napier, James. *Life of Robert Napier*. Edinburgh: William Blackwood, 1904.

Neumann, Josef. *Volkswirtschaftliche und soziale Bedeutung der technischen Entwicklung in der Schiffbauindustrie*. Leipzig, 1910.

Noble, C. B. *The Brunels, Father and Son*. London: Robden-Sanderson, 1938.

Noel-Baker, Philip. *The Private Manufacture of Armaments*. London: Victor Gollancz, 1936.

North, Douglass C. "Ocean Freight Rates and Economic Development 1750-1913." *Journal of Economic History* 18 (1958): 537-555.

———. "Sources of Productivity Change in Ocean Shipping, 1660-1850." *Journal of Political Economy* 76 (1968): 953-970.

Öhlrich, G. W. *Deutschland zur See*. Hamburg, 1849.

Orchard, J. Ewing. *Japan's Economic Position*. New York: McGraw-Hill, 1930.

Orcutt, H. F. L. "Tools and Gauges in the Modern Shop." *TIESS* 46 (1902-03): 248-274.

Overzier, Paul. *Der Amerikanisch-Englische Schiffahrtstrust*. Berlin, 1912.

Parris, Henry. "The Nineteenth-Century Revolution in Government: A Reappraisal Reappraised." *Historical Journal* 3 (1960): 17-37. Reprinted in Peter Stansky, ed., *A Victorian Revolution*. New York: New Viewpoints, 1973.

Parsons, Sir Charles A. "The Marine Steam Turbine from 1894 to 1910." *TINA* 53, pt. 2 (1911): 79-95.

Payne, Peter L. *British Entrepreneurship in the Nineteenth Century*. London: Macmillan, 1974.

Peirson, J. Gordon. *Great Shipbuilders, The Rise of Harland and Wolff*. London: A. H. Stockwell, 1935.

Pollard, Sidney. "Barrow-in-Furness and the Seventh Duke of Devonshire." *Economic History Review*, n.s. 8 (1955): 213-221.

———. "British and World Shipbuilding, 1890-1914: A Study in Comparative Costs." *Journal of Economic History* 17 (1957): 426-444.

———. "The Decline of Shipbuilding on the Thames." *Economic History Review*, n.s. 3 (1950): 72-89.

———. "*Laissez-Faire* and Shipbuilding." *Economic History Review*, n.s. 5 (1952): 98-115.

Pollock, David. *Modern Shipbuilding and the Men Engaged in It*. London: E. and F. N. Spon, 1884.

———. *The Shipbuilding Industry*. London: Methuen, 1905.

Rathbone, P. "Shipwrights' Trade Societies." National Association for the Promotion of Social Science. *Special Science Report*, 1860.

Rawson, Geoffrey. *Ships and Seamen*. London: Thornton, Butterworth, 1934.

Read, Edward James. *Fast Cruisers*. London, 1888.

———. *Our Ironclad Ships*. London, 1869.

———. *Shipbuilding in Iron and Steel*. London, 1869.

Rees, W. Stewart. *Some Merseyside Shipbuilders*. Liverpool, 1944.

Reid, J. M. *James Lithgow, Master of Work*. London: Hutchinson, 1964.

Renner, W. "Schiffbau und Schiffahrt auf den Grossen Seen." *Jahrbuch der Schiffbautechnische Gesellschaft*, 1909.

Rennie, Sir John. *Autobiography*. London, 1875.

Richardson. A. *Vickers, Son and Maxim*. London, 1902.

Richardson, Joseph. *History and Antiquities of Furness*. Barrow-in-Furness, 1881.

Richmond, Sir Herbert W. *Sea Power in the Modern World*. London: G. Bell and Sons, 1934.

Robb, A. M. "Ship-building." In Charles Singer et al., eds., *A History of Technology*. Vol. V. Oxford: Oxford University Press, 1958.

Roberts, Geoffrey. *Demarcation Rules in Shipbuilding and Shiprepairing*. University of Cambridge Department of Applied Economics, Occasional Paper no. 14. Cambridge: Cambridge University Press, 1967.

Roberts, W. P. "The Formation of the Society of Naval Architects and Marine Engineers." *Transactions of the Society of Naval Architects and Marine Engineers*, 1945 (Historical).

Robertson, D. H. *A Study of Industrial Fluctuation*. London: P. S. King and Son, 1915.

———, and Morison, D. C. "Shipping and Shipbuilding." *London and Cambridge Economic Service*, 1923.

Robertson, Paul L. "Shipping and Shipbuilding: The Case of William Denny and Brothers." *Business History* 19 (1974).

Robinson, C. N. "Naval Reinforcements in War Time—The Supply of Warship Material and Machinery." *The Naval Annual* 9 (1895): 149-166.

Robinson, Sir Robert S. *Report on Dockyards*. London, 1874.

Rolt, Lionel T. C. *Isambard Kingdom Brunel*. London: Longmans, 1964.

Ross, J. H.; Newton, A. P.; and Benians, E. A., eds. *Cambridge History of the British Empire*. Vol. VI, *Canada*. Cambridge: Cambridge University Press, 1930.

Rosenberg, Nathan. "Technological Change in the Machine Tool Industry, 1840-1910." *Journal of Economic History* 23 (1963): 414-443.

Ross, Colin P. J. *Produktionsbedingungen der Seewerke*. Berlin, 1911.

Rousiers, Paul de. *The Labour Question in Britain*. London: Macmillan, 1896.

Rousseaux, P. *Les mouvements de fond de l'économie anglaise, 1800-1913*. Louvain: Institut de recherches économiques, 1938.

Rowe, William H. *Maritime History of Maine*. New York: W. W. Norton, 1948.

Rusett, E. W. de. "On the Use of Hydraulic Riveting in the Construction of the 'Mauretania'." *TINA* 49 (1907): 308-311.

Russell, John Scott. *The Fleet of the Future, Iron or Wood?* London, 1861.

Russell, John Scott. "On the Education of Naval Architects in England and France." *TINA* 4 (1863): 163-176.

———. "On the Nature, Properties and Applications of Steam and on Steam Navigation." *Encyclopaedia Britannica*. 7th ed. Edinburgh, 1841.

———. "Reply to Statement on Iron Shipbuilding." *The Fleet of the Future, Iron or Wood?* London, 1863.

———. *Very Large Ships, Their Advantages and Defects*. London, 1863.

Russo, G. "Fifty Years Progress of Shipbuilding in Italy." *TINA* 53 pt. 2 (1911).

Rutherford, W. *The Man Who Built the 'Mauretania'*. London: Hillside Press, 1934.

Sanders, R. Y. "Foreign Trade and Shipbuilding." *Journal of the Royal Statistical Society* 87 (1924): 220-255.

Sanderson, Michael. *The Universities and British Industry, 1850-1970*. London: Routledge and Kegan Paul, 1972.

Saul, S. B. "The Engineering Industry." In D. H. Aldcroft, ed., *The Development of British Industry and Foreign Competition, 1875-1914*. London: George Allen and Unwin, 1968.

———. *The Myth of the Great Depression 1873-1896*. London: Macmillan, 1969.

Sawyer, John E. "Entrepreneurship in Periods of Rapid Growth: The United States in the 19th Century." Paper presented to the Conference on Entrepreneurship and Economic Growth, Cambridge, Massachusetts, November, 1954.

Schroedter, E. "The Ironmaking and Shipbuilding Industries in Germany." *Jahrbuch der Schiffbautechnische Gesellschaft*, 1902.

Schwarz, T. "Der Amerikanische Schiffbau in Letzen Jahrzehnt." *Jahrbuch der Schiffbautechnische Gesellschaft*, 1902.

———, and Halle, Ernst von. *Die Schiffbauindustrie in Deutschland und im Auslände*. Berlin: E. S. Mittler und Sohn, 1902.

Scott, J. D. *Vickers*. London: Weidenfeld and Nicolson, 1962.

Scott, W. R., et al. *An Industrial Survey of the South West of Scotland*. London: Board of Trade, 1932.

Scott, Walter. "Some Remarks on the Commercial Organization of Shipyards." *TNECI* 17 (1900-01): 129-146, 225-226.

Scott's Shipbuilding and Engineering Company. *Two Centuries of Shipbuilding*. London, 1920.

Shannon, H. A. "The First Five Thousand Limited Companies and their Duration." *Economic History* 2 (1932): 396-424.

———. "The Limited Companies of 1866-1883." *Economic History Review*, o.s. 4 (1933): 290-316.

Shields, John. *Clyde Built*. Glasgow: William Maclellan, 1947.

Shipbuilding Trades Committee. "Proposed Machinery." 1908.

Slade, Adolphus. *Naval Construction and Naval Promotion*. London, 1846.

Sloan, E. W., III. *Benjamin Franklin Isherwood, Naval Engineer*. Annapolis: United States Naval Institute, 1965.

Smellie, John. *Shipbuilding and Repairing in Dublin*. Glasgow: McCorquodale and Co., 1935.

Smiles, Samuel. *Men of Invention and Industry*. London: John Murray, 1884.

Smith, Adam. *An Inquiry into the Nature and Causes of the Wealth of Nations*. Ed. Edwin Cannan. New York: Modern Library, 1937.

Smith, Edgar C. *A Short History of Naval and Marine Engineering*. Cambridge: Cambridge University Press, 1938.

———. "Some Pioneers of Refrigeration." *Newcomen Society Transactions*, 1942-43.

Smith, E. W. Fraser. "The Future of British Engineering and Shipbuilding." *TNECI* 31 (1914-15).

Smith, J. W., and Holden, T. S. *Where Ships Are Born, Sunderland 1346-1946*. Sunderland: Thomas Reed, 1946.

Stephen, A. Murray. *A Shipbuilding History, 1750-1932*. London and Cheltenham: Ed. J. Burrow and Co., 1932.

Stigler, George. "The Division of Labor is Limited by the Extent of the Market." *Journal of Political Economy* 59 (1951): 185-193.

Strang, John. "The Improvement of Tidal Rivers." *Journal of the Royal Statistical Society* 16 (1857).

Sulley, Philip. *History of Ancient and Modern Birkenhead*. Liverpool, 1907.

Swan, Hunter and Wigham Richardson. 1906.

Swann, L. A., Jr. *John Roach, Maritime Entrepreneur*. Annapolis: United States Naval Institute, 1965.

Talbot, F. A. *Steamship Conquest of the World*. London: Heinemann, 1912.

Taylor, Arthur J. *Laissez-Faire and State Intervention in Nineteenth Century Britain*. London: Macmillan, 1972.

Terano, S., and Yukawa, M. "Development of Merchant Shipbuilding in Japan." *TINA* 53 pt. 2 (1911).

Thompson, Joseph L., and Sons. *One Hundred Years, 1846-1946*. Sunderland, 1946.

Thornton, R. H. *British Shipping*. Cambridge: Cambridge University Press, 1939.

Thurston, R. H. "Iron Manufactures in Great Britain." *Journal of the Franklin Institute*, 1871.

Tinbergen, Jan. "Ein Schiffbauzyklus?" *Weltwirtschaftliches Archiv* 34 (1931): 152-164.

Todd, G. "Joint Stock Companies, 1844-1900." *Economic History Review*, o.s. 4 (1932).

Tralau, Gustav. *Über den Standort der Schiffbauindustrie*. Rostock, 1927.

Tramond, Joannès and Reussner, André. *Eléments d'histoire maritime et coloniale contemporaine*. Paris, 1924.

Trebilcock, R. Clive. "British Armaments and European Industrialization, 1890-1914." *Economic History Review*, n.s. 26 (1973): 254-272.

———. "Legends of the British Armament Industry, 1890-1914: A Revision." *Journal of Contemporary History* 5 (1970): 3-19.

———. "A 'Special Relationship'—Government Rearmament and the Cordite Firms." *Economic History Review*, n.s. 19 (1966): 364-379.

———. " 'Spin-Off' in British Economic History: Armaments and Industry, 1760-1914." *Economic History Review*, n.s. 22 (1969): 474-490.

Tragelles, E. O. *Comparative Merits of Iron and Wood in Shipbuilding*. Cornwall Polytechnic Society, 17th Annual Report. Falmouth, 1849.

Tyler, David B. *The American Clyde*. n.p., 1958.

———. *Steam Conquers the Atlantic*. New York: D. Appleton-Century, 1939.

Usher, A. P. "Growth of English Shipping, 1572-1922." *Quarterly Journal of Economics* 42 (1928): 465-478.

Vickers, Son and Maxim. *Vickerstown Souvenir*. 1903.

Victoria County History of Durham. Vol. II. London: Archibald Constable, 1907.

Victoria County History of Lancaster. Vol. VIII. London: Archibald Constable, 1914.

Victoria County History of Yorkshire. Vol. II. London: Archibald Constable, 1912.

Wallace, Frederick W. *Wooden Ships and Iron Men*. London: Hodder and Stoughton, 1937.

Walton, Gary M. "Productivity Change in Ocean Shipping after 1870: A Comment." *Journal of Economic History* 30 (1970): 435-441.

Ward, David. "The Public Schools and Industry in Britain after 1870." In W. Laqueur and G. Mosse, eds., *Education and Social Structure in the Twentieth Century. Journal of Contemporary History*, no. 6. New York: Harper and Row, 1967.

Ward, John. "Presidential Address." *TIESS* 51 (1907-08): 1-46.

Waters, Sydney David. *From Clipper Ship to Motor Liner*. London, 1939.

Webb, Sidney, and Webb, Beatrice. *The History of Trade Unionism*. London: Longmans, Green, 1920.

———. *Industrial Democracy*. London: Longmans, Green, 1920.

Welch, J. J. "The Scientific Education of Naval Architects." *TNECI* 25 (1908-09): 177-198.

Wheeler, C. H. "Commerce of the Great Lakes." *Transactions of the Society of Naval Architects and Marine Engineers*, 1897.

Wilkinson, Ellen Cicely. *The Town that Was Murdered*. London: Victor Gollancz, 1939.

Wilmot, Sydney M. Eardley. *Our Fleet Today*. 2nd ed. London: Seeley and Co., 1900.

Wood, Walter. *The Battleship*. London: Kegan Paul, 1912.

Wright, Richard J. *Freshwater Whales*. Kent, Ohio: Kent State University Press, 1970.

Wurl, M. "Technical Education in Germany." *TNECI* 20 (1903-04).

Yarrow, Lady. *Alfred Yarrow, His Life and Work*. London: Edward Arnold, 1923.

Young, Allyn. "Increasing Returns and Economic Progress." *Economic Journal* 38 (1928): 527-542.

Young, Charles F. T. *Fouling and Corrosion of Iron Ships*. London, 1867.

Zimmermann, Erich. *Ocean Shipping*. New York: Prentice-Hall, 1921.

Index

Aberconway, Lord (Sir Charles McLaren), 54, 75, 99
Aberdeen, 56, 60, 63, 68, 163
Accounting practices, 78-80
Admiralty, 13, 74, 97, 100, 101, 134, 135, 142, 144, 149, 171, 202-229. *See also* Navy; Royal Naval Dockyards
African Steam Ship Company, 93
Aldcroft, Derek H., 3, 137, 138
Alexander Stephen and Sons, 51, 55, 73, 74, 76, 84, 85, 101, 121, 134
Alfred Holt and Company, 15
Allan Line, 103-104
Amalgamated Society of Engineers (A.S.E.), 154, 157, 158, 160, 161-162, 166, 184
American Shipbuilding Company, 42
Anglo-Persian Oil Company, 226, 227
Apprentices, ratio to journeymen, 154, 155
Apprenticeship, 137, 138, 139, 141-142, 154-156, 276nn10,11, 279n41; of entrepreneurs, 74
Aquitania, 18, 110, 147
Argentina, 217
Armaments manufacturers, 97-101, 214-222, 226-227
Armor, 202-203. *See also* Armaments manufacturers
Armstrong, Sir William, 120
Armstrong College, 141, 143, 144, 145
Armstrong-Mitchell, 52, 84, 97, 98, 100, 220
Armstrong-Whitworth, Sir W. G. and Company, 22, 51, 54, 57, 58, 63, 74, 80, 81, 82, 97, 98, 99, 115, 124, 142, 144, 146, 203, 213, 214, 219, 220, 221, 222, 270n33
Arrol, Sir William, 124
Associated Shipwrights' Society (A.S.S.), 155, 156, 157
Ateliers et Chantiers de Provence, 43
Atlantic Transport Line, 20
Austin, S. P., 73, 79, 81, 85
Austin and Hunter, 135
Australian Land Company, 23
Awning decked vessels, 21

Barclay, Curle, 51, 100
Barnaby, N., 208, 219, 221
Barrow-in-Furness, 49, 52, 53, 54, 55, 57, 58, 59, 62, 67-68, 91, 111, 112, 114, 122, 124, 163, 197, 262n3
Barrow Shipbuilding Company, 52, 56, 84, 101, 106, 111. *See also* Vickers, Son and Maxim
Barry, Patrick, 71, 72
Bath Iron Works, 128
Beardmore's, *see* William Beardmore
Beaufoy, Colonel Mark, 131
Belfast, 34, 48, 49, 51, 52, 53, 55, 56, 57, 58, 59, 62, 63, 66-67, 69, 84, 91, 96, 105, 110, 112, 128, 161-162, 163, 186, 197
Bell, Lowthian, 45-46
Bethlehem Steel Company, 41

Beverly, 68, 85
Bibby Line, 92, 105, 135
Bideford, 68
Biles, J. H., 214
Birkenhead, 51, 57, 58, 65, 84, 93, 97, 119, 168
Blair and Company, 89
Blyth, 62
Boer War, 223-224
Boilermakers, 154, 160, 166; Society, *see* United Society of Boilermakers and Iron and Steel Ship Builders
Bolchow-Vaughn, 99
Bone, W. J., 144
Bonuses, 198
Bowley and Wood wage indices, 173-174, 175-177, 179, 196
Brassey, Lord, 209, 223, 224, 265n40, 267n24
Brazil, 97, 219
Bristol, 52, 58, 89; Channel, 183
British Petroleum, 226
Browne, Sir Benjamin, 140
Brunel, Isambard Kingdom, 17, 131-132, 220
Burmester and Wain, 100
Burnett, J., 55
Burns, John, 223

Caird and Company, 51, 54, 55, 154
Cambridge, University of, 144
Cammell-Laird, 51, 60, 65, 66, 81, 84, 85, 97, 119, 144, 212, 213, 215, 222, 270n33. *See also* Charles Cammell
Camperdown, Lord, 224
Canada, 97, 98; shipbuilding, 12, 14
Capital, per worker, 126-127, 267n19; supply, 55-56, 62
Cardiff, 159
Cargo Fleet Company, 100
Cargo vessels, 16, 17, 20-22, 102, 134. *See also* Tramp steamers
Cart, River, 110
Cartels, *see* Industrial concentration
Caulking machinery, 120
Cavendish family, 101, 111, 265n40. *See also* Devonshire, Duke of
Cecil, Evelyn, 224
Central Marine Engineering Company, 89
Chamberlain, Joseph, 23
Chandler, Alfred D., Jr., 96-97
Chantiers de Construction Maritime, 43
Chantiers de France, 43
Chapman, 130, 132
Charles Cammell, 97, 98, 203, 214
Charles Connell and Company, 51, 75, 84, 177, 186-190, 193, 194
Charlottenburg, 145
Chatham, 72
Chile, 219
China, 203, 219
China Navigation Company, 93
Chiswick, 57
Churchill, Winston, 220
Clyde, 14, 26, 34, 38, 39, 40, 46, 47, 49, 51, 52, 53, 55, 56, 57, 58, 59, 60-62, 63, 66, 67, 68, 69, 70, 76, 80, 81, 84, 85, 89, 91, 109, 110, 120, 123, 143, 159, 161, 163, 167, 168, 174, 177, 184, 196, 197, 252-253. *See also* Glasgow; Scotland
Clydebank, 51, 55, 58, 97, 109, 110, 112, 123
Clydebank Company, 52. *See also* J. and G. Thomson; John Brown and Company
Coal supplies, 56-57, 61, 63, 66, 68
Coleman, D. C., 3-4, 74, 233
Conciliation, 161, 168
Connell's, *see* Charles Connell and Company
Cook, Welton and Gemmel, 85
Cork, 67
Cornell University, 145
Coventry Ordnance Works, 99, 222
Covered berths, 118-119, 128
Cowes, 68
Cramp, Charles, 40-41
Cranes, 116, 118, 119, 121, 123-124, 125, 126, 129
Cunard, Samuel, 61
Cunard Line, 19, 61, 103, 104, 105, 106, 119, 135, 223, 224, 225, 226
Currie, Donald, 223
Customers, relations with, 102-106

D. and W. Henderson, 81, 84. *See also* Henderson family
Dalmuir, 52, 99, 110, 123
Danzig, 145
Dartmouth, 68

David Colville and Sons, 101
David Rowan and Company, 89
Davies, Colonel, 99
Deadman, H. E., 144
Dee, 56
Demarcation disputes, 163-169
Denny, Peter, 93, 105
Denny, William, 61, 103, 134
Denny's, *see* William Denny and Brothers
Department of Science and Art, 139-140, 145
Devonshire, Duke of, 265n40, 267n24
Dick, Kerr and Company, 99
Dillon, Malcolm, 54
Dockyards, *see* Royal Naval Dockyards
Dr. Denny's Anti-Fouling Paint Company Ltd., 91
Doxford's, *see* William Doxford and Sons
Dreadnought, 204-205
Drillers, 157
Drilling machinery, 120, 121
Drogheda, 67
Dublin, 52, 58, 67, 71
Dumbarton, 35, 52, 55, 58, 61, 62, 63, 122, 145, 174
Duncan, Robert, 75, 85, 111, 112, 119
Dundee, 56, 60, 68, 71
Dundee Shipbuilding Company, 81
Dunn, James, 221

Earle's Shipbuilding Yard, 80, 84, 100, 213, 221
East Anglia, 58
Eastern Shipbuilding Company, 41-42
École d'Application du Génie Maritime, 146
Economist, 19, 284n44
Edinburgh, University of, 74
Edinburgh Agreement, 158, 160-161
Education Department, 145
Elder, John, 15, 61. *See also* John Elder and Company
Elder, Mrs. John, 143
Elder-Dempster and Company, 23, 95
Electric lighting, 121
Electrical machinery, 120, 121, 122, 123, 129
Elgar, Francis, 144, 209, 221-222, 262n45
Elswick, 63, 82-83, 98, 99, 115, 144, 162
Employers' federations, 67, 159, 161-162, 167, 168
Employment, 31, 33-34, 71, 77, 81-82, 191-192, 265n2; Clyde, 62, 81-82; North East Coast, 64; Royal Naval Dockyards, 169; fluctuations, 177, 178-180, 191-192. *See also* Unemployment
Engineering trades, 152, 154, 155, 157, 166
Engineers, *see* Amalgamated Society of Engineers
Entrepreneurs, background, 74-75
Entrepreneurship, 1-8, 84, 233-234, 266n10; American, 3, 5, 7; German, 2, 3, 5, 7
Experimental tank, 133-134, 206

Fairfield Shipbuilding and Engineering Company, 51, 81, 84, 122, 144, 177, 178-180, 184, 186, 188-189, 213, 215, 222, 270n33
Family connections, 72, 73-74
Faversham, 68
Federation of Shipbuilders and Engineers of England, Scotland, and Ireland, 159
Federation of Shipbuilding and Engineering Trades, 158, 167
Feinstein, Charles H., 30-31, 187-189, 193
Fincham, John, 132
Firth, B. A., 99
Firth, Thomas, *see* Thomas Firth
Firth of Forth, 56
Fisher, Admiral Lord, 204-205, 207, 209, 221
Fishing vessels, 56, 60, 61, 68, 85, 102
Fluctuations in output, 25-48, 84, 85; and mechanization, 115, 117, 118, 152
Fore River Ship and Engine Company, 128
Foreign operations, 97-98, 270n33
Foreigners, sales to, 26, 37-38, 250-251, 259n18, 260n22
Forwood Committee, 214
Fox, Robert, 132
France, 38, 202, 204, 222, 225; naval architecture, 130, 131, 142; shipbuilding, 12, 42-43, 44-46, 91-92, 129, 261n41, 265n2
Freight rates, 26-28

Froude, R. E., 133
Froude, William, 132, 133, 206
Furness, Lord, 40, 75, 81, 92, 96, 99-100
Furness Group, 100, 101
Furness-Westgarth, 89, 96
Furness, Withy and Company, 51, 84, 96, 121

Germany, 37-38, 202, 204-205, 224, 225; shipbuilding, 12-13, 28-29, 38-40, 43-45, 57, 91, 128, 129, 145, 227-228, 262n46, 265n2, 267nn8,19, 285n64
Gillmor, H. C., 120
Glasgow, 41, 45, 51, 55, 57, 58, 60, 61, 62, 67, 68, 101, 109, 114, 133, 134, 160, 161, 177, 197. *See also* Clyde; Scotland
Glasgow, University of, 74, 142, 143, 144, 145, 221-222
Glasgow Iron and Steel Company, 101
Gourlay, Andrew, 55
Grangemouth, 68
Graving docks, 118-119
Gray, William, 75. *See also* William Gray and Company
Grayson's, 65
Great Eastern, 17, 111, 131-132
Great Eastern Steam Ship Company, 105
Greenock, 51, 54, 55, 58, 62, 133, 143, 154
Greenwich, 41, 141, 143, 144
Grimsby, 56, 68
Grimsby Shipbuilding Company, 85

Hadfield's, 99
Hall, Henry, 72
Hamburg-Amerika Linie, 19, 39, 104, 147
Hamilton, J. J., 80
Harlan and Hollingsworth, 41
Harland, Edward, 66, 161, 278n33
Harland and Wolff, 19, 47, 51, 55, 58, 66-67, 74, 75, 81, 84, 85, 89, 91, 92, 95, 96, 99, 100, 101, 105, 110-111, 112-114, 121, 124, 128, 134-135, 147, 224, 268n30, 270n26
Hartlepool Engine Works, 96
Hartlepools, 49, 53, 57, 62, 63, 96. *See also* West Hartlepool
Harvey Steel, 202, 203
Haslar, 133
Hawes, Sir Benjamin, 220
Hawthorn, Leslie, *see* R. and W. Hawthorn, Leslie and Company
Hebbern, 63, 146
Henderson family, 73. *See also* D. and W. Henderson
Henry Withy, 79, 96
Hill, Sir Norman, 20
Hills, A. F., 161
Holidays, 177
Houlder Line, 23
Hours of work, 159-161, 170, 171, 172, 174, 177, 186, 279n41
Howdon, 51, 58, 90, 96
Hull, 56, 100
Humber, 49, 56, 60
Hunter, George B., 73, 100, 136
Hydraulic machinery, 120, 121, 122, 123, 129
Hylton, 104

India General Navigation and Railway Company, 94
Industrial concentration, 2-3, 96-102, 267n27
Inman Line, 19, 106, 111, 223
Innovation, 3, 29, 116, 117, 147-148, 206, 207
Institution of Naval Architects, 134, 143
Integration, vertical, 29-30, 90, 92, 96-102
Interlocking directorates, 73, 97, 98, 99
International Mercantile Marine, 38, 93, 95, 96, 224, 225
Inverclyde, Lord, 223, 224, 225
Investment, 34, 35, 70, 71, 72, 73, 76, 77-84, 265n2, 266n6, 267nn18,23,24, 270n31
Irish workers, 163; religion, 67, 264n37
Iron shipbuilding, 11-12, 14, 61-62, 72, 116, 153
Iron supplies, 62, 63, 268nn4,8
Irrawaddy Flotilla Company, 94
Irvine's Shipbuilding Company, 51, 84, 96
Italy, 37-38, 43, 97, 225

J. and G. Thomson, 52, 55, 78, 97, 98, 109, 123. *See also* Clydebank Company; John Brown and Company
J. L. Thompson and Company, 51

J. S. White, 84
Jamieson, A., 122
Japan, 38, 43, 203, 204, 219, 225
Jarrow, 29, 51, 54, 58, 63, 90, 96, 101
John, William, 144
John Brown and Company, 51, 55, 58, 80, 81, 82, 84, 96, 97, 98, 99, 100, 101, 109, 110, 126, 134, 203, 213, 214, 215, 222, 254, 270nn26,31,33. *See also* Clydebank Company; J. and G. Thomson
John Elder and Company, 52
Joint-stock companies, *see* Limited liability

Kindleberger, Charles P., 2, 92
Kirk, A. C., 61, 80
Knight, Robert, 46, 160
Krupp, 39; steel, 202, 203

Labor, control, 53-55; costs, 45, 47, 52-53, 55, 248; recruitment, 29, 51-52; unions, *see* Trade unions
Lagan, 51, 52, 66, 67, 69, 111
Laing's, *see* Sir James Laing
Laird, Macgregor, 93
Laird Brothers, *see* Cammell-Laird
Laissez-faire, 201-202, 227
Lamport and Holt, 95
Lanarkshire Steel Company, 101
Land costs, 57-58, 114-115
Landes, David S., 2-3, 5
La Platense Flotilla Company, 79, 94
Leith, 56
Leven Shipyard, *see* William Denny and Brothers
Levy, Hermann, 101
Lewis, W. Arthur, 2, 82, 106
Limited liability, 63, 70, 72-73, 75-76, 84, 266n13, 267n19
Liners, 18, 19-20, 56, 61, 63, 84, 85, 94-95, 101, 257nn24,31
Lithgow, William T., 93, 104
Liverpool, 52, 53, 60, 61, 65, 66, 67, 68, 69, 85, 89, 159, 164, 264n31
Liverpool, University of, 144
Liverpool Shipbuilding Company, 65
Lloyd's Register, 12, 21, 22, 23, 60, 119, 136, 144, 149
Loch Thom, 133
London, 52, 53, 60, 64, 89, 90, 114-115, 133, 164
London and Glasgow Company, 215, 222
Londonderry, 67
Lusitania, 110

Machinery, 71, 79, 80, 84, 108, 115-129, 166, 167
McIver, David, 223
Mackie and Thomson, 85
McLaren, Sir Charles, *see* Lord Aberconway
McMillan's, 122
Marine Engineering, 89, 153, 206, 211, 213, 268n2, 272n4
Marketing, 2, 56
Marwick, W. H., 75
Massachusetts Institute of Technology, 145
Materials testing, 135-136, 137
Maudslay's, 89
Mauretania, 17, 18, 104, 119, 123
Maxim Gun Company, 98
May, Charles, 120
Maywald, Karel, 30-31, 187
Mechanization, 29, 193
Mersey, 49, 51, 53, 58, 60, 64-66, 75, 90, 154
Metal trades, 152, 153, 154
Michigan, University of, 145
Middlesbrough, 52, 63, 168
Migration of workers, 157, 163, 178, 197
Millwall Iron Works, 12
Mitchell, Dr. Charles, 98, 100
Moran Brothers, 128
Morgan, J. P., 38, 93, 95, 225
Morton, R., 71
Murdoch and Murray, 85

Napier, David, 74
Napier, Robert, 61, 74, 80, 84, 89, 97, 99, 132, 213, 222
Napier family, 73
National Physical Laboratory, 134
Naval Architecture, 130-137; university chairs, 143-144
Naval Construction and Armament Company, 52, 112, 265n40. *See also* Barrow Shipbuilding Company; Vickers, Son and Maxim
Naval shipbuilding, 30, 31, 32, 59, 60, 61, 63, 66, 84, 85, 97-102, 205-222; in private shipyards, 211-222, 284n39;

tonnage, 215-216
Naval vessels, 16, 18; design, 202-205
Navy, 201, 202, 204-205, 222. *See also* Admiralty
Navy League, 19
Newcastle, 51, 52, 54, 55, 57, 58, 63, 68, 141, 143, 145, 146, 197, 252-253
New Shipwright Building Company, 71
Newshot Isle, 110
New York Shipbuilding Company, 128
New Zealand Shipping Company, 23
Noble, Sir Andrew, 220, 221
Nord Deutscher Lloyd, 39
Nordenfelt Gun and Ammunition Company, 98
Nordenfelt submarine, 112
North East Coast, 21, 34, 45, 51, 53, 55, 56, 60, 62-64, 69, 101, 146, 153, 159, 160, 197, 252-253
North East Coast Institution of Engineers and Shipbuilders, 135, 139, 146
North-Eastern Marine Engineering Company, 89
North Shields, 63
Northumberland Shipbuilding Company, 51, 96
Norway, 37-38
Number of firms, 86, 268nn32,2

Oil tankers, 22-23
Orient Line, 23
Oswald, T. R., 55
Output, 250-251, 258n2; of firms, 24, 86; for foreigners, 250-251; by regions, 252-253
Overtime, 160-161, 277n24, 279n41

Pacific Steam Navigation Company, 95
Palmer, Charles Mark, 40, 55, 63, 75, 90, 97
Palmer's Shipbuilding Company, 22, 29, 45, 51, 54, 74, 75, 79, 81, 84, 85, 96, 98, 99, 101, 106, 109, 121, 213, 215, 267n23, 271n52
Parsons, Sir Charles, 15, 74, 89, 136, 148, 267n24
Parsons Marine Steam Turbine Company, 89, 221, 267n24
Partnership, 70
Passenger liners, 16, 17-18
Peninsular and Oriental Line, 23, 222
Pensions, 198-199
Perret, J. R., 144
Piecework, 164, 165, 170, 171, 174, 182-186
Pile, William, 131
Pile family, 73
Pirrie, Lord, 66, 89, 92, 93, 95, 148, 224
Pneumatic machinery, 120-121, 122, 123, 129
Poplar, 57, 115
Port facilities, 57
Port Glasgow, 51, 52, 58, 62, 111, 119
Port Mulgrave, 90
Portsmouth, 132, 133, 142
Power for machines, 119-123, 129. *See also* Electrical machinery; Hydraulic machinery; Pneumatic machinery; Steam-powered machinery
Price, J., 45
Productivity, 186-190, 193, 194, 199
Profitability, 82-83, 99
Projectile Company Ltd., 99
Pusey and Jones, 40

R. and J. Evans, 65
R. and W. Hawthorn, Leslie and Company, 51, 81, 140, 222
R. Thompson and Sons, 84
Randolph, Charles, 15
Randolph and Elder, 75
Raydon, Potter, 85
Readhead and Sons, 51, 75
Reaney, Son and Archbold, 40
Reed, E. J., 133, 208, 209, 220, 222
Refrigerated vessels, 23
Rendel, George, 220
Repair work, 56, 57, 58, 63, 65, 67, 85, 89, 91, 100, 164, 174, 208, 213, 268nn3,5
Richardson, Wigham, 75. *See also* Wigham Richardson; Swan, Hunter and Wigham Richardson
Richardson, Duck and Company, 51, 85
Richardson, Westgarth and Company, 79, 96, 99
Ritchie, William, 66
Ritchie and McLaine, 66
Rivers Steam Navigation Company, 94
Riveters, 117, 152, 154, 162-163
Riveting machinery, 116, 117, 118, 120, 121-123, 129
Roach, John, 40
Robert Stephenson, 96

Robinson, Admiral Sir Spencer, 208
Robson, P. J., 160
Ropner and Sons, 51, 84
Rowan, F. J., 122
Royal Dutch Shell, 226
Royal Mail Steam Packet Company, 95
Royal Naval College, 141, 143, 144, 145
Royal Naval Dockyards, 31, 159-160, 201, 205-211, 221, 222, 284n39; efficiency, 206-207; labor force, 169-172; schools, 142, 143, 144, 145-146; wages, 169-170, 171, 172
Royal School of Naval Architecture and Marine Engineering, 143, 144, 146
Runciman, Walter Jr., 20
Russell, John Scott, 17, 44-45, 131, 132, 146
Russell and Company, 51, 85, 93, 104, 154
Russia, 97, 98, 204, 224, 225

Sailing ships, 14-15, 62, 63
Salaries, 198
Saltburn, 90
Sampson, J., 99
Saul, S. B., 4
Sawyer, John, 5
Scarborough, 63
Schwabe, G. C., 92
Scientific research, 3, 7, 130-137, 147-148, 149; naval, 206
Scotland, 35, 49, 60, 63, 69, 153, 154, 186, 197, 266n10
Scott, John, 12, 45
Scott's of Greenock, 11, 51, 55, 73, 84, 93, 133, 213
Semiskilled labor, 152, 153, 164, 165
Shaw Savill and Albion Line, 23
Shelter decked vessels, 21
Shipbuilding Employers' Federation, 67, 168, 197
Shipconstructive and Shipwrights' Association, 157, 158
Shipping, American, 9-11
Shipping lines, 26-28; and shipbuilding, 92-96; rings, 19, 94-95. *See also* individual lines
Shipwrights, 71, 117, 137, 151, 153, 154, 155, 156, 157, 161, 163, 164-165, 167, 169, 170, 171
Shipwrights Provident Union of the Port of London, 164. *See also* Associated Shipwrights' Society; Shipconstructive and Shipwrights' Association; United Kingdom Amalgamated Society of Shipwrights
Shipyard Standing Agreement, 162
Shire Line, 23, 95
Shops, 71, 72, 108-115
Short Brothers, 51
Sir James Laing, 51, 73, 79, 81, 84, 91, 92
Sir Raylton Dixon, 52, 75
Size of vessels, influence of increasing, 16, 17-18, 30, 116
Skill differentials, 166, 193-196
Skilled labor, 117, 120-121, 129, 151, 152, 153, 157, 161, 164, 165-166, 169, 181
Smith Dock Company, 63, 81
Society of Naval Architects and Marine Engineers, 41
Southampton, 55, 57, 64, 67, 85, 197
South Kensington, 143
South Shields, 51, 58, 63
South Wales, 45, 58
Spain, 38, 98
Spar decked vessels, 21-22
Specialization of output, by firms, 84-87, 267n27; by region, 59-61, 68, 70
Specialized ships, 18-23, 46
Speed, influence of increasing, 16, 17-18, 19, 116
Steam Engine Makers, 157, 166
Steam power, vessels, 14-16
Steam-powered machinery, 119-120, 121, 122, 129
Steam turbine, 15-16, 23. *See also* Sir Charles Parsons; Parsons Marine Steam Turbine Company
Steel Company of Scotland, 101
Steel industry, and shipbuilding, 6, 29-30, 36-37, 46-47, 56-57, 62, 66, 68, 90, 91, 99, 100-101, 214, 262n45, 263n20, 271n52, 285n64; cartels, 62, 101, 214, 263n20, 271n37
Steel shipbuilding, 14, 116, 259nn10,11, 263n20
Stephen, William, 71
Stephen of Linthouse, *see* Alexander Stephen and Sons

Stockton, 51, 63, 85, 89
Stosch, General von, 39
Subcontracting, 164-165
Subsidies, 38, 39, 42-43, 46, 222-226, 227-229
Sunderland, 51, 52, 55, 58, 85, 89, 104, 131, 135, 145, 146, 252-253, 257n23, 266n8
Suppliers, 88-92, 268n9, 269n10
Swan, C. S., 74, 100
Swan, H. F., 100
Swan, Hunter and Wigham Richardson, 47, 52, 55, 63, 73, 79, 81, 84, 85, 100, 101, 105, 119, 121, 123, 124, 128, 221. *See also* Wigham Richardson
Sweden, 97, 130

T. Royden, 65
Tay, 56
Tay Shipbuilding Company, 71
Technical change, and labor relations, 164, 166. *See also* Innovation
Technical education, 3, 4, 7, 29, 44-45, 137-150; American, 145, 146, 147, 148; French, 146; German, 145, 146, 147, 148; of entrepreneurs, 74
Technical staff, 136
Tees, 49, 51, 52, 61, 62, 63, 84, 89, 91, 168
Tellier, Charles, 23
Thames, 51, 52-53, 56, 58, 60, 64, 65, 70, 75, 89, 90, 91, 154, 184, 196, 197, 211
Thames Iron Works, 12, 64, 121, 123, 213, 214, 215
Thearle, S. J. R., 144
Thomas Firth, 98, 99, 222
Thompson, J. L., 72
Thornycroft, John I., 47, 57, 64, 81, 84, 85
Time horizons, 4-7, 29, 116, 233-234
Tod and M'Gregor, 109, 119, 123, 272n4
Tonnage measurements, 237-238
Tools, *see* Machinery
Torpedoes, 202-203, 204
Torquay, 133
Trade unions, 156-163; effect on wages, 161, 181-184; number, 159
Trades, irregular employment, 177, 178-180; multiplication of, 108, 115-116, 137, 151-154, 156, 164-165
Tramp steamers, 18, 20-23, 26-28, 46, 56, 59, 60, 61, 64, 84, 94, 95, 107, 257n33. *See also* Cargo vessels
Tranmere Bay, 97
Trebilcock, R. Clive, 226, 227
Turret ships, 257n23
Twigg, E. S., 199
Tyne, 49, 51, 52, 53, 57, 58, 59, 62, 66, 69, 70, 84, 90, 91, 97, 115, 159, 160, 163, 167, 168, 184, 196, 197
Tyne Pontoon and Dry Dock Company, 100
Tynemouth, 63
Tyser Line, 23

Unemployment, 160, 172, 177-178, 180, 181, 186, 242, 277n24, 279n41
Union Castle Line, 95
Union Iron Works, 128
Union Steam Ship Company of New Zealand, 105
United Kingdom Amalgamated Society of Shipwrights, 157
United Pattern Makers, 166
United Society of Boilermakers and Iron and Steel Ship Builders, 46, 117, 154-155, 157, 158, 162, 165, 166, 168, 184, 197
United States, shipbuilding, 9-13, 14, 28-29, 40-42, 43, 44, 46, 72, 91, 128, 129, 136, 145, 147, 227-229, 261n31, 262n46, 265n2
United States Shipbuilding Company, 42
Unskilled labor, 152, 153, 157, 163, 164

Value of output, 30-31, 32
Vernon's, 65
Vickers, Albert, 98
Vickers, Son and Maxim, 47, 52, 54, 55, 68, 81, 85, 91, 97, 98, 99, 112, 114, 116, 124, 213, 215, 221, 222, 270n33. *See also* Barrow Shipbuilding Company; Naval Construction and Armament Company
Vulkan Works, 128

Wages and earnings, 29, 34, 35, 36, 45-46, 53, 64, 160, 165, 169-170, 171, 172, 173-186, 190-200, 239-240, 243-247, 262n46; disputes, 161; real wages,

190-192, 193, 194, 199, 241; regional differentials, 196-197, 262n1
Walker, William, 199
Walker-on-Tyne, 58, 82-83, 98, 99, 115, 119, 146
Wallsend-on-Tyne, 52, 53, 55, 100, 146
Wallsend Slipway and Engineering Company, 100
Warships, *see* Naval vessels; Naval shipbuilding
Waterford, 67
Waterman Brothers, 154
Watts, Sir Philip, 144, 221
Wear, 49, 51, 52, 53, 55, 56, 58, 61, 62-63, 72, 84, 89, 91, 131, 154, 159, 163, 167, 196, 197
Weardale Steel Company, 100
Webb's Academy, 145
Welch, J. J., 144
Welding, electric, 123
Well decked vessels, 21
Westgarth, English and Company, 96
West Hartlepool, 21, 51, 52, 53, 54, 145, 186. *See also* Hartlepools
Whitby, 62, 63, 90
White, Sir William H., 140, 143, 144, 209, 214, 219-220, 221
Whitehead Torpedo, 203; Company, 270n33
White Star Line, 19, 20, 95, 104, 105, 106, 223, 257nn24,31
Whitworth, Sir Joseph, 98, 203, 222
Wigham Richardson, 75, 81, 100
William Allen and Company, 96
William Beardmore, 52, 81, 98, 99, 123, 222
William Denny and Brothers, 35, 36, 52, 73, 76, 79, 80, 85, 93-94, 105, 123, 133-134, 135, 174-176, 177-178, 179, 181, 186-194, 198-199, 239-241, 243-244, 248
William Doxford and Sons, 52, 79, 81, 84, 85, 257n23
William Gray and Company, 52, 54, 84
William Hamilton, 52
William Simons, 85
Wilson, C. H., 80
Wilson Line, 100
Wolff, G. W., 74, 92
Wood, John, 89
Wooden shipbuilding, 9-13, 14, 49, 56, 62, 70-72, 82, 156-157, 164, 256nn9,10,11
Workman, Clark and Company, 52, 67, 84, 85, 101, 124
Worshipful Company of Shipwrights, 103

Yarrow, Alfred, 74, 134, 142
Yarrow's, 57, 64, 81, 84, 85, 98, 114-115, 124

HARVARD STUDIES IN BUSINESS HISTORY

1. JOHN JACOB ASTOR, BUSINESS MAN, by Kenneth Wiggins Porter*
2. JAY COOKE, PRIVATE BANKER, by Henrietta M. Larson*
3. THE JACKSONS AND THE LEES: TWO GENERATIONS OF MASSACHUSETTS MERCHANTS, 1765-1844, by Kenneth Wiggins Porter*
4. THE MASSACHUSETTS—FIRST NATIONAL BANK OF BOSTON, 1784-1934, by N. S. B. Gras*
5. THE HISTORY OF AN ADVERTISING AGENCY: N. W. AYER & SON AT WORK, 1869-1949, revised edition, by Ralph M. Hower*
6. MARKETING LIFE INSURANCE: ITS HISTORY IN AMERICA, by J. Owen Stalson*
7. HISTORY OF MACY'S OF NEW YORK, 1858-1919: CHAPTERS IN THE EVOLUTION OF THE DEPARTMENT STORE, by Ralph M. Hower*
8. THE WHITESMITHS OF TAUNTON: A HISTORY OF REED & BARTON, 1824-1943, by George Sweet Gibb*
9. DEVELOPMENT OF TWO BANK GROUPS IN THE CENTRAL NORTHWEST: A STUDY IN BANK POLICY AND ORGANIZATION, by Charles Sterling Popple*
10. THE HOUSE OF HANCOCK: BUSINESS IN BOSTON, 1724-1775, by W. T. Baxter*
11. TIMING A CENTURY: HISTORY OF THE WALTHAM WATCH COMPANY, by C. W. Moore*
12. GUIDE TO BUSINESS HISTORY: MATERIALS FOR THE STUDY OF AMERICAN BUSINESS AND SUGGESTIONS FOR THEIR USE, by Henrietta M. Larson*
13. PEPPERELL'S PROGRESS: HISTORY OF A COTTON TEXTILE COMPANY, 1844-1945, by Evelyn H. Knowlton*
14. THE HOUSE OF BARING IN AMERICAN TRADE AND FINANCE: ENGLISH MERCHANT BANKERS AT WORK, 1763-1861, by Ralph W. Hidy*
15. THE WHITIN MACHINE WORKS SINCE 1831: A TEXTILE MACHINERY COMPANY IN AN INDUSTRIAL VILLAGE, by Thomas R. Navin*
16. THE SACO-LOWELL SHOPS: TEXTILE MACHINERY BUILDING IN NEW ENGLAND, 1813-1949, by George Sweet Gibb*
17. BROADLOOMS AND BUSINESSMEN: A HISTORY OF THE BIGELOW-SANFORD CARPET COMPANY, 1825-1953, by John S. Ewing and Nancy P. Norton

18. NATHAN TROTTER: PHILADELPHIA MERCHANT, 1787-1853, by Elva Tooker*
19. A HISTORY OF THE MASSACHUSETTS HOSPITAL LIFE INSURANCE COMPANY, by Gerald T. White*
20. THE CHARLES ILFELD COMPANY: A STUDY OF THE RISE AND DECLINE OF MERCANTILE CAPITALISM IN NEW MEXICO, by William J. Parish
21. THE RISE AND DECLINE OF THE MEDICI BANK, 1397-1494, by Raymond de Roover*
22. ISAAC HICKS: NEW YORK MERCHANT AND QUAKER, 1767-1820, by Robert A. Davison*
23. BOSTON CAPITALISTS AND WESTERN RAILROADS: A STUDY IN THE NINETEENTH-CENTURY RAILROAD INVESTMENT PROCESS, by Arthur M. Johnson and Barry E. Supple*
24. PETROLEUM PIPELINES AND PUBLIC POLICY, 1906-1959, by Arthur M. Johnson
25. INVESTMENT BANKING IN AMERICA: A HISTORY, by Vincent P. Carosso
26. MERCHANT PRINCE OF BOSTON: COLONEL T. H. PERKINS, 1764-1854, by Carl Seaburg and Stanley Paterson
27. THE MATURING OF MULTINATIONAL ENTERPRISE: AMERICAN BUSINESS ABROAD FROM 1914 TO 1970, by Mira Wilkins
28. FINANCING ANGLO-AMERICAN TRADE: THE HOUSE OF BROWN, by Edwin J. Perkins
29. BRITISH MERCANTILE HOUSES IN BUENOS AIRES, 1810-1880, by Vera Blinn Reber
30. THE BRITISH SHIPBUILDING INDUSTRY, 1870-1914, by Sidney Pollard and Paul Robertson